Materials Matter

Urban and Industrial Environments
Series Editor: Robert Gottlieb, Henry R. Luce Professor of
Urban and Environmental Policy, Occidental College

Materials Matter
Toward a Sustainable Materials Policy

Kenneth Geiser

The MIT Press
Cambridge, Massachusetts
London, England

This book was set in Adobe Sabon in QuarkXPress by Asco Typesetters, Hong Kong, and was printed on recycled paper and bound in the United States of America.

Library of Congress Cataloging-in-Publication Data

Geiser, Kenneth.
  Materials matter : toward a sustainable materials policy / Kenneth Geiser.
    p.   cm. — (Urban and industrial environments)
  Includes bibliographical references and index.
  ISBN 0-262-07216-5 (hc. : alk. paper) — ISBN 0-262-57148-X (pbk. : alk. paper)
  1. Materials—Government policy. 2. Materials—Environmental aspects. 3. Materials—Health aspects. I. Title. II. Series.
TA403.6.G45   2001
620.1′1—dc21

00-048966

In memory of my parents, Kenneth R. and Kathryn R. Geiser

Contents

Foreword

After several decades of debate among proponents of environmental improvement, there is convincing evidence that the dominant cause of environmental degradation—and hence the proper locus of remedial action—is the design of the major technologies of production. Thus, the modern automobile, which is a useful social and economic good, became a major source of photochemical smog when its engine was redesigned for high compression. In turn, this required the introduction of a new material—tetraethyl lead—as a fuel additive, which sharply increased human exposure to this toxic element. Once more redesigned, but this time to be driven by electric motors and powered by batteries recharged from solar sources, automobiles could again become less polluting.

It follows that if we are to pursue the goal of environmentally sustainable production, then the component factors of production—labor, energy, and materials—should conform to this mandate. The requirements, if far from being met, are well known: labor's working environment and the general environment should be free of anthropogenic toxic substances and conserving of natural resources; and energy sources should be renewable, that is, solar. However, as *Materials Matter* shows, the suitability of different materials to systems of sustainable production has been less explored. Indeed, past experience suggests that there has been little effort to deliberately design materials to suit *any* system of production, let alone one dedicated to environmental quality. Instead, new materials are often created not to specifically meet a stated need, but in the hope that the need will be found or created.

The most demonstrative examples of this designless approach can be found in the petrochemical industry, which produces a multiplicity of materials from a single basic source, such as crude oil or natural gas. For example, the manufacture of ethylene, which is heavily used to produce a large-scale plastic end product, polyethylene, yields propylene as a by-product. Since the manufacturing process is continuous, the propylene must be disposed of. However, the cost of manufacturing ethylene can be reduced by half if propylene is instead used as a raw material for the production of acrylonitrile. In turn, acrylonitrile can be polymerized into acrylic fiber, which is tough enough for use in outdoor carpets. This chain of chemical events is, of course, driven by economics: the profit margin of polyethylene in the massive and highly competitive market for household plastic film ultimately depends, to a degree, on the sale of acrylic outdoor carpet. On its face, such an item would be expected to represent a relatively small market. However, that problem was overcome when someone realized that green acrylic carpet could substitute for grass on such admirably large areas as baseball diamonds and football fields.

Now let us consider the suitability of the material that has emerged from this petrochemical odyssey, acrylic carpet, as an input to an environmentally sustainable production system, let us say, a baseball diamond. Let us suppose that in the course of planning a new major league baseball stadium, a meeting is held to consider the choice between alternative materials to cover the playing field: grass and acrylic carpet. And since the team owners, we shall assume, have decided to join the ranks of "green" corporations, the discussion includes the environmental suitability of the field covering. With the help of consultants, grass and acrylic fiber are compared with respect to durability, energy required in manufacturing, waste disposal problems, and initial and annual costs. Finally, the players' representative on the planning team brings up an issue that turns out to be definitive: on synthetic carpet, players experience uncomfortably hot feet, sprained big toes, and abrasions when sliding to make a difficult catch. Grass wins, as it has, in fact, in most of the new stadiums built in recent years.

Acrylic carpet—and plastics generally—exemplify a policy made explicit in a history (self-published) of the Hooker Chemical Company, a pioneer petrochemical company now extinct:

Rather than manufacturing known products by a known method for a known market ... the research department is free to develop any product that looks promising. If there is not a market for it, the sales department group seeks to create one.

It is this policy that has driven the petrochemical industry's remarkable growth since the 1940s. The industry has grown by invading *existing* markets—for soap, natural fabrics, paper, glass, and grass—and replacing them with detergents, synthetic fabrics, plastics, and acrylic carpet. And, as we now know, this economically driven process has invaded the ecosystem as well, so that detergents pollute surface waters; plastic film, intended to wrap food, turns trash-burning incinerators into dioxin sources; and a polyurethane mattress, when it smolders, suffocates the sleeper.

Modern environmentally destructive systems of production have arisen not from a failure of design, but from a principle of design that, since it is based only on technical feasibility and economic desirability, excludes from consideration the systems' impact on the global ecosystem on which productive enterprises—not to speak of their customers—depend. As we strive toward sustainable systems of production, it will be essential to incorporate in them materials, working conditions, and forms of energy that are—by design—intended to support the quality of the environment and the welfare of the people who live in it.

Barry Commoner
Director, Center for the Biology of Natural Systems
Queens College, City University of New York

Preface

The origins of this book, though I certainly did not know it at the time, go back 15 years ago to the dusty streets of Bhopal, India. It was 1985 and I had been invited to speak at a conference on chemical plant safety. The conference had been organized in memorial to the terrible night that methyl isocyanate gas had been released from the Union Carbide pesticide production plant in Bhopal, resulting in the deaths of thousands of sleeping neighbors of the plant. Upon the invitation of the conference hosts, I and two other participants agreed to spend a couple of afternoons in the streets of the neighborhoods nearest the plant, where we could listen to the accounts related by the victims of the accident. We sat behind small card tables, often in silence, as family after family came forward to tell us of the horror of that night and about the losses they could count in terms of friends and family members who were killed, neighbors who were disabled, personal health problems, and nightmares that would not end. It was a profound and moving experience. There was little that I or the others could offer except a willingness to sit and listen. When I asked these distraught and grieving people what they wanted me to do, they repeatedly pleaded that I go back to the United States and make certain such an accident could never happen again. Sitting there in the heat and grief, I made a silent promise that I would commit what energies I had in my life to trying to fill that request.

Back at my university, I continued my teaching and research on environmental policy. As that work matured, I was often struck by the way in which my students and colleagues seemed inclined to focus on environmental protection, rather than on the technologies of production that seemed to me to be so much at odds with natural systems. As I came to learn more about the toxicity and hazards of industrial materials, a question kept creeping into my mind that I often asked other faculty and business acquaintances—why were so many industrial materials toxic and hazardous? Sure, pesticides were toxic for a reason, but why were cleaning solvents, pigments, flame retardents, and plasticizers toxic? Was toxicity a result of the function that an industrial material was designed for, or was toxicity some kind of inadvertent property that had simply not been "designed out" when the material was first developed?

As it turned out, I would have lots of time to contemplate this question. Beginning in 1986, I and several environmental leaders in Massachusetts began a campaign to enact a state law that would directly address toxic chemicals. Concerned about the plight of local community residents (such as those in East Woburn) who feared that their drinking water had been contaminated by mismanaged toxic and hazardous wastes, we were convinced that proposed waste incinerators were not an acceptable solution to the increasing volumes of hazardous wastes. Seeking a better solution, we crafted a legislative bill that would encourage the state's manufacturing firms to reduce the volume and toxicity of their hazardous waste streams by reducing the use of toxic chemicals in their production processes. Following the passage of the law, I was invited by colleagues at the University of Lowell (later renamed the University of Massachusetts Lowell) to take over as the director of the university's new Toxics Use Reduction Institute, one of the three state agencies charged with implementing the law. The Massachusetts Toxics Use Reduction Program proved to be a remarkable experiment for testing our hypothesis that the use of many toxic chemicals could be reduced or eliminated by focusing the attention of industry managers directly on those chemicals in the design of their products and production processes. The results have been quite impressive. By 1998, the use of some 190 toxic chemicals in Massachusetts industry had been reduced by 33 percent and the generation of toxic chemical by-products had been cut nearly in half. Indeed, an independent evaluation of the experience revealed that after accounting for all expenses, Massachusetts manufacturers had actually saved money by reducing the use of toxic chemicals.

Still, the question about toxicity persisted. If we could dramatically reduce the use of toxic chemicals and save industry money in one state, then why were these chemicals used at all, and why had they not been eliminated earlier? Why had chemists and materials scientists produced so many toxic and hazardous chemicals, and why had those in manufacturing so willingly bought and used them, even though the chemicals were known to be dangerous? Why had the materials production industries produced such highly effective materials and so defiantly downplayed their hazards? And what about the environmental and public health activists? Had they settled too early, accepting pollution abatement and exposure control technologies rather than agitating for inherently safer and cleaner materials and processes? Was it even technically and economically possible to produce a safer menu of industrial materials? Finally, thinking beyond the immediate problems, was it possible for us to offer our children an array of highly functional chemicals that would be cleaner, safer, and less energy intensive than those that we had been offered by our parents? For several years I wrestled with these questions and engaged my colleagues in endless discussions about them. Finally, fed up with just talking about all this, I found the motivation that drove me to conceptualize, research, and eventually write this book.

I am quite grateful that Robert Gottlieb of Occidental College encouraged me to stop procrastinating and get down to the task of writing, and am even more thankful that MIT Press was interested in publishing the book. I am particularly appreciative that my colleagues at the university, in the Toxics Use Reduction Program, and at the Toxics Use Reduction Institute provided me with a year-long sabbatical during which I conducted most of the research. Over nearly 18 months of writing, a variety of people read drafts, offered advice, and assisted with references. For all of their help and support, I would like to thank Frank Ackerman, Paul Anastas, Nicholas Ashford, Scott Bernstein, David Berry, Halena Brown, Barry Commoner, Pat Costner, Greg DeLaurier, Louise Dunlap, Michael Ellenbecker, Dan Fiorino, Nadia Haiama, Elizabeth Harriman, Don Huisingh, Fran Irwin, David Kriebel, Sheldon Krimsky, Carl Lawton, Charles Levenstein, Gracia Matos, David Morris, John O'Connor, Kirsten Oldenberg, Joanie Parker, Amy Pearlmutter, Margaret Quinn, Mark Rossi, Lyle Schwartz, Neil Seldman, Ted Smith, Randall Swartz, Beverly Thorpe, Joel Tickner, Sukant Tripathy, Hans van Weenen, David Wegman, Bill Walsh, Iddo Wernick, and Rand Wilson.

By focusing directly on industrial materials, it has become clear that we are not facing an environmental crisis so much as an industrial and technological crisis. We have created an innovative and vibrant industrial economy that produces products galore, but it is little accountable for the environmental or health consequences of these commodities. It is a kind of two-dimensional system (cost and performance) when what is needed for a truly sustainable society is a more multidimensional system that is much more socially responsive. The development and management of industrial materials is critical to our ability to survive and prosper. However, we should not accept a materials system that creates tragedies like Bhopal, Love Canal, or Woburn. We must find a better way to ensure that the economy of the future is more respectful of nature and more accountable to all of us who wish to share in its material benefits.

Materials Matter

1
Material Incompatibilities

Materials flow through a system, which is made up of materials, energy, and the natural environment and is governed by man-made institutions such as production, consumption, technology, transportation, and government. What is significant about the interaction between the parts is that they compose a material system; they function like a system and have to be treated by policy-makers as a system.
—National Commission on Materials Policy, 1973

From an environmental perspective, materials do matter. Some materials are exceedingly hazardous to make and use and, once discarded, pollute and contaminate the environment, while other materials are made safely and degrade naturally once disposed. Consider the silk "drop line" that is produced by a common spider. The substance is made from pure protein and water in a gland below the spider's abdomen. It is strong, elastic, resilient, and easily decomposed when discarded. Compared ounce for ounce with steel, the silk drop line is five times stronger, and compared with our strongest plastics, it is able to absorb several times the impact force without breaking. However, unlike steel or plastic, the spider manufactures the drop line at ambient temperatures, under normal pressure, without the use of toxic chemicals, and with no hazardous wastes left over. The feat is enough to incur the envy of any materials scientist.[1]

Every day we use scores of products made from a broad array of materials. Many are naturally occurring substances, mined from the earth or harvested from the land, while others are synthetic materials manufactured in complex chemical cracking and conversion processes. The seemingly endless supply of products that we purchase are assembled from a wide range of these substances. Refrigerators today may contain more than seventy different materials and automobiles are assembled from hundreds of unique substances. Every year the world's industrial enterprises pump out a torrent of products that enrich our lives, support our health, ease our work, and entertain and amuse us. Yet many of the materials in the same products that so satisfy us also create risks to our health and the environment. As we mine, synthesize, process, distribute, use, and, finally dispose of these materials, we generate worrisome threats to the sustainability of the ecological systems upon which we depend.

We do not need these threats. We could enjoy a rich and rewarding supply of products with substantially less impact on the environment and on our health. The enormous wastefulness of advanced consumer economies could be redirected to using and recycling materials more efficiently. By reusing materials in continuous, closed loops, we could significantly reduce the environmental burden of consumer wastes. By paying closer attention to the efficiencies of material use, we could extend the use of materials and better manage their flow though our economies. Of even more significance, we could redesign the physical and chemical properties of our materials and reengineer their uses to create safer and less problematic substances that could be used in more sensitively managed operations.

The premise of this book is quite simple. If we paid closer attention to the materials that we produce, we could pay less attention to the impacts of those materials once they are released to the environment and people are exposed to them. Instead of investing in complex technologies for managing toxic pollutants and hazardous wastes and negotiating complicated institutional systems for permitting environmental releases and enforcing standards of human exposure, we could try to produce safer materials and use them more carefully. While there are probably several criteria that could be used to define "safer materials used more carefully," this book will focus on two rather broad and encompassing factors: toxicity and dissipation. By designing less toxic materials and using them in processes that are less likely to dissipate them into the environment, we could go a long way to creating sustainable materials systems.

In doing this we could learn from the way in which nature produces and uses materials. By more consciously modeling our materials and their uses on processes of nature, we would be more likely to fit our materials needs into the ecological systems by which the planet operates. This trend is already under way. There are government and industrial programs organized to promote more efficient and less wasteful production and consumption processes. Product and materials recycling programs thrive in some countries and some industries. There are research scientists studying how nature makes and uses materials, and there are technologies that perform useful functions with little pollution, material consumption, or energy requirements. These modest experiments point the way to a more sustainable economy, one that more lightly touches the earth's systems. But we need more. To move toward sustainable materials systems, we will need a much more extensive commitment. We will need greater attention—public, private, and governmental attention—to the materials that we use today and those we could be using in a safer future.

1.1 Materials and the Environment

The U.S. economy consumes a vast amount of materials. This includes metals, various nonfuel minerals, wood- and plant-based materials, and a host of synthetic chemicals. A century ago, 161 million metric tons of materials flowed through the national economy; by 1995 the figure was well over 2.8 billion metric tons. This translates into nearly 10 metric tons of raw materials moved per person per year. Over the course of the twentieth century, the rate of materials consumption has steadily increased, with consumption of more than half of the materials occurring during the last 25 years of the century. This includes a 29-fold increase in the consumption of nonfuel minerals, a 14-fold increase in use of metals, and an 82-fold increase in the use of fossil fuel-based synthetic chemicals.[2]

During the past century, we also witnessed a remarkable change in the composition of the materials we use. Traditional materials such as wood, natural fibers, and agricultural materials have been replaced by new, syn-

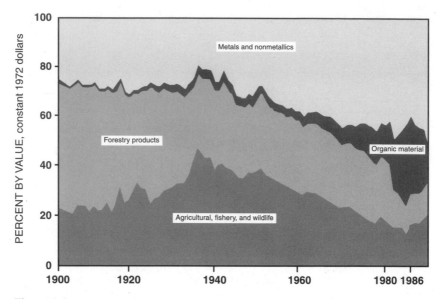

Figure 1.1
U.S. consumption of materials since 1900. Source: U.S. Bureau of Mines, *The New Material Society: Material Shifts in the New Society*, Vol. 3, U.S. Department of the Interior, Washington, D.C., 1991.

thetic materials such as metal alloys, composites, and plastics. Figure 1.1 shows that almost half of the materials consumed in 1900 were based on renewable resources such as wood and plant- and animal-based materials, while by 1990, the consumption of these resources had declined to less than 8 percent. Since World War II, the use of nonrenewable resources such as petroleum-based materials has grown substantially.

Historically, the United States has been a major consumer of the world's resources. With about 5 percent of the global population (270 million people) and 7 percent of the its land area, the United States consumes nearly one-third of the world's nonenergy materials. This is changing. Growth of the world's population and the increasing social and economic development of the rest of the planet's people is slowly eroding this dominance. Today, consumption of materials in the rest of the world is growing at nearly twice the rate as that in the United States.[3]

There are well over 700,000 chemical substances identified in the scientific literature, although fewer than 70,000 are used in industrial production. Of these, the U.S. Environmental Protection Agency (EPA) estimates that 15,000 nonpolymeric chemicals are produced or imported into the United States in amounts of more than 10,000 pounds per year and just over 2800 chemicals are produced or imported in excess of 1 million pounds per year. Each year the materials industries add another 1000 new chemicals to this long list.[4]

The production, use, and disposal of these materials generates substantial costs to the environment. These costs show up in the damage caused during the extraction of materials, in the pollution generated and energy consumed in producing and using materials, and in the disposal of waste materials. The amount of polluting materials emitted into the air is estimated to have doubled and in some cases tripled over the past century, peaking around the 1970s. For instance, sulfur dioxide emissions increased from about 10 million tons in 1900 to about 30 million tons in 1970. Likewise, emissions of volatile organic compounds (VOCs) rose from 7 million tons to over 25 million tons during the 1960s. Both have decreased since the 1970s, unlike nitrogen oxide emissions, which rose to a peak in 1970 and have not decreased since.[5]

In 1997, more than 21,000 industrial facilities in the United States reported releasing 2.58 billion pounds of toxic chemical wastes to the air, water, land, and disposal facilities. Over half of these releases occurred as air emissions, while another 218 million pounds of chemicals were discharged into the nation's rivers, lakes, and bays.[6] These are the figures reported by some of the nation's largest chemical processing firms to the EPA for its Toxics Release Inventory (TRI), but these reports include only a portion of all of chemicals released by literally thousands of other facilities that are not covered by this inventory. Add to these numbers the huge amounts of less toxic wastes generated by commercial businesses and domestic activities, and the impact on the environment is staggering.

The nation produces just over 209 million tons of municipal solid wastes a year, or about 4.3 pounds of waste per person per day. Of this volume, over 16 percent is burned in incinerators and 57 percent is buried in landfills, with the remaining 27 percent is recycled or composted. The country's industries generate some 40.7 million tons of hazardous wastes a year. This works out to about 300 pounds for every man, woman, and child per year. Of this hazardous waste, nearly 76 percent is disposed of on land, while 10 percent is recovered and 9 percent is incinerated. It is expected that over the next decade the total waste stream in the United States will increase by over 20 percent.[7]

Once municipal and industrial wastes are released to the environment, they can cause significant hazards. The landfilling of wastes produces volatile emissions to the air and potent liquid leachates that seep into unprotected groundwater. The incineration of municipal wastes converts solid and liquid materials into hazardous air emissions and toxic ash. Even more potential damage is caused by the extraction and processing of raw materials in the nation's hundreds of mines, smelters, refineries, pulp mills, and chemical synthesis plants. The four primary materials processing industries—metals, chemicals, paper, and plastics—generate 71 percent of the country's toxic air releases, and these airborne industrial discharges can be carried by global air currents to the far reaches of the planet and far up into the atmosphere. In addition, these same four industries create the largest industrial energy demand, which further adds to the planet's pollution. Mining and smelting are estimated to consume between 5 and 10 percent of the world's generated energy supply.[8]

1.2   Problem Materials

Many of the materials we use today have been produced and disposed of for decades with little evidence of concern. Yet, some materials have proven to be quite harmful either because of direct human exposure or, more indirectly, because of their effects on ecosystems. With mounting evidence of significant impacts, various national governments have moved to sharply regulate and even ban the production and use of the most hazardous of these substances. The United Nations International Registry of Potentially Toxic Chemicals lists some 600 substances that have been banned or severely restricted by at least some governments.[9] Pesticides such as DDT, mirex, chlordane, and heptachlor have been banned or heavily controlled in many industrialized countries. Even quite common industrial chemicals such as carbon tetrachloride, chlorofluorocarbons, trichloroethane, and asbestos have been phased out or restricted to only the most limited uses. Efforts to regulate or phase out the use of these chemicals have often been quite complex, costly, and contentious. As an illustration of these struggles, we can consider here three brief case studies on tetraethyl lead, polychlorinated biphenyls, and chlorofluorocarbons.

Tetraethyl Lead

Tetraethyl lead (TEL) was first introduced commercially during the 1920s as an additive to gasoline to improve performance and reduce the audible "knock" during combustion. The chemical was the product of a research consortium set up between the General Motors Corporation, Du Pont Chemical, and Standard Oil of New Jersey. For General Motors, the additive opened a new era in automobile comfort and convenience, and for Du Pont and Standard Oil, the chemical represented a windfall because of its potentially large market.[10]

The toxicity of TEL was well recognized at the time of its first commercialization. German chemists had warned Thomas Midgley, the chief research chemist at the consortium laboratory, about lead poisoning hazards in the manufacture of TEL. Even with this warning, the executives at General Motors and Du Pont decided to proceed, assuming that they could control the hazards in production and that the dilute nature of the additive in the finished gasoline product would not create a significant environmental hazard. During the first 2 years of production, several workplace fatalities associated with exposure were reported, but they attracted little public attention. Then rather suddenly, in October of 1924, five workers died and many others became ill at Standard Oil's Elizabeth, New Jersey, production facility and media coverage of the conditions at the plant first brought the hazards of TEL to public attention.

The reaction was intense, with New York City, Philadelphia, and several states initiating actions to ban the sale of leaded gasoline. Although the U.S. Bureau of Mines issued a report downplaying the occupational and environmental hazards of TEL in 1925, the Surgeon General of the U.S. Public Health Service (USPHS) called for a temporary ban and organized a conference on the hazards of leaded gasoline. While the noted public health advocate, Alice Hamilton, and others argued at the conference that leaded gasoline was both an environmental and an occupational hazard, the conference focused solely on the occupational issues and concluded that a ban on leaded gasoline was not warranted if "its distribution and use are controlled by proper regulation."[11]

Between 1926 and 1977 production of TEL by Du Pont and that by a new joint venture with General Motors and Du Pont, the Ethyl Corporation, rose from 1000 tons a year to over 233,000 tons a year. Although the corporations continued to defend TEL, medical research began to show that many Americans had elevated levels of lead in their blood and that the lead came from gasoline. Still, the decision to phase out TEL was not based on environmental contamination, but rather on the decision by General Motors to comply with the 1970 federal Clean Air Act by installing catalytic converters to control vehicular emissions. The converters selected by General Motors were easily compromised by the lead in gasoline and therefore General Motors finally concluded that leaded gasoline would need to be phased out. Du Pont and the Ethyl Corporation were reluctantly convinced during the early 1980s to abandon their resistance to regulations on TEL and by 1984 the use of leaded gasoline in the United States was discontinued.

Polychlorinated Biphenyls

Polychlorinated biphenyls (PCBs) are a group of aromatic hydrocarbons that have been used as dielectric fluids in electrical transmission equipment, transformers, and capacitors; as oil stabilizers in some heat exchangers and hydraulic systems; and as plasticizers. These chlorinated compounds have been highly valued because they are chemically inert (resistant to oxidation, acids and bases, and other chemical agents), heat resistant, nonflammable, and electrically nonconductive. PCBs were first produced commercially by Monsanto Chemical in 1929 under the trade name, Aroclor, with little suspicion of their potential hazards. While PCBs were also produced in Germany, France, and Japan, Monsanto remained the only U.S. manufacturer of PCBs until 1977, when the company closed down its primary production facilities. Between 1929 and 1977, approximately 1.4 to 1.7 billion pounds of PCBs were produced in the United States.[12]

Monsanto closed out its PCB production because of increasing evidence of health effects in wildlife and mounting government pressure to legally prohibit production. The first scientific evidence of PCBs entering the environment came from two Swedish studies in the mid-1960s.[13] The studies that followed demonstrated that PCBs were released to the environment when electrical equipment or hydraulic systems leaked or were incinerated or improperly discarded. PCBs are highly persistent and resist natural degradation; once released into the environment, they tend to accumulate in the fatty tissue of living organisms.

PCBs were first found in the United States in 1969 in oysters harvested from Florida's Escambia Bay. The following year 146,000 chickens had to be destroyed in New York because they were contaminated with high levels of PCBs. PCBs were soon identified in both wildlife and humans. Ospreys living along the Massachusetts and Connecticut coasts were found to have high PCB concentrations in their body tissue. In 1968 a major human exposure occurred in Kyushu, Japan, where 1300 people became ill from consuming rice oil accidentally contaminated with PCBs. Symptoms that included eye disturbances, skin lesions, and adverse neurological effects were blamed on the accident. Soon studies of nursing mothers in the United States found significant concentrations of PCBs in their breast milk, and several leading public health professionals began to advocate phasing out the use of PCBs. Although Monsanto withdrew PCBs as a plasticizer in 1970, company scientists debated the health studies and argued that the chemicals were effectively safe if used properly in contained systems.[14]

By the 1970s, professional associations and government agencies began to respond. Although acute occupational hazards of PCBs had first been observed in the 1950s, the American Conference of Government Industrial Hygienists now moved to set voluntary threshold limit values for Aroclor. In 1971 the U.S. Food and Drug Administration (FDA) set a residue level for PCBs in food at 5 parts per million. With increasing evidence of both human and environmental risks, a ban on the production of PCBs was inserted into the Toxics Substances Control bill being considered by the U.S. Congress in 1976. Recognizing the inevitability of this legal prohibition, in 1976 Monsanto announced its decision to cease production.

Chlorofluorocarbons

In 1987 the Montreal Protocol to the International Convention for the Protection of the Ozone Layer set out a plan for the phaseout of the production and use of chlorofluorocarbons (CFCs). CFCs are a family of chlorinated fluorine compounds that are generally nontoxic, nonflammable, and chemically inert. CFCs were first developed by Du Pont's Thomas Midgley during the early 1930s and marketed as a refrigerant called Freon. In the United States, Du Pont Chemical became the major supplier, marketing CFCs as substitutes for industrial solvents suspected as carcinogens and as safe substitutes for the flammable gases used as refrigerants. Because CFCs appeared nonhazardous and chemically stable when they were first introduced, there was little concern about their environmental effects.[15]

Initial concern over CFCs did not originate from direct evidence of harm, but rather from a hypothesis put forward by two research scientists, Mario Molina and Sherwood Rowland. In an influential article in the magazine *Nature*, Molina and Rowland hypothesized that once released to the atmosphere, CFCs could float to the upper stratosphere and react with the sensitive ozone layer that protected the planet from ultraviolet radiation.[16] At the time there was no evidence of ozone damage, in part because there was no continuous monitoring of the ozone layer, so many were skeptical of the ozone threat, particularly those who manufactured CFCs.

Still, the Molina and Rowland hypothesis raised the first public concern over CFCs. The initial focus was on CFCs used as aerosol propellants. First Oregon and then the EPA banned the sale and use of CFCs as propellants. Internationally only Sweden, Norway, and Canada followed the United States in this ban. Most countries waited for further studies. In 1977 and again in 1980 the U.S. National Academy of Sciences prepared reports assessing the potential for CFC destruction of the ozone layer. Then in May of 1985 the British Antarctic Survey published a report providing the first clear evidence of a weakening trend from 1979 to 1985 in the ozone layer over Antarctica. Subsequent studies validated this evidence and identified a similar weakening in ozone concentrations over the Arctic as well.

The large CFC manufacturers remained skeptical of the link between CFCs and ozone layer loss, but increasingly government and the scientific communities came to a consensus that "man-made chemicals are responsible for much of the ozone loss."[17] With broad government and scientific support, the Montreal Protocol was signed in 1987, and after the announcement that Du Pont could produce a functional substitute, the protocol was activated in 1989 with the ratification by eleven nations that represented two-thirds of global CFC use. A timetable was worked out at a London meeting in 1990 that set a schedule for the phaseout of all CFC production and use in the industrialized nations by the year 2000.

1.3   Toward a Safer Materials Management System

These three brief cases have several common features. Each demonstrates how innovative research and corporate entrepreneurship led to the development and commercialization of a highly effective industrial material. In each case a darker side of the material came to be recognized and a protracted struggle led to the eventual phaseout of the material's use. Each case reveals the same underlying conflict in social values: product functionality and economic performance conflict with human safety and environmental protection. Such conflicts are common in the history of industrial materials. Performance and cost drive a search for increasingly sophisticated materials, but health and safety and concern for the environment raise cautions and restrain the enthusiasm with which materials are adopted.

The cases are also different. PCBs and TEL were of concern because of their toxicity. Toxicity is a property of a material that is determined by its chemical structure. Many chemicals are quite toxic and many others are less so. When living organisms are exposed to a toxic material in volumes or under conditions that are likely to inflict harm, the substance is said to be hazardous. Thus toxicity is determined by the chemical composition of a substance, while the hazardousness of a chemical is determined by the manner in which the substance is used. Hazardous human exposures to toxic materials can occur in occupational settings where workers produce, use, or dispose of toxic substances and in domestic settings where consumers use products containing toxic chemicals. Both people and other living organisms can be exposed to hazards when toxic wastes are released to the environment, when accidents occur, or when products containing toxic chemicals are discarded. Tetraethyl lead was identified first as an occupational health hazard, and from its earliest production there was evidence of mortalities associated with workplace exposures. Only later was it recognized that the burning of leaded gasoline was likely to distribute small amounts of lead throughout the environment from automobile exhaust. This is more like the problem of CFCs.

CFCs were never identified as hazardous to human health. Indeed, they were marketed as a nontoxic substitute for the ammonia used in refrigeration and the chlorinated solvents used throughout many workplaces. It was not human toxicity that drew Molina and Rowland to hypothesize about environmental damage, it was atmospheric dissipation. The volatile nature of CFCs meant that during manufacture and use a certain amount of CFCs would be dissipated into the atmosphere and slowly float up to the ozone layer, where it could stimulate chemical changes. The world's most concerted effort to phase out a dangerous industrial material was focused, not on a material toxic to humans, but on one that could threaten the sensitively balanced chemistry of the planet.

Dissipation is a term drawn from physics to describe the conversion of concentrated amounts of high-quality materials into widely dispersed substances that are of lower quality because they are so difficult to recapture and reuse. Dissipation of this sort is not readily recognized as hazardous, but when chemicals—even quite benign chemicals—overload sensitive parts of the environment, ecological systems can be compromised and the effects can be life threatening. When extensive dissipation is combined with high levels of toxicity, as occurs in the application of pesticides to agricultural fields or the runoff of hydrocarbons from roadways and parking lots, the combination of factors increases the significance of environmental harm.

In studying the hazards of industrial materials, it is important to consider both the potential toxicity of the substances and the potential for dissipation during their production, use, and disposal. We have long recognized the threats posed by toxic chemicals. There is a substantial literature in the fields of public and occupational health that analyses the complex interaction of chemical structure, human exposure, and biological response. The specialties of epidemiology, toxicology, and pharmacology today provide the core scientific underpinnings of this literature, but the much earlier writings of Alice Hamilton, Ellen Swallow, John

Andrews, Harriet Hardy, and many others identified the basic principles of material toxicity decades ago.[18]

Awareness of material dissipation is a more recent phenomenon. Although concern over the loss of the nation's high-quality natural resources appears in a broad body of literature from the turn of the century on, it is only in the past 30 years that much study has been carried out on the effects of wasted materials widely dispersed into the environment. Much of the analysis in this area has been conducted by chemical, civil, and environmental engineers. During the past decade a new specialty called "industrial ecology" has developed that is focused on tracking materials flows, reducing materials wastes, and creating opportunities for more intensive use of materials during their life cycle.[19]

The cases of tetraethyl lead, polychlorinated biphenyls, and chlorofluorocarbons reveal the process by which industrial materials are adopted and the long and protracted process by which their hazards are recognized and addressed. A more comprehensive review of the history of industrial materials reveals many other similar struggles, although the way in which each conflict arises and is addressed is determined by the characteristics of the specific material and its uses. Nevertheless, the processes are always lengthy, costly, and often contentious, and they are likely to be continuously repeated in the future unless we change the strategies we use to manage industrial materials.

In the United States and most other industrialized countries, these conflicts are institutionalized in the structural relationship between private and public organizations. We rely on private corporations to invent, distribute, and market the materials used to manufacture products. Government agencies are empowered to regulate environmental and human exposures to those materials. These agencies base their regulatory policies on scientific studies of the health and ecological effects of each substance. With nearly 70,000 substances used in industry, this would be an enormous task even if we had enough scientific studies to rely on, but we do not. Most of today's industrial materials are used with an incomplete understanding of their health and environmental effects. Of the chemicals produced and imported into the country in quantities of over 1 million pounds a year, a recent EPA study found that only 7 percent had a complete set of the basic health and environmental screening tests, while 43 percent lacked even one of the most basic studies.[20] This only accounts for the largest-volume chemicals used in the country; we have far less data on the thousands of other substances manufactured and used in much smaller quantities. However, the dilemma is not simply caused by an absence of evidence. Even if we had the data, the task of writing regulations for thousands of chemicals and monitoring and enforcing those regulations would be well beyond the means of any government. Instead, we rely on a lot of good will among chemical users, a lot of concern over liability for chemical damage, a lot of professional denial, and a lot of just plain ignorance.

Fifty years ago, when there were far fewer firms producing far fewer products and using chemicals that had been around for some time, this dilemma may have been less worrisome. However, the scale of production today, the rapidity of industrial development around the world, and the kind of environmental concerns that are arising—climate change, endocrine disruption, biodiversity loss—suggest that the conventional approach to the management of industrial materials is inadequate. The current strategy alone is too uncertain, too burdensome, and too costly to guarantee the level of safety we should desire.

The problems of the current system of materials management are clear: It focuses on one substance at a time; it is dependent on a vast amount of scientific study; it requires a large investment of public and private resources; and it addresses the issue of a material's safety long after most materials have been on the market. This approach focuses too much on identifying and ensuring a nonhazardous level of exposure and not enough on developing a safer system of materials. From this perspective, it appears that a more efficient and effective approach would be to focus on the materials and not on the exposure. Manufacturing less-toxic materials and using them in less-dissipative processes would ensure more safety. Safer materials that are used more intensively and in more contained processes would provide for a more sustainable future.

This is a very exciting time in the development of industrial materials. Our scientific and technical knowledge means that chemicals that were once available only through serendipitous discovery can now be easily designed for quite tailored and imaginative uses. These new materials could be as safe and ecologically sound as they are effective and inexpensive. They could be managed in ways that reduce the wastefulness and dissipation that currently prevails, without compromising the material quality of our daily lives.

1.4   The Objectives of This Book

It is this prospect of a more sustainable materials system that underlies the purpose of this book. In pursuing this goal, the text draws on a large literature on resource management, environmental pollution, and hazardous chemicals, although it differs from much of that work by placing the central focus on industrial materials rather than on the environmental effects of materials or on the economics of resources and wastes. This is a focus that has been less developed until relatively recently.[21] Specifically, the book has three objectives: first, to examine the history of industrial materials in order to identify how today's materials were developed and what efforts were made to respond to their environmental and human health impacts; second, to examine potential future routes for materials development that might be more conducive to health and environmental protection; and, third, to consider what private and public policies could most effectively guide such developments. The book looks to both history and the future in its search for a means of ensuring safer and more effective materials for future generations.

The term *materials* is used quite broadly and extensively in this text. The conventional concept encompasses all those substances, chemicals, and compounds that make up the earth. This book focuses on those materials that are produced and used in human societies. Often the term *materials* is used in this particularly anthropogenic sense (i.e., related to humans), as in "our materials." In particular, the book focuses on industrial materials, which means those materials that are commonly used by industries to make the goods and services that support consumer economies. Materials such as food and drugs and energy materials used as fuels are not covered. Forest and wood products are also excluded, but primarily to reduce the scope of the research. Industrial materials may be products, process chemicals, or wastes. No distinction is made between those materials that end up in consumer products and those that are used to make products but do not show up within the finished products. Thus, industrial materials include raw materials, feedstocks, process chemicals, intermediates, recycled materials, materials as industrial pollutants, materials in production wastes, materials in products, and materials in discarded products.

A large number of materials exhibit toxic properties and there are many different forms of toxicity (carcinogens, teratogens, neurotoxins, ecological toxins, etc.). The poisonous effects of these substances is typically determined by the nature of the exposure, or the dose. Still, not all substances are equally toxic, and the potency of toxic materials varies a great deal. Here the term *toxic material* is used to refer to substances with relatively high degrees of potency in at least some form of toxicity.

In seeking a more sustainable system of materials, the analysis attempts to link the development and use of materials with the international search for more sustainable forms of development.[22] Sustainable development involves economic activities that meet current social needs without threatening the capacity of future generations to meet their own needs. By trying to link these two subjects, the book builds an argument that the materials systems of the past were not sustainable and that the vision of sustainability is a useful metaphor for redirecting our patterns of materials development and use in the future. In doing so, the text recognizes that the materials of the future must continue to meet the conventional criteria of high performance and low cost, but adds to these objectives a commitment to human safety and environmental protection. Specifically, the argument tries to address the dual problems of toxicity and dissipation with two strategies: detoxification and dematerialization.

The term *detoxification* arises out of toxicology. Here the term describes the reduction of the toxic characteristics of materials used in products and processes. This could be accomplished by reducing the volume of toxic materials used in a process or product, or by substituting more benign substances for toxic chemicals, or by changing the toxicity of materials through chemical changes that reduce or eliminate their toxic properties. *Dematerialization* is a term that arises out of the recent work in industrial ecology. It means increasing the intensity of service derived from each unit of material used. This could involve recycling and reusing materials, designing products that use fewer materials, or substituting nonmaterial services for material-intensive products. Moving toward a more sustainable system of materials will require various economic and corporate strategies, but in terms of the materials themselves, these two strategies—detoxification and dematerialization—offer avenues for achieving a safer and more environmentally protective future.[23]

This book attempts to build a foundation for those who are promoting a more sustainable materials future. The subject is very broad, and there was no expectation that the text could cover all of the relevant factors or details. In order to properly focus the book, certain somewhat arbitrary boundaries have been set on the subject, and these need to be acknowledged before proceeding because they are true limits of the book. First, the book focuses centrally on industrial materials and then primarily on metals and chemicals. A more expanded coverage would include nonmetallic minerals, wood, fuels, and perhaps materials used in agriculture, because these materials also contribute to the great material wealth of our national economy and to environmental damage. While it would be useful to consider these other materials, this would have greatly expanded the text.

Second, the subject describes only the experience in the United States. This is a particularly difficult limit because for many years Europe led in materials development and today materials supplies and problems are certainly global in nature. Nevertheless, materials policies are still largely national in scope, and there is at least some justification in focusing on the United States because this country remains one of the most dominant players in shaping materials policies throughout the international economy. Finally, the book focuses specifically on the environmental and public health issues of the development and use of materials. Remaining true to the broadest goals of a sustainable society would require consideration of other factors, such as social equity and justice. Again, while it is regrettable, these aspects have been given little space here.

The book is divided into four parts. The first presents a history of industrial materials and the efforts made by government and private officials to respond to concerns raised over environmental and public health issues. This history provides a perspective on how industrial materials developed and a background for considering how people tried to identify and respond both to the dissipation of these materials and to their toxicity. The second part provides an overview of the primary federal policies developed to address industrial materials and then reviews the performance of those policies as to their effectiveness. This leads to a proposal for an alternative approach that integrates detoxification and dematerialization as policy strategies. The third part considers several avenues of materials development and use, including the familiar ones of recycling and reuse and use of renewable materials, as well as the possibilities of advanced and engineered materials and biobased materials. It assesses the likelihood that these approaches will lead to more sustainable results. The final section sketches out in more detail a set of policy strategies that would lead toward less toxicity in the menu of industrial materials and less dissipation in how they are used.

The story begins in the middle of the nineteenth century. An understanding of the origins of industrial materials and what was known and done offers a beginning for considering how to better direct the materials systems of the future. The period that lasted from the 1860s to the 1980s was an exciting time for industrial materials. During this period, thousands of materials and production processes were invented and patented. Many technologies for extraction and synthesis of materials were invented and refined, and most of the major production operations of manufacturing were established and optimized. The majority of commercial products that we enjoy today were invented and commercialized, and the technologies of waste treatment and pollution control were developed and adopted.

The history of industrial materials continues with a more focused consideration of the way in which the awareness of the environmental and public health aspects of these materials emerged and tended to shape their use. The second part of the book turns more centrally to the evolution of federal policies directed at industrial materials and assesses the effectiveness of these policies. With this background, it is then possible in the final two parts to consider the possibilities of a new and different approach to the development and use of industrial materials in this new century.

I

Developing Industrial Materials

2

Developing Industrial Materials

The Industrial Revolution was merely the beginning of a revolution as extreme and radical as ever inflamed the minds of sectarians, but the new creed was utterly materialistic, and believed that all human problems could be resolved given an unlimited amount of material commodities.
—Karl Polanyi

By convention, archaeologists label early periods of human development by the names of materials that were of significance during the time. Thus, the Stone Age is followed in about 3000 B.C. by the Bronze Age, and the Bronze Age is replaced by the Iron Age around 1000 B.C. While these materials helped to characterize human society during each of these historic times, they also make up a large share of the artifacts that we use today to understand those distant cultures.

Somewhere around 3000 B.C. humans developed the kiln and learned to smelt and use metals. There is evidence of iron axes and implements in use by 1000 B.C., and hydraulic cement was used in the construction of the early Roman Forum. Metallurgy, cut stone, clay pottery, and products made from plant and animal materials characterized much of the first millennium A.D. These materials persisted well into the seventeenth century, and the materials used in 1700 differed fairly little from those used centuries before.

The development of science and the expansion of trade during the eighteenth century, and most of all, the development of modern industry during the nineteenth century changed all of this. Innovations in the use and understanding of materials that occurred during this time laid the foundation for the development of the metallurgical, ceramic, petrochemical, and polymer materials of today. The eighteenth century saw the development of two branches of early chemistry: organic materials—

those based on carbon compounds, such as hydrocarbons—and inorganic materials—those based on noncarbon chemistries, such as metals. New methods were developed for converting minerals into usable inorganic materials, such as sulfuric acid, potassium carbonate, alum, and artificial soda for use in the production of soap, glass, pottery, and leather. But it was the development of the organic chemicals and particularly the coal-tar chemicals, the petrochemicals, and the polymers of the nineteenth and twentieth century that literally transformed the material basis of human society. These organic materials are of a fully different generation than materials that had been used before and are quite recent additions to the human experience.

The development of industrial materials in the United States has a unique history that is heavily structured by the resources of the continent and the social and political conditions of their exploitation. First, the land was rich in resources—forests for lumber; minerals for metals; and coal, gas, and oil for fuel and feedstocks. Second, although U.S. materials inventors lagged behind European contemporaries for nearly two centuries, the United States finally achieved leadership in innovation during the twentieth century. Third, the emergence of the large, integrated, private corporation increasingly set the institutional structure for innovation and production of materials. Fourth, the growth in materials consumption was determined by two factors: the need for physical infrastructure and consumers hungry for commercial products. Fifth, government has been a continuing handmaiden, promoting materials development with laws, trade protections, and resources for assistance and investments. Finally, wars were critical in reshaping materials markets and spurring innovations. The result has been a near revolution in one century in materials production and consumption. Indeed, many of the most common materials used today go back no more than one human generation, and there are many people still living who can remember a world without them.

2.1   European Advances in Materials for Production

Until the middle of the twentieth century, most of the major developments in industrial materials began in Europe and arrived in the United States as intellectual imports. Industrial chemistry was born in England in the middle part of the nineteenth century and matured in Germany around the turn of the century. French, Swiss, and German professors helped to pioneer the early sciences of chemistry and physics, but it was the business entrepreneurs of England that put this knowledge to its most practical industrial uses.

Some substances have been produced in modest quantities since antiquity. These include the mining of salt and sulfur and the production of gypsum and quicklime by the heating of minerals. Vitriol, alum, and sulfuric acid production had begun in the late eighteenth century, and by the turn of the century saltpeter, borax, and sal ammoniac (ammonium chloride) were being imported from the East. Sulfuric acid had traditionally been manufactured by burning sulfur and absorbing the effluent gas in glass flasks of water, but in 1746 John Roebuck demonstrated increased efficiencies by enclosing the whole process in lead-lined, brick chambers. The alkali industry developed in England during the early part of the nineteenth century with the manufacture of soap, glass, and gunpowder. Initially, soda was derived by leaching the ashes of sea plants and potash from wood ashes. After a French druggist named Nicolas Leblanc invented a process for chemically converting salt to salt cake, sodium carbonate, and caustic soda, British entrepreneurs set up alkali production factories in Lancashire, Glasgow, and Tyneside. Salt for the Leblanc process came from brine and the sulfur came from the sulfide residues of local copper, lead, and iron mines.[1]

The modern organic chemical industry arose during the 1860s in the textile-weaving valleys of England. The earliest developments involved dyestuffs, such as "Perkin mauve," a deep purple dye invented by William Henry Perkin using aniline derived from benzol, a distillate of coal tar. During the following decade, scores of British firms were established to manufacture dyes, but by the 1870s the German chemical industry, propelled by the development of new dyes such as alizarin, synthetic indigo, and the azo dyes, grew even more rapidly. The commercial production of these German dyes, which obviated the need for thousands of acres of madder root and indigo production, marks the first large-scale substitution of synthetic chemicals for a naturally grown organic chemical. By the end of the century German producers dominated the global organic chemicals market, while the British industry largely dominated in inorganic chemicals.

2.2   Origins of the Early Materials Industries in America, 1630–1860

The roots of the American chemical industry go back to John Winthop, Jr., the son of the noted governor of the Massachusetts Bay Colony. In 1635, the younger Winthrop set up a crude chemical processing operation in Boston in the back of his pharmaceutical shop to manufacture alum for curing leather hides and saltpeter for the production of gunpowder.[2] While early colonists had made glass, pottery, and brick; evaporated seawater to produce salt; and burned wood ashes to make soap, these were traditional craft production processes. The Winthrop operations were the first to create chemical changes through processes based on an understanding of chemistry. Winthrop went on to set up the first chemical stock company in America and then in 1638, to open a salt works on Boston's North Shore, where in 1648 he was granted a charter to manufacture gunpowder.[3]

Much of the nation's early production relied on the basic materials of the earth, forest, and agricultural fields. Alcohols were derived from sugar fermentation; fibers came from cotton, wool, and flax; cellulose originated from wood; and resins were tapped from tree sap. Charcoal formed during the slow burning of logs provided organic chemicals such as acetic acid (vinegar), acetone (a solvent), and methanol (wood alcohol). Because of its high heating characteristics, charcoal became an important intermediate in the processing of other materials, particularly in the smelting of raw minerals to refine them into metals of commercial value.

The American colonies contained a large variety of natural mineral resources. Iron ore was first found at Roanoke, North Carolina, in 1585 and a lively iron smithing industry arose there by the early part of the seventeenth century. Bog ore was dredged from the bottom of New England ponds during the 1630s. The first iron furnace was opened at Saugus, Massachusetts, in 1645.[4] Lead deposits were first found in the iron ores of Falling Creek near Jamestown in 1621, and in 1682 the French explorer Robert La Salle found lead deposits in the Mississippi Valley. A copper lode was discovered in Massachusetts in 1632, and the first commercial mining of copper ore was begun in Connecticut by John Winthrop who was by then governor of that colony. Ore from this mine and others that opened in New Jersey was shipped to England because by an agreement with the Crown, the smelting of copper was prohibited in the colonies. The British were quite eager to extract the minerals found in the colonies, but so jealously did they guard their prerogative to process metals in England that the Parliament passed the Iron Act in 1750 to reserve all metal finishing for English mills.[5]

A bituminous coal mine was opened in Richmond, Virginia, in 1750 to provide fuel for local blacksmiths, and both bituminous and higher grade anthracite coal were discovered in Ohio and Pennsylvania. The abundance of wood in the colonies inhibited the use of coal as a domestic fuel, but in England, where the forests were nearly devastated, coal found a strong market for domestic heating and as fuel for firing metal smelters. By the mid-1750s large amounts of coal were being shipped from Virginia and Maryland to Britain.

After the American Revolution, independent domestic materials industries began to emerge. The first iron works in the western part of the colonies was established at Bourbon, Kentucky, in 1791. By that time lead ores were being smelted in crude wood fire pits in the Mississippi Valley. A gold mine was opened in North Carolina in 1793, and during the 1820s small gold deposits were discovered in Virginia, Georgia, and Alabama. The first zinc metal was produced at the Washington Arsenal in 1835. The discovery of phosphate rock near Charleston, South Carolina, and later in Florida and Tennessee made the United States a leader in phosphorus production and phosphate fertilizers.[6]

Of the metals, iron had the most profound effects because it became the cornerstone upon which the nation's physical infrastructure could be built. From bridges to buildings to rail lines to vehicles, structural iron was critical. The pig iron and iron casting industry developed all along the East Coast, but by the 1850s the industry had begun to concentrate in northwestern Pennsylvania. Initially, the smelting of iron had required charcoal, which could be manufactured wherever there were trees, but Pennsylvania offered anthracite coal, which could achieve a higher and longer lasting heat. The key was coke. As charcoal is produced from the roasting of wood, coke is manufactured through the roasting of coal. However, unlike charcoal, coke retains its structural characteristics during firing so that air can flow easily through the mineral being smelted, increasing both the temperature and the volatilization of impurities. At the time, coke was produced in a simple domed masonry oven referred to as a "beehive oven" owing to its shape. Because coking was a batch process that required coal to be baked for some 48 hours at a time, a coke manufacturer built batteries of several hundred beehive ovens, each connected through a horizontal flue to a common smokestack. By 1845, the H. C. Frick Coke Company had begun to build large batteries of coke ovens near the coal mines of the Pennsylvania hills, and when iron ore was discovered on Lake Superior's south shore, the Jackson Mining Company began shipping iron to Pennsylvania.[7]

Early in the nineteenth century the chemical industry began to develop from the efforts of many small entrepreneurs such as Charles Goodyear (see box 2.1). As the industry grew, Philadelphia became its center. A druggist named John Harrison began the production of sulfuric acid there in 1801 and soon added white lead. The Du Pont Company was consolidated in 1802 and opened a gunpowder factory on the banks of the Brandywine River. There, Lamont du Pont invented a new blasting powder based on sodium nitrate and sold under the trade name Mammoth Powder. The New York Chemical Company, specializing in vitriol, alum, acids, dyes, and paints opened in 1823, and Eugene Ramiro Grasselli opened a chemical company to manufacture vitriol and alum under his own name in Cincinnati in 1839. By 1850 there were 170 chemical firms in the country selling $5 million worth of products.[8]

Mineral prospecting was primarily a trial-and-error process in which many unsuccessful claims were set and a few fortunate strikes were located, often by mere serendipity. The gold found at Sutter's Mill in California in 1848 was discovered quite by accident when a carpenter named James Marshall, who had been engaged to build a sawmill, decided to dredge out the tailrace and spotted the glittering minerals in the scouring of the current. This discovery began a series of adventurous prospecting rushes that opened up the West and revealed its great mineral wealth. Gold was discovered in Colorado in 1859 and in Montana in 1862, and silver was discovered in Nevada in 1859 and in Idaho in 1864. Between 1849 and 1850, more than 100,000 men traveled to California to pan for gold. In the first year alone, nearly $10 million worth of gold was panned, sieved, or mined from the streams of the Sierra Nevada.[9]

Box 2.1
The Invention of Rubber

With only a meager knowledge of chemistry, much of the early invention of materials occurred though a crude form of empiricism that depended on multiple trials and many errors. The invention that made natural rubber a household product provides a good example.

Since the eighteenth century it had been recognized that the sap derived from the South American *Hevea brasiliensis* tree could produce a moldable resin called "gum rubber." Both British and American entrepreneurs had tried to develop gum rubber for waterproofing textiles, and in 1824 Charles Macintosh, a Scottish merchant, had succeeded in producing a popular raincoat from the material. However, the further use of gum rubber was limited by its undesirable habit of softening in warm weather and becoming brittle in cold weather.

Charles Goodyear, a hardware merchant from Philadelphia, had spent years mixing natural rubber with various compounds in an effort to overcome the temperature sensitivity of gum rubber. In 1837 he succeeded in acquiring a contract with the federal post office for rubberized mail pouches, but his efforts went bankrupt when the heat in his warehouse melted the products. Undeterred, Goodyear continued his experiments. Yet, it was an accidental spill of rubber mixed with sulfur on a hot kitchen stove in 1839 that finally revealed to Goodyear a process that stabilized the rubber. By 1841 Goodyear had developed this sulfur-based process, which he dubbed "vulcanization" after Vulcan, the Roman god of fire and metal working, and scaled it up for industrial production. Unfortunately, Goodyear's poor management skills kept the business near continual bankruptcy.

Source: Charles Morrow Wilson, *Trees and Test Tubes: The Story of Rubber*, New York: Henry Holt and Company, 1943.

The Civil War interrupted much of this early prospecting by diverting men into military service, but the war also greatly expanded markets for metals and chemicals. Copper, iron, and lead production were greatly expanded by both the Union and the Confederacy, and the material needs of the war spurred the production of gunpowder, fertilizers, pharmacueticals, leather, and textiles.

2.3 The Early Period of Industrialization, 1860–1914

The period that followed the Civil War was marked by a rapid expansion of the national economy. For the nation this was a period of intellectual ferment, scientific discovery, and technical invention. The steam engine emerged as a source of transportation and industrial power. Hydro-electric power was first generated commercially in 1877 at Niagara Falls, and during that year Nicholas Tesla developed the rotating magnetic field that made long-distance transmission of power possible. Two years later Thomas Edison invented the first incandescent light bulb, and very rapidly electricity became essential to the nation's emerging economy.

Between 1860 and 1910 the U.S. population nearly tripled, and with it grew an increasingly affluent domestic market for industrial products. During the 1880s the value added to the economy by manufactured goods first began to exceed the value added by agricultural products. By 1890, the United States surpassed Great Britain in the volume of its industrial output and became a recognized world leader in manufacturing. In large part, this heated growth was a result of the ample supply of the natural resources of the land. The basic industrial materials—coal, copper, lead, iron ore, and lumber—were abundant, easily accessible, and cheap. Since the county lacked the larger markets of Europe and lagged behind Britain and Germany in technological sophistication, the supply of natural materials was a critical factor in advancing U.S. industrial capacity.

Iron remained the primary material. By 1860 there was a large and well-established iron industry, with ironworks spread throughout 20 states, although nearly 60 percent of the nation's iron output came from more than 125 blast furnaces in Pennsylvania. By contrast, the steel industry was much smaller, with only 13 establishments producing less than 12,000 tons of steel a year.

Steel is an iron–carbon alloy derived by smelting pig iron to drive off impurities. In 1856, Henry Bessemer was granted a British patent for a pneumatic production process that forced compressed air through molten iron in order to raise its temperature high enough to oxidize contaminants such as silicon and magnesium. The first successful American steel mill using the Bessemer process was established in Michigan in 1864, just in time to meet the rapidly rising demand for steel railroad rails. The introduction of the Bessemer steel process greatly increased the speed and scale of steel production, and output jumped from 42,000 tons in 1870 to 1.2 million tons in 1880, exceeding the production of wrought iron in that year. Like iron, steel required coal and coke. In 1880 there were 186 companies producing nearly 3.3 million tons of coke; within a decade there were 253 firms producing 12 million short tons.[10]

By the 1890s the United States had become the world's largest producer of Bessemer steel. However, the Bessemer process was not unchallenged. In 1868, a British engineer, William Siemens, developed a regenerative steel production process called the "open hearth" process (because the hearth was visible during firing), which increased the heat by recycling back into the molten iron the superheated gases driven off from the initial heating. Unlike the Bessemer process, the open hearth process liberated phosphorus from the molten iron and could accept a reasonable amount of scrap iron as feedstock. This was particularly attractive in the United States, where the iron ore was rich in phosphorus and scrap iron was plentiful and cheap. In 1888 the first open hearth steelmaking process was established in Pennsylvania and the process was rapidly adopted throughout the country. During the first decade of the 1900s, the open hearth process became the nation's dominant steelmaking process.[11]

The early chemical industry increased in number of firms and volume of production during this period. Between 1850 and 1914 the total number of manufacturing establishments grew from 170 to 633 and the value of products grew from $5 million to $222 million.[12] Still, the emerging glass, textile, and consumer products industries relied on British imports for inorganic chemicals such as acids, bleach, and caustic soda, and on German imports for dyestuffs and other organics. The larger European markets and the low transatlantic transportation costs kept the price of British and European imports low enough to inhibit the development of a chemicals industry in the United States. In inorganic chemicals, the British Leblanc technologies were well established and the Britain-to-America freight rates were cheap since shippers could usually rely on a return cargo of grain. Only after the 1880s did the inorganic chemicals industry assume more accelerated growth, aided in part by the protection of government tariffs enacted by Congress in 1897. Slowly, the dependence on imports for inorganic chemicals diminished, although reliance on Germany for dyestuffs continued right up to World War I.[13]

The Invention of New Materials and Processes

Technological development was driven by changes in the economy. The growing consumer market and a tight labor market impelled industries to seek technological advances that increased labor productivity. Product innovations and process improvements created opportunities to manufacture cheaper goods with fewer employees. Two major technological advances that occurred around the turn of the century—the by-product recovery oven and the electrolytic cell—not only drove the development of new materials, but more important, promoted significant process improvements and the development of the field of chemical process engineering.[14]

Until the 1900s, the development of the U.S. organic chemical sector was limited by the widespread use of the beehive coke oven. Since its introduction, coal distillation in the beehive oven had become critical, not only for making coke for the iron and steel industry, but also for making a gaseous fuel as a by-product. This gas, called "town gas," found a ready market in the lamps used to illuminate city streets and homes. However, the beehive ovens were particularly inefficient. For every ton of coal converted to coke, a ton of coal was burned as fuel. A significant amount of heat was lost as by-product; worse still was the loss of potential resources as the impurities in the coal were volatilized. During the mid-1800s European engineers became particularly interested in the constituents of this coal gas and they developed a new coking oven called the "by-product recovery oven," which was rapidly adopted in Europe during the 1870s and 1880s. Unlike the beehive ovens, the by-product recovery ovens produced a coke from the more ubiquitous and lower grade bituminous coals of the European continent and captured for later use the volatile tars, ammonia, and aromatic hydrocarbons (benzene, xylene, toluene, and naphtha) emitted during combustion.

The recovery of these by-products proved most fortuitous in Europe. By recovering the by-products of coke production, the recovery ovens provided the foundation for the development of the European organic chemical industry. Since they had significant investments in batteries of beehive ovens, U.S. investors were slow to adopt the by-product recovery ovens, and over the years it is estimated that roughly $20 million a year was lost in potential revenues. Only when the European models were fully adopted in the United States after 1910, largely for the production of municipal gas, did coal gas become a feedstock for the emerging domestic organic chemical industry.[15]

The by-product recovery oven was equally important in the development of alkali chemicals. Alkali chemicals such as sodium carbonate, sodium hydroxide, and potash are inorganics that are more caustic than acidic. During the mid-nineteenth century, the conventional Leblanc process for caustic soda production was replaced by the Solvay process. This process was named for two Belgian brothers, Ernest and Alfred Solvay, who invented a procedure for producing soda by reacting sodium chloride first with ammonia and then with a carbon-based material such as limestone or carbon dioxide. The process required an inexpensive source of ammonia and here the Solvays turned to the town gas companies and their by-product recovery ovens, for it was possible to manufacture ammonia cheaply from the coal distillation by-products. In 1881 two Americans convinced the Solvays to build an ammonia/soda ash plant in the United States. With the Solvays as partners, they formed the Solvay Process Company (later the Semet-Solvay Company) and built their first plant near Syracuse, New York. It was here in 1892 that the Solvay Process Company built the first by-product recovery ovens in the United States, which were used to produce ammonia. These ovens also produced an excess gas that was used to illuminate lamps in the plant. Again, this by-product became as important as the coke. By 1898, the New England Gas and Coke Company had built a battery of four hundred by-product recovery ovens at Everett, Massachusetts for the production of municipal illumination gas.[16]

The other important technology was the electrolytic cell. The use of electricity and the development of electrolytic reduction was critical to the development and refinement of several materials. Copper plating by electrolysis was first tried in the 1840s and afterward there arose a small electroplating industry in Europe and the United States. However, advances in both electrochemistry and electroplating were inhibited by the low voltages available from existing dynamos and voltaic batteries. The development of hydropower during the late nineteenth century provided the electricity needed for electric furnaces and larger scale electrolytic processes. In 1886 Charles Martin Hall of Oberlin, Ohio, developed the country's prototype electric furnace for smelting ore when he discovered that an electric current sent through a bath of alumina and molten cryolite could liberate the aluminum. In 1888, Hall founded the Pittsburgh Reduction Company (later, the Aluminum Company of America) and began production of an aluminum for applications in artificial limbs, bathtubs, cookware, tableware, and lightweight mess kits.

Electrochemistry opened new frontiers in inorganic chemistry by using energy to break apart natural chemical compounds. In the late 1880s E. A. LeSeuer established the Electrochemical Company in Newton, Massachusetts, to produce caustic soda from salt brine using an electrolytic cell, and in 1892 he opened the nation's first electrolytic alkali plant at Rumford Falls, Maine.[17] Bleaching powder (chlorine) was first produced by the electrolytic cell by the Mathieson Alkali Company in 1896 and by Herbert Dow, who had set up a plant at Midland, Michigan, to exploit the salt brine deposits there. Dow began licensing the electrolytic cell in 1895 and 2 years later he founded the Dow Chemical Company with a product line that soon included bleach, caustic soda, sulfur chloride, carbon tetrachloride, chloroform, and sodium chloride. Hooker Chemical opened a similar chlorine production facility at Niagara Falls in 1906. These electrochemicals required cheap electric power, which was available from the hydropower plants at Niagara Falls and from coal- and wood-fired plants on the Great Lakes. Several firms—Grasselli, Kalbfleisch, and Nichols—began producing manufacturing-grade sulfuric acid during the 1880s, although it was not until the New Jersey Zinc Company adopted the contact catalytic process in 1899 that American firms competed well with European sources.[18]

During the latter part of the nineteenth century, the development and diffusion of chemistry and engineering knowledge had advanced enough that some innovations in materials were developed by building on an understanding of other innovations. This was true in the development of plastics, which moved from cellulosic to synthetic chemicals through the invention and substitution of increasingly more versatile and inexpensive polymers. John Wesley Hyatt, an Albany printer, is credited with inventing the first commercially viable cellulose-based plastic, which he called Celluloid. Hyatt was experimenting with a process developed earlier by a British inventor, Alexander Parkes, when he discovered a transparent film had formed from an accidental spill of a common printing staple, collodion. By trial and error, Hyatt found that dissolving cellulose nitrate in ether and ethyl alcohol gave the material a moldable quality that made it useful for coatings, lacquers, and photographic emulsion. Hyatt patented his process in 1870, and Celluloid soon appeared in combs and brushes and cheap jewelry. Eastman-Kodak adopted Celluloid as the base material for photographic film in 1884, but its high flammability (Hyatt's production facility burned to the ground in 1875) stalled its further development.[19]

Further development of plastics involved a search for safer polymers through incremental improvements in the natural polymers (cellulose) that formed the backbone molecular structure. This led to the development of cellulose acetate, which was used in the production of rayon fiber and became a staple of fine textiles, although it could not take dyes easily. The development of a fully synthetic commercial polymer occurred with the invention of Bakelite in 1907 by a Belgian-born chemist named Leon Baekeland. Baekeland set out to develop chemical coatings and had a strong interest in the reaction between phenol and formaldehyde, which seemed to form a large molecular backbone similar to cellulose. As he pursued the possibilities of a phenol–formaldehyde substitute for shellac, Baekeland discovered that one experimental resin had formed a dense, hard ball, which he further refined to form an easily produced plastic. Recognizing the commercial significance of his invention, Baekeland established the General Bakelite Company in 1910 and opened a production facility at Perth Amboy, New Jersey. The new plastic rapidly became an economic success in products ranging from combs to radio insulators.[20]

As materials production developed, process innovation became as important as product innovation. Herbert Dow was a master at process improvements. Dow had set up the Dow Chemical Company in order to produce chlorine for the American bleach market, but for the next 2 years the operations were plagued by frequent explosions in the electrolytic cells and serious leaks of electric current. During this period Dow and his associates carried out scores of experiments, not to invent a novel material, but to rectify and optimize a production process so that an existing material could be produced in a stable process and at a marketable price.[21]

Innovation was an important engine of materials development. With so many inventors often working on the same processes, it was critical to develop a sophisticated patent system for clarifying ownership of intellectual property, resolving conflicts between entrepreneurs, and ensuring the opportunity for a hefty return on speculative capital. By the later years of the nineteenth century, the U.S. Patent Office and its counterparts in Europe had become the critical starting gate for the enormous waves of innovations that propelled the expanding materials markets.

The Maturing of the Materials Industries

By 1910 American chemical production—measured in dollars or tons—was greater than the combined output of England and Germany. Yet this industry remained largely an inorganic chemical industry. In 1914, the total share of organic chemicals (such as plastics, coal-tar products, and nitrogen compounds) was about one-quarter of the total value of the chemical industry as a whole. Until World War I, Germany remained the leader in organic chemicals. Although Charles Goodyear is credited with inventing the sulfur vulcanization process to make natural rubber stable (see box 2.1), England took the early lead in the production of vulcanized rubber and the output there grew by 6000 percent between 1840 and 1900. The first rubber bicycle tires were manufactured in England during the 1840s, and by the close of the century John Boyd Dunlop and Andre and Deoudard Michelin were producing rubber-based pneumatic tires for the emerging automobile industry.[22]

The petroleum industry developed and matured during this period. Pennsylvania oil was discovered in 1859, and production rose from 450,000 barrels in 1860 to 3 million barrels in 1862.[23] In an effort to stabilize oil prices, John D. Rockefeller organized the Standard Oil Trust in 1882 to consolidate all oil refining and distribution in Pennsylvania. In 1889, a German chemist named Herman Frasch discovered that refining crude oil in the presence of copper oxide could separate out the sulfur that plagued some midwestern oil fields and this rapidly brought online oil supplies from Ohio and Indiana. Texas oil was discovered at Beaumont in 1901, and for some time the country had plentiful sources from at least three regions to draw upon. Initially, kerosene used for illumination was the primary petroleum product, but oil was also used to make gasoline, naphtha, fuel oil, lubricants, and a special petroleum jelly called Vaseline. Natural gas was a by-product of oil extraction, but because it was volatile and difficult to transport, it was often burned off (flared) at refineries, even though a more expensive town gas with similar properties was being manufactured from coal in many cities. By 1905 the gasoline-powered internal combustion engine had beat out steam and electric engines as the dominant power source for automobiles and thereafter gasoline production became the driving force for the pumping and refining of oil.

This period also witnessed a major expansion in the mining and minerals industry. In 1859 Henry Comstock found silver in the Washoe, Nevada, area east of Lake Tahoe, and the discovery of rich deposits of both precious and base metals continued up through the end of the century. Gold and silver were discovered in Colorado during the 1850s and copper and lead during the 1870s. A highly profitable ore smelting industry arose around Denver, Pueblo, and Leadville for smelting silver and lead. Silver, lead, and copper ores were discovered in Utah in the 1860s and a smelter was established at West Jordan in 1864.

By 1880 there were two leading mining centers for iron ore—one in Missouri and the other in the Mesabi/Lake Superior region. The lower phosphorus content of the Mesabi ore was more attractive to the steel interests, and the opening of the Soo Canal made Lake Superior ore readily available to the Pennsylvania smelters. In the Pennsylvania mountains, the H. C. Frick Coke Company set up nearly 1000 beehive coke ovens to turn coal into coke. The availability of iron ore from the Lake Superior region, the cheap Pennsylvania anthracite coal, and the massive amounts of available coke made Pittsburgh the center of the steel industry, and by the end of the decade the region boasted four large rolling, plate, and wire mills.[24] As with chemicals, the iron and steel industry grew with assistance of federal tariffs such as those established in 1871 to protect the railroad steel rail market from British producers.

The first government permits for copper mining on Michigan's Upper Peninsula were issued in 1844, and in that year the Pittsburgh and Boston Company opened the Cliff Mine on the north shore of the peninsula. Copper was discovered in Idaho in 1885 when a prospector who had spent years prospecting for gold on the Coeur d'Alene River stumbled across an outcropping of galena while trying to retrieve a strayed burro.[25] The Utah Copper Company was organized in 1903 and constructed a large smelting and processing plant on the shores of the Great Salt Lake. The St. Joseph Lead Company uncovered the rich lead ores of southeastern Missouri and production there rose rapidly from 20,000 tons in 1871 to 260,000 tons by 1900. Arsenic trioxide was first produced as a by-product of smelting gold and silver ores at Everett, Washington, in 1901. By the turn of the century, the United States had become the world leader in zinc production, a position it held for the next 70 years. Production of cadmium from zinc became commercially viable when the Grasselli Chemical Company began marketing cadmium as a by-product of its zinc production at its Cleveland smelter in 1907.[26]

Long before the Treaty of Guadalupe Hildago gave the Colorado River Basin to the United States, gold, silver, and copper had been identified in the plateaus and mountains of this great, arid region by the Jesuits and other early Spanish explorers. In 1863 Henry Wickenberg opened a gold mine in Arizona along the Hassayampa River and 3 years later he sold the mine to Benjamin Phelps. Silver was extracted from the Silver King Mine opened in the area around Globe, Arizona, in 1875. The first bars of Arizona copper came from the mines at Morenci in 1875, but it was the investments of Phelps, Dodge and Company at the Clifton mines and at the Copper Queen mine at Bisbee that generated the largest supply of copper from Arizona.

The Northwest also proved rich in minerals. Gold was found in the rivers and creeks of Washington and Oregon and lead was discovered at Coeur d'Alene, Idaho, in 1885. The Bunker Hill and Sullivan Mining Company was incorporated at Coeur d'Alene in 1887 and quickly commenced construction of one of the nation's largest smelters. Gold, silver, and copper were also discovered near Butte, Montana, during the 1860s and 1870s. By 1900, the Anaconda Mine in Montana was producing nearly 50,000 tons of copper each year. As the new century opened, the United States was a leading producer of coal, copper, iron ore, zinc, phosphate, molybdenum, lead, tungsten, and many other minerals.

This is a period marked by significant industry restructuring. During most of the nineteenth century the mining and chemical industries were composed primarily of small, family-owned and managed businesses. Improvements in transportation and the integration of the rapidly growing domestic market following the turn of the century resulted in the growth in size and dominance of some firms at the expense of many others. During the 1880s the Illinois Steel Company was formed as a merger of several small mills, and the Carnegie Steel Company grew through the outright purchase of other companies. The Mesabi iron ore mines were merged into the Lake Superior Consolidated Iron Mines Company in 1893. Between 1902 and 1904, three du Pont cousins merged their operations and consolidated the explosives market to form the predecessor of the Du Pont Corporation. By the second decade of the century, corporations like Carnegie, Standard Oil, and Du Pont dominated the rapidly growing steel, oil, and chemicals markets.

2.4   The Interwar Period, 1915–1945

The years from 1915 to 1945 marked a second great period of industrial development and witnessed rapid growth in the materials industries. If the period up through 1915 was focused on the development of structural materials and inorganic chemicals, the period that followed World War I was focused on organic chemicals. The great wealth of natural resources remained a critical factor in this growth, but increasingly important was a growing sophistication in research and technological development. Although there were important new innovations, the focus of research and development rested primarily on adapting and improving upon European technologies.

World War I stimulated industrial expansion. The constricted flow of foreign raw materials during the war and the surge in materials demands greatly extended the mining of metals such as copper, chromite, and tungsten. War demands pressed the iron, copper, and petroleum industries for rapid expansion. The demand for iron and steel was unprecedented. In 1916 alone, U.S. mills produced 39 million tons of pig iron valued at $633 million. A barrel of high-grade Pennsylvania crude oil jumped from $1.88 to $4.00 between 1914 and 1918, and copper doubled in price between 1914 and 1916.[27] Du Pont responded to war needs with new factories for the production of munitions, and Dow increased its production of phenol for explosives.

The war revealed the vulnerability of America's dependence on other countries for many materials—some quite critical to national security. The most important of these were the German coal-tar chemicals required for the manufacture of munitions, dyes, fertilizers, and medicines. Critical to the war effort was the need to develop explosives, and the most effective explosives were based on trinitrotoluene (TNT) and picric acid, both of which could be best produced through the distillation of coal. At the beginning of the war, the Semet-Solvay Company began a crash construction program of coal by-product recovery ovens to produce the organic chemicals needed to manufacture explosives for European sales. By 1917 the company was under contract to the U.S. government to build and operate an enormous integrated explosives plant at Split Rock, New York, where a significant amount of Allied munitions were manufactured.[28] Following the war, U.S. firms turned to these by-product recovery ovens for the production of coal-based chemicals such as dyestuffs, ammonia, and pharmaceuticals, and the domestic organic chemical sector grew significantly, aided in part by government tariff programs.

The war sharply increased the demand within the country for industrial chemicals and permitted U.S. producers to gain greater access to international markets. The total value for national output increased from $221 million in 1914 to $730 million in 1919, while the number of producers rose from 633 to 1053.[29] Hostile relations with Germany presented an opportunity for the U.S. government to expropriate German intellectual property and openly violate German patents. The government established the Alien Property Custodian's Office, which confiscated German properties in the United States for resale to U.S. businesses and worked through the Chemical Foundation to parcel out confiscated patents at nominal fees to U.S. companies.[30]

The government also set up chemical factories to exploit European technologies, including an ammonia processing plant at Muscle Shoals, Alabama, based on a German process developed by Fritz Haber and Carl

Bosch. Haber and Bosch had worked in secret to develop a synthetic process for fixing nitrogen from air that could then be used to make ammonia without coal. Because the Haber–Bosch process had proven so effective in producing cheap gunpowder from synthetic nitrates and nitric acid during the war, the Allied forces were eager to learn the procedures and adopt the process for their own industries.

By the close of the war, U.S. companies had a wealth of new product ideas and substantial financial resources for investments in research and development. Between 1919 and 1936, well over 300 corporate research and development laboratories were established in the chemical industry.[31] Du Pont, among others, heavily recruited German scientists. Many young Americans enrolled in German and Swiss universities for training in chemistry and physics and returned with well-crafted skills and ideas. During these years, new chemical engineering programs were established at the Massachusetts Institute of Technology, Columbia University, and the University of Wisconsin.

Nylon, one of the most successful early synthetic resins, emerged from one of these new laboratories at Du Pont Chemical in 1934. Early work had been pioneered by a German chemist named Hermann Straudinger. Wallace Carothers, a contemporary of Straudinger, had come to Du Pont to establish a high-quality research department There, a research team headed by Carothers developed a synthetic rubber, which was named Neoprene by Du Pont and marketed as a substitute for the natural material. However, Carothers was more deeply interested in creating polymer fibers that could compete successfully with cotton and rayon. One afternoon in 1934, one of Carothers' assistants pulled a stirring rod out of a beaker containing a reaction of dibasic acids and diamines and found it trailed a long, ductile fiber that shortly hardened into a silk-like fiber. This strong thread was named Nylon by Du Pont and within 3 years it was being sold as woven fabric, stockings, climbing rope, and industrial-grade fabrics.[32]

This was also a period of increasing industrial integration. Sales of cars became a pivot for the chemicals and automobile industries as the demand for cars spurred a demand for both iron and steel and chemicals such as lacquers, paints, rubber, and plastics. Automobile use also increased the demand for petroleum. Prior to the internal combustion engine, oil was primarily used for illumination and lubricants. The demand for gasoline drove the development of oil drilling and refinery technology. Simple distillation of oil had been improved during the 1920s with the development of thermal "cracking" (heating in a vertical tube) to break up the oil into various fractions and separate out the lighter fractions (primarily gasoline) for the fuel market. New catalytic techniques for fractionating oil began replacing thermal cracking in petroleum refining during the late 1930s. The search for higher octane gasoline with an antiknock additive resulted in the formation of the Ethyl Corporation for the manufacture of tetraethyl lead. The Koppers Company, a builder of coke ovens, began forming partnerships with public utilities to build coke ovens that sold gas to the municipal utilities and coke to local industries while Koppers retained the coal tars for the production of paving and roofing materials, dye intermediates, and aromatic hydrocarbons. The growth of the consumer product market, particularly in radios, records, cameras, and photographic films, also spurred the growth of the chemical market. The chemical industry was itself becoming a large market for chemicals used as intermediates and catalysts in the processing of finished chemical products.

This period was also marked by substantial corporate consolidation. The war had demonstrated the importance of large-scale industries capable of continuous operations and well-organized research divisions, and this required big companies, corporate administrative organizations, and stable markets. The days of the family-owned, privately financed companies were over. In 1920 the Semet-Solvay Company, the largest alkali chemical producer, merged with several other inorganic chemical manufacturers to form Allied Chemical and Dye Corporation, which at the time became the nation's largest chemical company. By 1930 Anaconda and Kennecott Copper controlled 34 percent of the world's copper production, while in petroleum five large companies led by Standard Oil and Royal Dutch Shell supplied 35 percent of the world's oil.[33]

Recognizing the problems created by rapid fluctuations in the price and supply of oil and kerosene in the United States, John D. Rockefeller had set up the Standard Oil Trust Agreement in 1882 and the Trust soon grew to control a significant proportion of the domestic and international petroleum markets. The iron and steel industry had likewise been concentrated into U.S. Steel by J. P. Morgan, and the aluminum industry was organized by the Mellon family into a virtual monopoly called the Aluminum Company of America (Alcoa). In the European chemical industry, cartels such as the German Nitrogen Syndicate and I. G. Farben arose in Germany, and in Britain the United Alkali Company and Imperial Chemical Industries (ICI) emerged to dominate their national markets.

However, in the United States, such monopolies were viewed with suspicion and skepticism. In 1911 federal authorities challenged the Standard Oil Trust under the Sherman Anti-Trust Act and after lengthy litigation finally forced it to break up into several large companies.[34] By the 1920s, as the chemical industries began to seriously expand, the lessons of the Standard Oil Trust were well acknowledged. Thus, instead of monopolistic cartels, the U.S. materials industry grew by mergers and acquisitions into a set of large, highly competitive corporations. In addition to some five hundred mergers in the industry during the 1920s, this involved large-scale diversification and consolidation through stock ownership and holding companies.[35]

Up until World War II, the chemical industry was dominated by Allied Chemical and Dye, Du Pont, and Union Carbide and Carbon. Du Pont built up its nitrocellulose-based products, including paints, varnishes, artificial leathers, and rayon and became the largest producer of synthetic ammonia and synthetic nitrates. Through a loosely affiliated collection of process divisions, Union Carbide grew to dominate the markets for aliphatic chemicals, liquid gas, ferroalloys, and carbon products. A second tier of rapidly growing companies arose to fill available market niches. These included American Cyanamid, Monsanto, Dow, Kodak, and Merck. Dow expanded markets for calcium chloride, magnesium chloride, and magnesium and developed new silicone chemistries that led to the establishment of a joint venture with Corning Glass in 1942 for silica-related products.[36]

World War II

Just as World War I had revealed the U.S. dependence on German organic chemicals and provided both the commercial market and the government support for the mining and chemicals industries, World War II played an equally impressive role in promoting the domestic materials industries. The war revealed how critically the nation's security rested on access to materials. The Japanese advances in the Pacific were particularly damaging. In a short time the United States lost its access to a quarter of its supplies of chromite, three-quarters of its imports of tin, and almost all of its supplies of natural rubber.[37] As wartime needs depleted some of the nation's richest mineral reserves, formerly unprofitable mineral deposits became more valuable. Military requirements demanded enormous amounts of copper, aluminum, tungsten, manganese, and magnesium for steel production, shipbuilding, aircraft, and electronic devices.

As the war intensified, the nation imported lead, zinc, copper, antimony, and mercury from Mexico; copper and nickel from Canada and Chile; nickel from Cuba; vanadium from Peru; petroleum from Venezuela; and bauxite from Dutch Guiana. At home, the government used an inflated "premium price policy" to encourage production of marginal copper, lead, and zinc ores. Technological developments helped produce domestic bauxite from low-grade Arkansas ores, alumina from clay and alumite, and magnesium from dolomite. Indeed, new high-grade copper ore was discovered near Tucson, Arizona; a new cobalt/copper supply was found in Oregon by the U.S. Bureau of Mines; and private operators opened a new deposit of tungsten in Idaho.[38]

During World War II, the federal government became an aggressive partner in developing the chemical industry. Following the Japanese blockade of exports from the natural rubber plantations of Malaysia, it established the Rubber Reserve Program, a crash program among the leading rubber and chemical companies, to develop a synthetic rubber substitute. In Germany, I. G. Farben had developed a synthetic rubber called Buna-N made with 25 percent styrene contained in a butadiene chain. The Rubber Reserve Program called upon the nation's four leading rubber manufacturers and Standard Oil to set up new operations for the production of synthetic rubber based on styrene–butadiene. Because of its early experience in styrene production, Dow Chemical was put in charge of the program and U.S. production of styrene, which had started at 2 million pounds per month in 1941, was ratcheted up to close to half a billion pounds by the end of 1944.[39] Through its enormous procurement capacities, the federal government drove the development of domestic petroleum refining. An insatiable demand for high-octane aviation fuel forced the petroleum industry to adopt and integrate new catalytic cracking and alkylation technologies in its refineries.

The war called for new hardware and sophisticated technologies in communications, transportation, and weapons. The development of synthetic rubber was followed by rapid advances in polymers, petrochemicals, and metal alloys. The war also spurred the development of the materials sciences. Research in solid-state physics, crystallography, physical metallurgy, ceramics, and the polymer sciences was heavily funded by military services during the early 1940s, and these initiatives became important to the development of the new field of materials science and engineering.

The Growth of the Petrochemical Industry

Work on petroleum-based chemicals, or petrochemicals, had begun in both Europe and the United States during the late 1920s. Prior to World War II, oil and chemical companies in the United States had begun to explore the various hydrocarbon products that could be derived from refinery offgases and natural gas feedstocks. Many organic chemicals can be derived from different feedstocks: coal, alcohol, oil, or natural gas. Germany, with ample coal reserves, continued to rely on coal and coke-oven by-products to support its organic chemical industry, but the abundance of oil and natural gas and the large oil refining capacity in the United States meant that the organic chemical industry was increasingly directed toward petroleum as its basic feedstock. Thus, the by-product recovery oven was replaced by the oil and gas refinery as the principal technology for converting fuel minerals into industrial materials. As the leading petroleum firms invested in large cracking and reforming facilities in order to improve fuel quality and yield, they discovered that they could readily produce valuable chemical raw materials as well. Sixty years earlier it had been the by-products of coal coking that became the raw materials for commercial chemicals; now it was the by-products of oil and gas production that emerged as the central source of chemical feedstock.[40]

While the oil and natural gas companies began to open chemical divisions to exploit the by-products of oil and gas refining, the chemical companies began to expand into petroleum. In these expansions, Standard Oil of New Jersey and Shell were leaders in the petroleum industry, while Dow and Union Carbide led in the chemical industry. Methanol, ethylene, propylene, olefins, and butadiene were the most significant by-products of oil refineries. Glycols, aldehydes, esters, amines, and ketones were the first chemicals produced from natural gas and the by-products of refinery offgases.

Union Carbide and Carbon and the Mellon Institute of Pittsburgh joined forces as early as 1913 to develop an inexpensive process to make ethylene and its derivatives. George Curme, a Mellon Institute chemist, found ethylene as a residual by-product in making acetylene from natural gas. Recognizing the opportunities, Curme began work on the production of ethylene dichloride, ethylene oxide, ethylene glycol, and ethanol. The work was slow and frequently interrupted, but by 1921 the company had set up a plant in the Kanawha Valley of West Virginia where ethylene derivatives could be made from natural gas. In 1927, these efforts led to the development of one of the first truly successful petrochemicals, Prestone, an ethylene glycol-based antifreeze additive.

In 1931 Shell Oil opened a chemical research laboratory at Emeryville, California, and the company was soon producing isobutylene, isopropyl alcohol, and methyl ethyl ketone from butyl alcohol at its Martinez petroleum refinery. In 1941 Dow Chemical built the first fully integrated petrochemical plant at Freeport, Texas, where the proximity to oil refineries, gas reserves, salt brine, and cheap energy was ideal for the production of chlorine, ethylene glycol, ethylene dibromide, and magnesium. Throughout the 1940s other firms followed Dow's lead in setting up large, integrated petrochemical plants on the Texas Gulf Coast.[41]

Wartime demands for fuel and chemicals spurred the steady growth of petrochemicals during the 1940s. Butylenes became indispensable for high-octane fuels and synthetic rubber. Ethylene and benzene were critical for styrene production, and demand for toluene for ordnance expanded rapidly. The spread of petrochemicals prior to the war had been modest, but following the war, the industry expanded rapidly. The destruction of much of the German and Japanese industrial capacity and the historical dependence Germany had on coal permitted U.S. firms with more experience in oil and gas chemistries to leap ahead in petrochemical development. In 1950, half of the total U.S. production by weight of organic chemicals was based on oil and natural gas; by 1960 this figure had reached 88 percent.[42]

Petrochemicals transformed the chemicals industry. By the close of the war, Union Carbide was the nation's largest producer of petrochemicals, with leading products in plastics, pesticides, and agricultural chemicals. Du Pont launched a series of new products, including a tough and non-stick coating called Teflon, a durable structural material called Kevlar, and a thin surface film called Tyvek. The corporation replaced its rayon line with new synthetic fibers like Dacron and Orlon and its cellophane line with Mylar. Dow Chemical transformed its war production facilities in Texas to produce styrene, polystyrene, ethylene, ethylene dibromide, ethylene glycol, and magnesium chloride from brine, oil, and natural gas.[43] The development of petrochemicals encouraged oil companies to capitalize on their by-products by expanding into commodity chemical production, and by the 1960s oil companies held large market shares of several basic petrochemical products, particularly polymers such as polypropylene and polyester.

2.5 The Postwar Period, 1946–1990

The close of the war marked the beginning of the third major period of development in the U.S. materials industries. Victory and the transition to peacetime cleared the road to a robust consumer economy. Electric power, electronic appliances, and telephones, as well as indoor plumbing and central heating became available to most of the population between 1940 and 1970. New home construction, automobile ownership, camper vans, barbecues, and televisions characterized an economy built on shopping, leisure, and convenience. The demand for consumer products led to a shift in both the volume and the composition of the nation's consumption of materials.[44]

Fuels and inorganic chemicals no longer dominated the nation's materials industries because the country now led in most organic chemicals as well. Consumer products became the leading market and plastics the flagship material. This period was characterized by a substantial increase in process development and product innovation and in the restructuring of the industry to face global markets. The great materials resources of the United States no longer determined its economic advantage. Although the two world wars had done much to deplete the highest-grade ores, this was less a factor than was the transition to petrochemicals and polymers. Near East, Latin American, and North Sea oil and gas reserves became as productive as the aging oil fields in the United States, and these new sources all came onto the global market at increasingly diminishing costs. For a time following the war, the United States dominated the world chemical industry, but this did not last. The rebuilding of the German and Japanese chemical industries occurred quite rapidly. German and Japanese technical knowledge had also developed during the war and the new, state-of-the-art plants built to regenerate those economies were soon successfully competing with the older U.S. and British facilities.

Following the war, the United States assumed a world-dominant position in iron and steel production, but this also did not last. The basic oxygen furnace, which relied on high-pressure oxygen injection, was developed in Austria and adopted in the United States during the 1950s, and during the next decade the electric arc furnace designed specifically to accommodate high levels of iron and steel scrap was introduced. While these two technologies substantially improved steelmaking productivity, their smaller scale and lower capital requirements meant that they were equally likely to be adopted in other countries. By the close of the 1970s, smaller scale steel operations and so-called "minimills" throughout the world were siphoning off much of the U.S. steel export market, and big U.S. steel companies were restructuring and downsizing.

The fastest-growing sections of the U.S. chemical industry after 1945 were petrochemicals, plastics, and pharmaceuticals. Each was dependent on new materials research as well as careful analysis of the driving forces of the consumer markets. Increasingly, the frontiers of the industry began moving from process improvements for producing bulk chemicals to the development of fine and specialty chemicals designed for specific applications by unique customers.

The petroleum and chemical companies that had built chemical plants for the government during the war now negotiated to take those plants over and expand them into enormous multiproduct petrochemical facilities. By now there were some sixty important petrochemicals, and most of these materials were being produced by a growing number of firms. Specialized chemical engineering contractors helped to set up many of these plants and in doing so transferred knowledge and technology among corporations. The result was intense competition and overcapacity in several commodities.

Petrochemicals and Plastics

Petrochemical plastics led the way. Much of the early scientific development of polymers had been conducted in Europe prior to the war (see table 2.1). The basic theoretical work on polymers and long-chain macromolecules was done during the 1930s by the German chemist, Hermann Straudinger. Polyvinyl chloride (PVC), polyvinyl acetate, and polystyrene were all first commercialized in Germany during the 1930s, and polyurethanes were patented in Germany in 1937. The critical work on high- and low-density polyethylene was conducted by British researchers at ICI using work initially begun by Wallace Carothers at Du Pont. Polyester was developed in 1941 by researchers at a small British firm and its production was soon shifted to ICI. Following the war, Karl Ziegler at the Max Planck Institute developed metallic catalysts for producing high-density polyethylene, and Guilio Nata at the Milan Polytechnic Institute demonstrated their use in producing polypropylene.

In the United States, Union Carbide and Dow refined these processes and propelled petrochemical plastics to commercial success after the war. The production of thermoset plastics from phenols grew steadily, but it was the growth of thermoplastic resins manufactured from olefins that ignited the industry. The explosion in demand for petrochemical fuels (gasoline, heating oil, and natural gas) meant that there was an abundance of inexpensive olefins from which ethylene and propylene could be readily made. By the 1950s, easily operated injection molding machines became available and cheap commodity thermoplastics such as polystyrene and polyvinyl chloride rapidly transformed consumer products. Largely because of low costs, ease of processing, and various functionality improvements, plastics began to be substituted for conventional materials in toys, kitchenware, containers, automobile parts, domestic appliances, furniture, tools, and electronic products. Between 1939 and

Table 2.1
First commercial production of selected industrial chemicals

Date	Chemical	Producer
1901	Phenol	F. Raschig (Germany); Hoffman-LaRoche (Swiss)
1903	Carbon tetrachloride	Griesheim-Elektron (Germany)
1908	Trichloroethylene	Wacker (Germany)
1913	Ethylene	Griesheim-Elektron (Germany)
1913	Ammonia	BASF (Germany)
1916	Acetic acid	Wacker (Germany)
1916	Ethylene oxide	BASF (Germany)
1916	Acetaldehyde	Hoeschst (Germany)
1917	Acetone	Hoeschst (Germany), Weitzman (U.K.), Standard Oil of New Jersey (U.S.)
1920	Vinyl acetate	Shawinigan Chemicals (Canada)
1923	Methanol	BASF (Germany)
1923	Butanol	BASF (Germany)
1930	Vinyl chloride	Wacker (Germany)
1930	Polystyrene	I. G. Farben (Germany)
1931	Polyvinyl chloride	I. G. Farben (Germany)
1934	Styrene–butadiene	I. G. Farben (Germany)
1938	Nylon-6,6	Du Pont (U.S.)
1939	Polyethylene	Imperial Chemical (U.K.)
1943	Butyl	Standard Oil of New Jersey (U.S.)
1950	Polytetrafluoroethylene (Teflon)	Du Pont (U.S.)
1953	Polyester	Du Pont (U.S.)
1957	Polypropylene	Montecatini (Italy)

Source: Peter H. Spitz, *Petrochemicals: The Rise of an Industry*, New York: Wiley, 1988; and Keith, Chapman, *The International Petrochemical Industry*, Oxford: Basil Blackwell, 1991.

1946 the annual production of synthetic resins and cellulosic plastics rose from 247 million to 1.2 billion pounds, while polystyrene production went from 1 million pounds in 1939 to 150 million pounds in 1947.[45]

Du Pont dominated the industry in the production of synthetic textile fibers. During the late 1930s the firm had marketed nylon as a replacement for its own cellulose-based rayon, and now, after the war, Du Pont produced polyester and acrylic fibers as competitors to nylon. In 1950, the threat of government antitrust litigation forced Du Pont to license its nylon patent to Monsanto and shortly thereafter Monsanto joined with American Viscose to develop acrylic fibers as well. By the middle of the decade, American Cyanamid also began production of synthetic fibers and soon the textile industry was being transformed with polyesters, polyamides, acrylics, and polyolefins. Between 1949 and 1969, natural fiber (cotton, wool, silk) consumption remained constant and rayon and acetate fiber consumption rose about 700,000 pounds, while total synthetic fiber production increased 3.5 billion pounds.[46]

By the 1960s the U.S. petrochemical industry had become a vast industry, with some fifteen major corporations producing a wide range of intermediary and consumer products. Prices for petrochemical plastics dropped consistently, and as prices fell the market for plastics continued to expand, although not always to the benefit of the large suppliers (see box 2.2).

The Challenge of the Global Market
The reemergence of the German chemical industry, the growth of the Japanese industry, and the rise of materials industries in the industrializing countries meant that the years after 1960 required U.S. materials industries to restructure in terms of organization, products, and partnerships. Many U.S. firms opened plants in other countries, and the greater part of bulk chemical production gradually shifted overseas. The oil companies, notably Shell and Exxon, were the first to expand their petrochemicals to international-scale production, but Du Pont, Dow, Monsanto, and others soon followed, with most of the first overseas investments located in Western Europe. By 1970 the direct foreign investments by U.S. chemical companies amounted to $13 billion or nearly one-third of all their domestic investments.[47] Even more important than these foreign investments in firmly altering the way in which U.S. materials suppliers thought about world markets was the oil crisis of the mid-1970s.

Although concern about the future of petroleum supplies had already surfaced in the 1960s as the demand for natural gas began to outstrip supply, the oil embargo of 1973 revealed the dependence of the U.S. economy on foreign suppliers. The immediate effect of rising oil prices on

Box 2.2
The Development of Polyvinyl Chloride

Many commercially viable materials have originated as recycled wastes from the production of other materials. Vinyl chloride and its derivative, polyvinyl chloride, provide a classic case. During the 1920s Union Carbide chemists began to explore polymers based on vinyl chloride. The corporation's ethylene oxide plant generated large quantities of ethylene dichloride wastes that could be used to produce the monomer, vinyl chloride, but the polymer was rigid and brittle. During the early 1930s, a Goodyear scientist, Waldo Semon, successfully introduced "plasticizers" such as tritolyl phosphate into the polymer to provide a more moldable and flexible product. Goodyear began to market this material under the name Koroseal for wire and cable insulation.

Both Union Carbide and Goodyear found that adding heat stabilizers and lubricants improved the performance of polyvinyl chloride, and the material was soon finding use in tubing, wall coverings, draperies, and insulation. New uses of polyvinyl chloride emerged throughout the 1930s, including vinyl coats and wall coverings. Rigid sheets were introduced in 1938 and flexible sheets in 1939. Even the rigidity of polyvinyl chloride proved of value as fabricators began to produce pressure and sewer drainpipes from the material. Still, domestic production of polyvinyl resins grew slowly, with less than 1 million pounds produced in 1939. The war promoted polyvinyl chloride production with new applications in underwater cable insulation and flameproofing of garments. Production of vinyl chloride monomer began to rise and at the close of the war U.S. production was up to 120 million pounds per year.

The technology of the vinyl chloride monomer changed little up until the 1950s, having settled into two processes: the direct reaction of acetylene with hydrogen chloride and the thermal decomposition of ethylene dichloride. During this period the largest producers of vinyl chloride monomer—Union Carbide, Goodyear, Dow Chemical, Allied Chemical, Goodrich, Diamond Alkali, and Monsanto—also produced the polymer, polyvinyl chloride. In 1955, Dow Chemical began to sell the monomer independently and a lively merchant market for polyvinyl chloride developed, with engineering firms producing simple, off-the-shelf processing technologies for less sophisticated producers.

The entry of so many producers into the market and the ever-expanding demand for polyvinyl chloride drew the price of vinyl chloride down. Even with total vinyl chloride production reaching 2 billion pounds in 1965, demand continued to rise. The result was falling prices and increasing investments in plant capacity. By the middle of the 1970s it was clear that the low price of vinyl chloride monomer, the large number of producers, and the highly competitive market with no dominant producer meant that there was little hope of substantial profits for any of the participants, even though the market remained on a steady upward trend.

Source: Peter Spitz, *Petrochemicals: The Rise of an Industry*, New York: Wiley, 1988.

petrochemicals was to increase prices and profits, but as prices rose, concerns developed that the cost of petrochemical products might soon surpass the price of products made from the traditional materials that they had so recently replaced. This proved to be less significant than the new competition that arose as oil-rich countries around the world began to develop their own petrochemical industries, with new plants constructed in Saudi Arabia, Algeria, and Mexico.

By the close of the 1970s, production in the U.S. chemicals industry had leveled off and the industry was characterized more by stability and maturity than by growth. The global market proved to be leaner and more competitive than the traditional markets of America and Europe, and there were many more competitors and opportunities than existed just after the war. This required more focused companies, and many of the leaders divested themselves of less profitable processes and products. Indeed, in the petrochemical markets during the 1980s, many firms found themselves with an overcapacity in production facilities; these were often closed or sold off. New partnerships and international relations emerged and national companies increasingly reconfigured themselves into multinational companies with facilities and markets throughout the world.

The U.S. mining industries, which ran at peak performance following the war, began a long-term decline starting in the 1960s. Today the nation imports a major share of its industrial metals from foreign mines and smelters. Over half of metals such as tin, tungsten, platinum, cobalt, chromium, zinc, and nickel now come from foreign sources. Still, the U.S. minerals industry remains prosperous. The value of processed materials from the mining industries stood at $413 billion in 1997, with the value of raw nonfuel mineral production estimated at $39.5 billion, which represents a modest 2 percent growth over the previous year.[48]

For the past two decades the frontier of industrial materials development and production in the United States has been in high-value materials such as fine chemicals, industrial solvents, engineered plastics, superalloys, advanced composites, and electronic materials. New catalysts and new process intermediates have improved product yield and lowered costs. Indeed, the United States has the largest chemical industry in the world. In 1996, U.S. chemical industry sales amounted to $372 billion compared with $216 billion for Japan and $495 billion for all of

Europe. On a value-added basis, the chemical industry accounts for about 11.3 percent of U.S. manufacturing and produces about 1.9 percent of the U.S. gross domestic product.[49] However, the greater part of the nation's industrial materials now comes from abroad. For several decades the United States has been a net importer of metals and various nonmetal minerals. Today a large share of the bulk chemicals of production are also produced offshore in other parts of the world, where newer plants and lower production costs lead to more competitive products. The nation remains a major world supplier of industrial materials, but many of its markets are determined more by international conditions than by factors under domestic control.

2.6   The Nation and Its Materials

The growth of the U.S. materials economy has depended on the success with which new materials, discovered or invented, have been converted into commodities for production. The infrastructure for managing this conversion—the materials industries—grew and developed along with the discovery and invention of these new materials. Huge industries arose in mining, metals, petroleum, and chemicals.

This brief history has demonstrated that many factors have helped to shape and promote the emergence of the nation's rich materials economy. Among the more significant have been the natural resources available for exploitation, the development of new technologies, the growth of industrial enterprises capable of transforming new materials into commercial products, the growth of a consumer market eager to use the products of industrial processes, government development and trade policies, and the requirements of war. The natural endowment of the North American continent provided a huge reservoir of high-quality fuel and nonfuel minerals and ample water supplies and rich soils for forest and agricultural materials up through the middle of the twentieth century. The diminished condition of these resources since then has rendered them a less significant factor.

Technological innovation has been critical to the development of industrial materials in the United States. It started out in modest and clumsy ways, relying on simple trial and error, but its development is now supported by a dense network of scientific studies and technological advances. However, specific technologies such as the Bessemer steel oven, the Leblanc soda process, the beehive oven, the electrolytic cell, the Buna-N synthetic rubber process, the Haber–Bosch nitrogen fixation process, and the catalytic oil refinery have at times acted either to spur or to retard the development of the nation's industrial materials.

A political and legal environment emerged to support the establishment and growth of limited-liability corporations that eventually amassed large pools of investment capital. The entrepreneurial skills and productive labor of those who emigrated to the United States provided a pool of social resources for inventing and manufacturing materials. As the economy grew, consumer wealth grew as well and with it, an ever-expanding market for physical infrastructure and material commodities. The federal government intervened often to assist in the exploitation of minerals, and the development of chemistries and government trade and tax policies protected and promoted the growth of the private corporations in the materials industries. Finally, wars, with their dislocative consequences, and the nation's military victories have resulted in intensive exploitation of national materials, access to global materials resources, significant domestic and confiscated research knowledge, and decades of expanding global markets.

For nearly a century the United States has been a world leader in the production and development of materials. The nation started with a rich supply of forest and agricultural materials, including wood; crops; and animal fur, hides, and oils. For some time the emerging economy was a net importer of inorganic chemicals. As the nation developed, the continent's minerals were discovered and tapped and iron, lead, copper, coal, and oil laid the foundation for an industrial economy. Gradually, the county built an inorganic chemicals industry with a strong export potential, but lagged in the development of organic chemicals. Wars and infrastructure development soon eroded the traditional minerals base, but permitted the development of new metals such as aluminum, zinc, and magnesium. By good fortune, the nation's oil and gas endowment meant that when organic chemicals were finally developed, they would be petrochemicals and these chemicals became the foundation of industrial production. Finally, plastics based on petrochemicals are now replacing not only the traditional materials but some of the newer metals as well. One hundred and fifty years of materials development has exceeded its pioneers' wildest expectations.

At the close of the twentieth century, the United States was no longer an unchallenged leader in materials development. European and Asian corporations and firms in many industrializing countries produce a wealth of materials, many of which are exported to the United States. The rich natural resources of the nation have been tapped and in some cases heavily drained. The great technical advances in petrochemicals and polymers provided a host of structural and military products and consumer commodities at diminishing costs. However, today many of those materials are made into products throughout the world, often at lower costs. The materials economy is a global economy, and the future of materials development in the United States rests on how well its industries perform in that market and how carefully its resources are used.

3

The Economy of Industrial Materials

Man cannot create material things—his efforts and sacrifices result in changing the form or arrangement of matter to adapt it better for the satisfaction of his wants—as the production of material products is really nothing more than a rearrangement of matter which gives it new utilities, so his consumption of them is nothing more than a disarrangement of matter which destroys its utilities.
—Alfred Marshall

One hundred and fifty years of economic development has provided a prodigious wealth of industrial materials. The nation's economy has become a gigantic materials processing system. It produces million of tons of industrial materials and, almost as rapidly, it consumes and discards them. A recent study of the country's materials use estimated that nearly 2.5 billion metric tons of nonfuel materials moved through the economy in 1990. Over 70 percent of these materials were used in construction. Excluding these materials, this still means that some 785 million tons of industrial materials were consumed in 1990. This is equivalent to just over 3 tons of industrial materials per person per day. Of the industrial materials, 330 billion metric tons were minerals and 112 billion metric tons were metals, with another 112 billion metric tons of nonrenewable organic chemicals (e.g., solvents, fibers, and plastics).[1]

The forces that drove the development of this great treasury of materials over the past century and a half still drive its continuing enrichment. Today, the country relies not only on its own resources but also on material imports from throughout the world. Invention, innovation, and technological development still provide the impetus for new materials and new production processes. The structure of the materials industries and the role played by the dominant corporations still determine both the pace of change and the resistance to it. The consumer market is now global, but remains unsatiated. Government policies still stimulate, protect, and challenge materials suppliers. War may be less a factor than at any time during the past century, but military preparedness remains an important factor in the development of advanced and strategic materials.

Before proceeding to consider how people have tried to respond to the environmental and public health issues presented by the development of these industrial materials, it is useful to briefly reflect on the economic results of this history. The industrial materials economy of today is a highly integrated complex of processes and firms, largely coordinated by market forces that set prices, drive demand, and determine the relative success of various competing materials. As an engine of economic growth, it is a well-oiled machine.

3.1 Materials Flows and Materials Systems

To elaborate on these points, it is useful to look at the present economy of industrial materials in terms of global systems. A system is a functionally integrated collection of processes that is set apart from its wider context and is therefore identifiable as a coherent unit. A system may be open or closed. The planet's energy system is considered open because it receives energy from the sun and releases heat back to the atmosphere. In contrast, the global materials system is considered closed because, with the minor exception of gains from meteorites, materials neither enter nor leave the system. Geologists theorize that all of the materials of the planet exist in a continual state of very slow motion. Those on the surface move about in well-recognized cycles. The hydrological cycle is a physical example, with water changing phase from ice to liquid to vapor during its cycling. Nitrogen, phosphorus, and sulfur cycles provide other examples. The same is true for the carbon–oxygen cycle where carbon fixed in plants during photosynthesis is returned to the atmosphere when plants die and decompose and where the oxygen released during photosynthesis is used in the fermentation of the decomposing plants.

When materials are mobilized, transformed and reconcentrated by natural processes, geologists view this system as a materials cycle. In such a cycle, quantifiable amounts of materials flow among different environmental media (air, water, and land) and are stored or sequestered in particular accumulations, which are called "inventories" or "stocks." At any given time a certain amount of material is flowing through the various media, while most materials are sequestered in comparatively fixed (although still moving) stocks.

The flow of materials in a materials cycle is subject to certain natural laws. Among these are the thermodynamic principles put forward by physicists during the nineteenth century.[2] The first principle holds that the total energy exerted in a physical transformation is neither increased nor decreased by the process, but rather is conserved. The second principle asserts that the "quality" of the energy, that is, its availability for future use, will always be diminished by each transformation. Albert Einstein's brilliant equivalency of energy and mass permits these principles to be applied to matter as well as energy. Therefore it follows that materials are neither lost nor gained, but conserved in all (non-nuclear) transformations, and that their "quality" (availability for future use) will always be diminished by such transformations. When the quality of energy or matter moves from higher to lower states, it is said to be "dissipated."

Resource economists use these same concepts by analogy to track the flow of materials from natural sources, through human economies, and back again to the natural environment. In this view, the human economy is a subsystem of the larger global systems. It is an open subsystem because materials and energy are exchanged between the economy and the environment. The model presented in figure 3.1 describes the flow of materials through this open economic subsystem. This concept is used to describe the global system of materials flows, where it serves as the

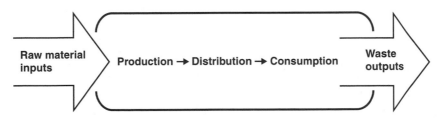

Figure 3.1
Materials balance model.

basis of a simple, linear *materials balance model.*[3] In an open subsystem, a materials balance requires that the amount of inputs equal the amount of outputs. This is the same model that serves as the basis of the economic input/output models that economists use to describe national economies.

Employed on a smaller scale, the materials balance model can be used to describe an individual industrial production plant. In a conventional industrial facility, raw materials are brought in to be processed and the production results in two types of output: finished products and so-called "by-products." These by-products may be recycled back into production, sold as co-products, released as pollutants, or sent offsite as wastes. From an engineering perspective, this model is attractive because over any given period of time, the flows of materials must balance; that is, the inputs must equal the outputs (both products and by-products) plus or minus any materials which at the time are still in the system (the inventories or stocks).

In considering the U.S. economy as a materials processing subsystem, the sources of inputs include the raw materials extracted from the land and its flora and the raw materials and products imported from other countries. Within the subsystem these raw materials are processed into materials that are then consumed directly or used to manufacture the many finished products that supply the consumer (primary) market, either domestically or abroad. Once the useful life of a product is finished, it is discarded. Some products and their materials may then be reclaimed, remanufactured, and reused in what is often called a "secondary" market. But ultimately nearly all materials and products reach final disposal, where they are discarded as dissipated wastes.

Figure 3.2 shows materials flows in the United States. This simple materials balance approach can be used to track the stocks and flows of materials from their sources as extracted resources, through their processing into products, into the use of those products and finally to their disposal as wastes. In the sections that follow, each of these stages is described and reference is made to the most common environmental and human health impacts of materials at each stage of their economic life.

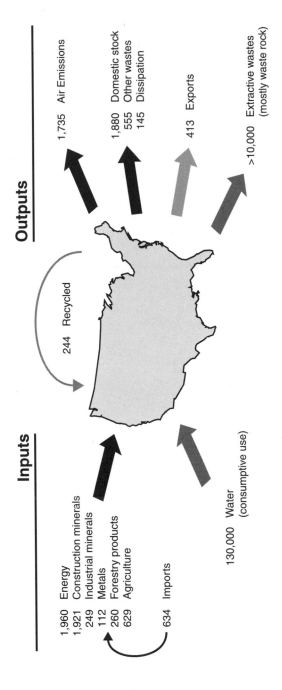

Figure 3.2
U.S. materials flows. Source: Iddo K. Wernick and Jesse H. Ausubel, "National Material Flows and the Environment." *Annual Review of Energy and the Environment*, 20, 1995, pp. 463–492.

3.2   Sources of Materials

Because thermodynamic principles dictate that matter can neither be created nor destroyed, usable materials must be extracted from the resources readily available on the planet. These resources are sometimes referred to by resource economists as "natural capital." They include renewable, replenishable, and nonrenewable resources. Renewable resources are those materials that are naturally re-reproduced, such as plant- and animal-based materials. Replenishable materials are those, such as air and water, that can be depleted but, given sufficent time, can be restored to homeostatic condition by natural processes. Nonrenewable, or depletable materials are those materials, such as high-quality minerals, which have a finite inventory and which, once used and discarded, cannot be replenished.[4]

Minerals make up the foundation of the earth's crust. Fewer than 200 minerals compose the vast bulk of this structure. Table 3.1 identifies the twelve elements that make up over 99 percent of the earth's crust. These minerals can be extracted from the earth by mining or pumping. They may be organic fuels (oil, gas, coal, etc.) or primarily inorganic nonfuels. The nonfuel minerals can be either metals (iron, copper, nickel, silver) or

Table 3.1
Twelve most common elements in the Earth's crust

Element	Percentage by weight
Oxygen	45.20
Silicon	27.20
Aluminum	8.00
Iron	5.80
Calcium	5.06
Magnesium	2.77
Sodium	2.32
Potassium	1.68
Titanium	0.86
Hydrogen	0.14
Manganese	0.10
Phosphorus	0.10
Total	99.23

nonmetals (stone, clay, gypsum). The metals are typically divided into ferrous (iron) and nonferrous (aluminum, copper, cadmium, chromium, cobalt, arsenic, lead, zinc, etc.) types. Some metals, such as silver, platinum, and gold, are precious nonferrous metals that are valued for their decorative qualities as well as their scarcity.

Mineral specialists refer to the total amount of known and unknown supplies of minerals as the resource. This differs from the concept of a reserve base, which is the demonstrated resources that meet minimum physical criteria for mining and production, and the concept of a reserve, which is that portion of the reserve base that could be economically extracted or produced at the time of determination. Because of this definition, the concept of a reserve is quite dynamic. Technological advances in extracting and processing minerals as well as increases in the price of a mineral tend to expand the amount of available mineral reserves. Table 3.2 shows the estimated U.S. reserves and reserve base of the most common nonfuel minerals and compares these with the total world supplies. Today it is estimated that lands of the United States contain about 14 percent of the world's copper reserves; 11 to 12 percent of the global lead, zinc, silver, and gold reserves; and 10 percent of the world's iron ore and sulfur and phosphate rock.[5]

Nonfuel minerals in the United States are obtained by mining mineral reserves, importing mineral ores, or reclaiming minerals from secondary markets. Table 3.3 identifies several of the most common minerals consumed in this country. In terms of production, the U.S. share of the world market for most metals has been declining over the past several decades. Still, the United States remains the western world's largest producer of refined aluminum, copper, and lead.[6] Table 3.4 shows U.S. and world production of the most common large-volume nonfuel minerals.

Hard-rock mining is carried out by excavating underground shafts or by stripping away surface rock and gouging out open pits. Over 80 percent of the minerals extracted from mines in the United States come from open pit mines. In general, open pit mining is cheaper, more productive, and safer than underground mining. Health and safety is a constant issue in underground mines. Since 1900 some 20,000 U.S. miners have been killed in coal mines alone. Both underground and open pit mines produce significant amounts of waste material called "spoil" or "overbur-

Table 3.2
U.S mineral reserves and reserve base, 1996[a]

	Reserves	Reserve base	Percent of world reserve	Percent of world reserve base
Thousand metric tons				
Copper	45,000	90,000	14.5	14.8
Lead	8000	20,000	11.6	16.7
Molybdenum	2700	5400	49.1	45.0
Zinc	16,000	50,000	11.4	15.2
Nickel	23	2500	0.05	2.3
Thousand metric tons of ore				
Iron	16,000,000	25,000,000	10.6	10.8
Sulfur	140,000	230,000	10.0	6.6
Phosphate rock	1,200,000	4,400,000	10.9	13.1
Metric tons				
Gold	5600	6100	12.2	8.6
Platinum group metals	250	7800	0.4	1.2
Silver	31,000	72,000	11.1	17.1

Source: U.S. Geological Survey, *Mineral Commodity Summaries*, Washington, D.C.: U.S. Government Printing Office, 1997, and National Mining Association, *Facts About Minerals, 1997–1998*, Washington, D.C., 1998.
[a] The reserve base consists of the in-place, demonstrated resources that meet minimum physical criteria for mining and production. Reserves are that portion of the reserve base that could be economically extracted or produced at the time of determination.

den." By volume and weight, the amount of solid mining and milling wastes associated with the extraction of metal ores typically far exceeds the quantities of material obtained. This is particularly true in the mining of coal, copper, gold, and platinum. It is estimated that each ton of refined aluminum involves the removal and processing of 4 tons of ore, while several thousand tons of gangue (the rock in which the mineral is found) and overburden must be moved to produce 1 ton of gold or any of the platinum metals.[7] The huge volumes of mining wastes can present significant local environmental consequences, and the mining of metals often produces toxic by-products. This is quite common for those metals that occur as sulfates, which can generate large amounts of liquid runoff that is high in sulfuric acid.

Table 3.3

Production and consumption of selected U.S. minerals, 2000 (thousand metric tons)

	Produced by mining	Primary production	Secondary production	Imports	Exports	Apparent consumption
Aluminum	—	3800	1400	4000	1700	7500
Cadmium	—	1800	—	600	600	2220
Chromium	0	—	103	429	23	522
Cobalt	0	—	3.1	8.2	1.5	11.5
Copper	1660	1870	640	2240	1295	3090
Gold	0.3	0.3	0.2	0.2	0.3	—
Iron ore	57,000	—	—	13,000	6500	68,400
Iron and steel	—	44,900	—	31,300	5000	109,000
Iron and steel scrap	—	—	70,000	6000	5000	68,000
Lead	520	350	1050	30	70	1750
Magnesium metal*	—	na	80	30	105	174
Mercury	na	—	0.4	0.1	0.1	0.4
Molybdenum	44.1	—	—	11.9	38.4	18.4
Nickel	0	—	—	143	35.6	137
Silver	1.8	2.3	1.7	2.8	2	4.5
Tin	0	—	17.8	45	6	59.7
Tungsten	0	—	—	3.3	0.2	12
Zinc	775	235	135	1015	601	1630

Source: U.S. Geological Survey, *Mineral Commodity Summaries, 2000*, Washington, D.C.: U.S. Government Printing Office, 2000.
Notes: —, Not reported; na, not available; *, excludes magnesium compounds derived from seawater.

Table 3.4
U.S. and world production of selected nonfuel minerals, 1996 (in metric tons unless otherwise specified)

Material	World production	U.S. production	U.S. percent
Metals			
Arsenic	42,100	0	0
Aluminum	20,700	3580	17
Cadmium	18,900	1530	8
Chromium	12,200,000	0	0
Copper (primary and secondary)	12,500,000	2,340,000	19
Gold (kilograms)	2,250,000	318,000	14
Iron (ore)	1,020,000,000	62,100,000	6
Lead	2920	436	15
Magnesium (primary)	341,000	133,000	39
Mercury	2890	Withheld	—
Nickel	1,080,000	1330	—
Silver	15,200	1520	10
Tin (smelter)	207,000	11,000	5
Zinc (primary and secondary)	7,530,000	366,000	5
Industrial Minerals			
Asbestos	2,290,000	9550	—
Cement	1,480,000,000	80,800,000	5
Gypsum	99,700,000	17,500,000	18
Phosphate rock	141,000,000	45,400,000	6
Silicon	3,200,000	412,000	13
Soda ash	30,400	10,200	34
Sulfur (all forms)	52,400	12,000	23

Source: U.S. Geological Survey, *Minerals Yearbook, Metals and Minerals, 1996,* Washington, D.C.: U.S. Government Printing Office, 1998.

Petroleum and natural gas are recovered by drilling into underground reservoirs, although oil can also be obtained by processing buried tar sands and saturated shales. Obtaining some oil and gas requires no more than tapping the reservoir because the trapped petroleum is under severe seismic pressures. As oil recovery matures, the seismic pressures are reduced and the oil must be pumped. Environmental impacts come from the drilling of wells (including the possibilities of a "blowout"), the discharge of pumping wastes, and the volatilization of hydrocarbons during the extraction, storage, and transportation of the product, all of which is complicated further when wells are placed offshore to extract oil from the seabed.

Market-driven mineral extraction leads to a continuous depletion of the highest quality and most accessible mineral resources. In the early 1900s copper ore ran at about 2 to 3 percent copper, but by the 1980s it averaged less than 0.7 percent, with the leanest Arizona ores running at just 0.4 percent copper. Lake Superior iron ore which at one time was the richest in the world, running about 55 percent iron, now seldom peaks over 35 percent iron.[8] Were it not for continued exploration and the discovery of new reserves, this depletion would result in diminishing stocks and ever-rising prices.

3.3   Processing of Materials

The processing of extracted ores separates, reduces, and refines the materials into intermediates, which serve as the fundamental building blocks for industrial materials. The production of metals from mineral ores is fairly straightforward, typically requiring crushing and heating the ore. Mineral ores are refined in large smelters in which heat is used to drive off impurities. Metals are typically found in compounds involving oxygen (oxides) or sulfur (sulfides) and other elements. Acquiring pure metals from the compounds found in high-grade alluvial ores requires little more than heating (smelting) or leaching with other chemicals to remove the oxygen or sulfide atoms. Obtaining the metals found in more heterogeneous ores involving relatively useless gangue minerals such as silica or calcium carbonate requires crushing and grinding the ores before physically separating out the metals by processes that rely on gravity, magnetism, or flotation.

Smelting is usually very energy intensive and quite polluting. For instance, the smelting of iron generates large amounts of slag rich in silica, phosphorus, and sulfur. Two to three tons of iron ore are required to yield a ton of iron metal. Prior to smelting iron ore, it is sintered (with heat) into nonporous nodules in a process that produces by-products such as sulfur dioxide and various airborne particulates. The furnace combustion in both smelting iron and making steel produces particulates, which are released as air emissions. In copper smelting, nearly a ton of sulfur dioxide gas may be produced for every ton of copper obtained from sulfide ores. Vigorous efforts at air pollution control throughout these processes can capture up to 95 percent of the possible air emissions, but this results in a substantial amount of sulfur-laden wastewater.

Petroleum also requires refining to produce fuels, lubricants, and the intermediates used in the manufacture of petrochemicals. Oil delivered to refineries is distilled by adding heat to separate the various components (fractions) on the basis of differences in boiling point. During the 1940s it was discovered that this thermal cracking could be greatly accelerated by adding suitable catalysts (acids) in what is now called "catalytic cracking." The heated oil is pumped into the lower ends of vertical fractionating towers. The lighter volatiles that collect at the top of the tower are pumped off first and then with the addition of more heat, the midrange and heavier fractions are pumped off from lower and lower levels in the tower. Thus gases, gasoline, and water are pulled out of the top vapors; kerosene is pulled off lower, gas oils (for diesel and furnace oils) are removed even lower; and lubricating oil stocks (for greases and waxes) are separated from the petroleum tars and asphalts at the bottom.

The products recovered through cracking are still fairly crude (each fraction can contain from 5 to 100 different compounds) so that additional distillation and stripping processes are required to further separate the products. For instance, the gasoline achieved through fraction tower distillation is not adequate to meet the performance requirements of transportation fuels, so the gas oil fraction is exposed to high temperatures and moderate pressures to break down, or "crack," the large gas oil molecules into smaller ones suitable for gasoline. Similarly, marketable lubricating oils are produced from crude oil fractions through a series of steps that include redistillation to drive off "lighter ends," solvent dewaxing to lower the viscosity, and filtering the oils through clays to capture various impurities.

Oil refineries have been quite polluting and notoriously hazardous. Heating hydrocarbon vapors raises the risk of exothermic reactions which, if uncontrolled, can destroy an entire facility. Hydrocarbon gas leaks from overpressuring and weakened seals and valves create air releases. Aqueous desalting effluents contain oil, salts, and sulfides. Hydrogen sulfide, phenols, and caustics appear in refinery process waters. Alkaline sludges, spent acids, oil-soaked clays, and filter cakes from pollution control equipment show up as hazardous wastes from refineries.

Organic Chemicals

Petroleum fractions provide the organic chemicals that are the required intermediates for producing a wide variety of products, including antifreeze, dyes, glues, paints, pharmaceuticals, cosmetics, solvents, inks, lubricants, and pesticides. The vast majority of hydrocarbon organic chemicals are derived from just three refinery streams: methane, olefins, and the aromatics. These make up the high-volume, low-price "first-order" intermediates that are the building blocks for hundreds of "second-order" petrochemical products. Figure 3.3 is a diagram of several of the most common petrochemical production lines. While the range of finished petrochemical products is large, the total volume is dominated by only a small number. Well over half of the world's total final output by weight of petrochemicals involves plastics and resins, and most of these are derived from ethylene and propylene. If synthetic fibers and rubber are included, this figure rises to nearly three-quarters.[9]

Methane is fairly readily produced from raw oil or gas streams, but most of the other hydrocarbons require substantial processing to obtain the desired products. Ethylene and propylene make up the largest share of olefins. Ethylene surpasses all other petrochemicals both in volume and in diversity of commercial use. Roughly half of all ethylene is used

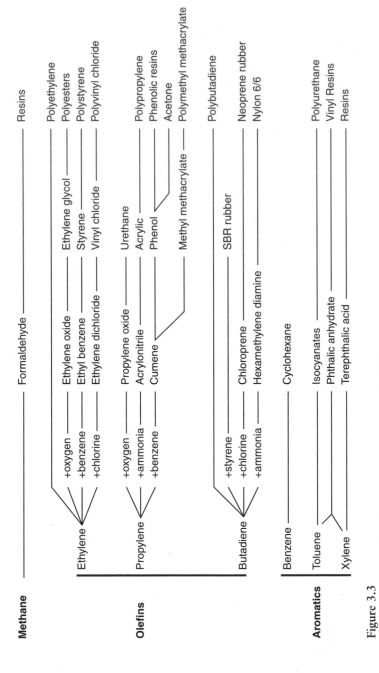

Figure 3.3
Principal routes of synthesis of common petrochemicals.

in the production of polyethylene, with lesser amounts used to make ethylene oxide, ethylene dichloride, and ethylbenzene. About half of the ethylene oxide produced is directly converted to ethylene glycol to make solvents and automotive antifreeze, with the rest used to make polyester (polyethylene terephthalate) fiber. Ethylbenzene is largely used to make styrene. Ethylene reacted with chlorine produces a broad range of industrial chemicals, including solvents and intermediates such as ethylene dichloride, which can be directly converted into vinyl chloride. The chlorinated solvents, such as carbon tetrachloride, perchloroethylene, trichloroethylene, and methylene chloride have large markets because of their high solvency, low flammability, and high vapor density.

The most common aromatic hydrocarbon is benzene, which is produced along with two co-products: toluene and xylene. Until World War II, benzene was primarily used in the rubber industry. The war created a heavy demand for toluene for use in the manufacture of explosives, and thus created a surplus of benzene, which found new uses as an additive in high-octane gasoline and in the production of pesticides, adhesives, paint removers, detergents, and plastics. Today, depending upon market demand, toluene or benzene can be a finished product. Half of all benzene produced is used in making styrene, with the remainder used to make phenol and adipic acid, an intermediate used in the production of nylon. About half the toluene made is reconverted to make benzene, with a quarter going into gasoline to improve octane. The xylenes are used in gasoline, solvents, polyesters, and other synthetic resins.

Manufacturing organic chemicals requires substantial amounts of heat, pressure, and chemical reaction. The required processes are among the most energy intensive in the economy, particularly those used to produce bulk chemicals such as ethylene, benzene, and polyethylene. Historically, these processes have been highly polluting; however, newer petrochemical complexes are heavily integrated and their waste generation is far less.

Polymers

Polymers are a broad class of organic chemicals used to make resins, fibers, films, adhesives, detergents, paints, coatings, rubbers, textiles, and thousands of components, casings, and domestic products. They are made of long chains of bonded hydrocarbon molecules that contain a pattern of repeating units called "monomers." In nature these polymers are the assemblages that form the structural properties of wood, cotton, wool, muscle, starch, and proteins. Synthetic polymers have structures similar to natural polymers, but involve a host of different monomers and assembly processes not found in natural materials. Table 3.5 lists some of the most common polymers and their production volumes.

Almost all of today's polymers are made from petroleum derivatives, the most common being ethylene and propylene. Polymers are produced synthetically when the petrochemical monomers mixed with various reagents are induced to bond together or assemble into desired repeating sequences. This can occur in a simple "step growth" process in which the assembly occurs slowly with the assistance of various auxiliary sub-stances such as water, methanol, or hydrogen chloride, which are then released as by-products, or in a more rapid chain reaction process that has little need for supporting chemicals.

Synthetic polymers are broadly classified as plastics, fibers, or elas-tomers and are differentiated by their modulus, or stiffness, with elas-tomers being the least stiff. Plastics are further differentiated by whether they are thermosetting in that their shape is determined in their first molding and is thereafter fixed, or thermoplastic, in that they can be remelted and reshaped. Roughly 90 percent of plastics are thermoplas-tics. Polymer resins are produced by hundreds of so-called compounders who tailor-make performance-grade resins that are sold in pellet form to thousands of plastic product fabricators. The fabricators heat the pellets to create a viscous, moldable fluid, which is shaped into final products by molding, extrusion, or blowing in specially designed product-forming machines. Commodity plastics are characterized by high production vol-umes, wide ranges of functionality, and low resin costs. The four most important commodity plastics are polyethylene, polypropylene, poly-styrene, and polyvinyl chloride.

Metals
For centuries iron was civilization's most important metal, but today alu-minum and a range of metal alloys have even greater commercial value. These metals are used to make thousands of products ranging from

Table 3.5
U.S. production, import, and export of selected polymers, elastomers, and resins, 1996 (millions of pounds)

Chemical	Production	Imports	Exports
Acrylic and monoacrylic fibers	460	60	190
Acrylonitrile–butadiene–styrene Resins	1477	189	364
Epoxy resins	662	43	374
Ethylene–propylene elastomers	318	30	84
High-density polyethylene resins	12,373	1229	1715
Low-density polyethylene resins	7784	219	1538
Nylon fibers	2798	306	150
Nylon resins	1120	155	na
Phenolic resins	3476	97	125
Polybutadiene elastomers	562	94	160
Polycarbonate resins	na[a]	90	325
Polyester resins	3826	625	374
Polyester resins (unsaturated)	1557	61	151
Polyolefin fibers	2563	52	123
Polypropylene resins	11,991	285	1480
Polystyrene	6065	232	583
Polyurethane elastomers	na	15	31
Polyvinyl chloride resins	13,220	373	1225
Rayon fibers	270	48	29
Rubber, natural	na	275	20
Styrene	11,875	591	2299
Styrene–butadiene elastomers	1015	108	293
Urea–formaldehyde resins	2147	33	51
Vinyl chloride	14,955	Negligible	2164

Source: Stanford Research Institute, *Chemical Economics Handbook, Manual of Current Indicators*, Palo Altto, Calif.: February 1999.
[a] na, Not available.

structural members to transportation vehicles, containers, tools, motors, turbines, electronic components, and household appliances. Several metals, such as zinc, chromium, copper, silver, and gold, are used in electroplating to place a protective or decorative coating on formed products ranging from jewelry to hubcaps. Metal plating is also the heart of electronic printed-wiring board production. Metal compounds are widely used as catalysts in chemical production; in coin production; in printing and photography; and in paint, ink, and dye pigments because of their brilliant colors.

The formation of metals involves rolling out plates or foils, drawing wire, casting in premade forms, or stamping cutouts from sheets. Most of the processes require heat and pressure and produce metal wastes and scrap that can often be directly recycled back into production. The performance of most of the common metals—iron, steel, copper, lead, and aluminum—has been enhanced by mixing small amounts of other substances into the molten metal during its processing. These additives alter the crystalline structure of the metal as it cools. The resulting alloys may be stronger, lighter, harder, more durable, more moldable, less corrosive, or more conductive. The most important additives used to make alloys are chromium, cobalt, manganese, molybdenum, nickel, silicon, titanium, tungsten, and vanadium.

Steel is a carbon–iron alloy and is by far the most common ferric alloy. The hardness of steel can be increased by introducting molybdenum, manganese, or tungsten, while nickel–chromium is added to confer stainlessness, and copper is added to reduce corrosion. Hard carbide cutting tools are created by alloying steel coatings with titanium. Copper is alloyed with zinc to form brass and with tin to make bronze. Both of these alloys are stronger than copper. Brass is cheaper than pure copper, while bronze is more corrosion resistant.

The formation of metal alloys consumes energy and creates hazardous wastes, but the products are often quite stable and harmless. Metal plaing also creates heavy metal waste streams, although these wastes can often be recycled. Metals used in dissipative products such as paints and pigments are more likely to erode as particulates and contaminate environments. Several metal compounds, such as lead, arsenic, and mercury, are lethal enough to provide active ingredients in pesticides and herbicides.

Inorganic Chemicals

The inorganic chemicals are a heterogeneous collection of compounds and include the salts, acids, and bases. The largest use of inorganic chemicals is in manufacturing and chemical processing, so inorganic compounds are less likely to show up in finished industrial products. Nonmetallic minerals provide the feedstock for many of the most important substances; these include sodium and potassium salts, calcium and sodium carbonate, sodium hydroxide, chlorine, ammonia, nitric acid, sulfur and sulfuric acid, and phosphorus and phosphoric acid.

Sodium and potassium chloride are produced by mining brine or evaporating seawater. The electrolysis of sodium chloride brine yields sodium hydroxide, which is largely used in the pulp and paper industry and in the production of chlorine and various organic chemicals. Most chlorine (70 percent) is used in the manufacture of organic chemicals, such as vinyl chloride, ethylene dichloride, glycols, and chlorinated solvents and methanes, and about 15 percent is used in the pulp and paper industry.

Nearly 90 percent of all sulfur produced globally is converted to sulfuric acid. Nearly all of this sulfuric acid is produced by the so-called "contact process," which involves the reaction of sulfur dioxide and air over a hot catalyst (e.g., platinum). The agricultural sector consumes over half of the sulfur produced, but the low cost of sulfuric acid makes it very attractive in chemical processing, manufacturing, and pharmaceuticals. Sulfuric acid is applied in high concentrations to phosphate rock to manufacture phosphoric acid, which is used to make agricultural fertilizers and phosphates for detergents. Ammonia production requires a combination of nitrogen and hydrogen gas and high temperatures. While the main use of ammonia is for fertilizers, it is also used to manufacture explosives, some polymers, and many refrigerants, and nitric acid is generated from the oxidation of ammonia.

Processing of inorganic chemicals can consume substantial amounts of energy. The wastes from such processing often include quite corrosive acid or alkaline wastewaters or mists and fumes. If the chlorine and sodium hydroxide markets are not balanced, the chlor-alkali processes can produce large volumes of either chlorine or caustic wastes. While these wastes pose rather immediate hazards, many inorganic chemicals also demonstrate significant long-term hazards as well.

Integration of Production

Most industrial chemicals are produced in complex, highly integrated processes that generate many related products. Often this occurs because the chemicals have the same raw material feedstock. For instance, ethylene is a fundamental intermediate used in the production of second-order chemicals such as ethylene glycol, ethylene oxide, ethylbenzene, and ethylene dichloride. These in turn are used to produce petrochemical products, such as styrene and vinyl chloride, which are then used to produce polymers such as polystyrene and polyvinyl chloride. But these finished products can also be produced using other starting feedstocks and other routes of synthesis.

Organic chemicals provide a good example of materials that can be processed (synthesized) through several different synthetic routes. As noted earlier, the German organic chemical industry developed around the distillation of coal, while the U.S. industry turned to petroleum during the 1930s. This need not have been the chosen direction. Semet-Solvay had been making coal-tar chemicals in the United States since 1910 and Du Pont had a line of coal-based solvents and was quite reluctant to become dependent on the oil companies.

Metals are also likely to be found in interrelated groups that offer integrated production opportunities and a wide range of product diversification. Table 3.6 lists some of the by-products and co-products associated with various primary metal mining operations. Aluminum, which is found as an oxide in bauxite ores, often occurs with other oxides such as those of iron, silicon, and tin as well as gallium, a metal that has found new uses in the electronics industry. Copper, zinc, and lead are often found as sulfides in the same ores. Depending on local conditions, these ores may also include cadmium, cobalt, arsenic, molybdenum, bismuth, silver, selenium, tellurium, and thallium. In order to produce pure copper, zinc, or lead, these impurities must be stripped away through smelting and refining. When the market is right, these metal impurities are recovered as commercially valuable co-products. This is particularly true of silver, which has a high market value for jewelry, coins, decorative plating, and electronic components. Copper mining has become a major commercial source of sulfur because some smelters today are capable of producing a ton of sulfur for every ton of copper. Indeed, some copper

Table 3.6
Common mineral by-product groups

Copper	Nickel	Zinc	Platinum
Antimony	Cobalt	Animony	Iridium
Arsenic	Manganese	Cadmium	Osmium
Cobalt	Lead	Gallium	Palladium
Gold	Antimony	Germanium	Rhodium
Manganese	Bismuth	Lead	Ruthenium
Selenium	Selenium	Silver	Niobium
Silver	Silver	Thallium	Tantalum
Tellurim	Tellurium	Lithium	Zirconium
		Rubidium	Hafnium

Source: Adapted from Robert U. Ayres, "Materials and the Environment," in Michael Bever, ed., *Concise Encyclopedia of Materials Economics, Policy and Management*, New York: Pergamon, 1993, pp. 141–151.

or nickel mines would not be profitable without the processing of such by-products.

The history of integrated production systems is often a history of by-product management because the profitable utilization of waste materials is a frequent determinant of a successful process. The alkali sector provides a good example. During the nineteenth century, the sal ammoniac industry in Britain merged with the soda industry because a by-product of the sal ammoniac industry, sodium sulfate, became an essential raw material in the Leblanc process. This sulfuric acid–soda industrial complex produced two new by-products: hydrochloric acid and calcium sulfide, both of which caused serious environmental damage. Public pressure and the opportunity for increased financial returns forced the industry to reclaim these by-products for further processing. The calcium sulfide could be reused in the manufacture of sulfuric acid. The hydrochloric acid was captured and treated with lime to form chlorine, which could be used for bleaching in the textile industry. This proved to be a great financial boon because as the newer Solvay process increasingly took over the market for caustic soda, the Leblanc process turned to chlorine production for its economic survival. Thus, chlorine, which previously was the less valued by-product, became the prime product of the older industry. Over the years, chemical engineers have pursued lengthy investigations to determine economically retrievable uses for industrial plant effluents and mine wastes that would otherwise arouse public concern. Where these analyses produced marketable commodities, integrated production processes and product diversification emerged.[10]

Integrated materials production has also created webs of materials dependent on fuel production. The production of fuels creates many of the by-products that are the intermediates for processing industrial chemicals. As indicated earlier, charcoal was used as a fuel for the early smelting of metals, and the production of charcoal generated gases that were used to make methanol and formaldehyde. Coal and coke soon replaced charcoal as a primary processing fuel in the refining of iron and copper, and the by-products of coke production provided the initial basis for the early organic chemical industry. Thus the production of iron, steel, and copper as well as municipal gas often financially underwrote the production of industrial chemicals. The transition to oil and natural gas as the primary fuel for the nation simply offered a new source of by-products. The fractional distillates generated at refineries to produce fuel oil and gasoline have become the source of the intermediates for hundreds of petrochemicals. The demand for gasoline and natural gas provides a substantial subsidy and a primary influence on the supply of petrochemicals. Indeed, it could be said that petrochemicals and plastics are a by-product of the nation's commitment to the automobile. As long as cars run on hydrocarbons, there will be a steady supply of cheap organic chemicals for petrochemical products.

3.4   Use of Materials

Industrial materials are used throughout the U.S. economy. Among the largest users by material weight are the machinery, fabricated metals, chemicals, transportation equipment, and electronics sectors. Among these, the automobile manufacturing industry is one of the largest consumers. Since the 1960s this one sector has used approximately one-fifth of the steel produced annually in the United States. In addition, the industry has consumed about one-tenth of the aluminum and copper, one-third of the zinc, and one-half of the lead. Although iron and steel continue to account for more than two-thirds of a typical new car's

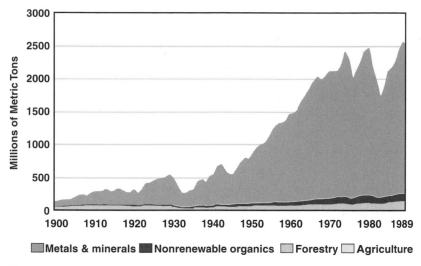

Figure 3.4
U.S. consumption of raw materials, 1900–1989. Source: Donald G. Rogich and Staff, "United States and Global Material Use Patterns," unpublished paper, U.S. Bureau of Mines. Washington, D.C., September, 1993.

weight, plastics, composites, and aluminum together account for another 10 percent.[11]

The volume of materials that flow through the national economy grew substantially over the past century. Indeed, the growth of material consumption dramatically outstripped the growth rate of the population. As demonstrated in figure 3.4, the total consumption of nonfuel materials (construction, mining, and industrial materials) was 17 times greater in 1989 than it was in 1900. The most accelerated growth during this period occurred between 1950 and 1975, when the nation was rapidly building its current infrastructure of suburbs, highways, airports, and skyscrapers. Since the 1970s, growth in consumption has been more modest and more dramatically influenced by business cycles.[12]

This growth pattern has not been true for all materials. Figure 3.5 traces the growth in the consumption of several large-volume materials since 1890, and suggests that consumption of specific materials follows a common economic cycle based on demand. During the early years, the consumption of a material is fairly modest, until at a certain point growth in demand begins to pick up. This rapid growth spurs more

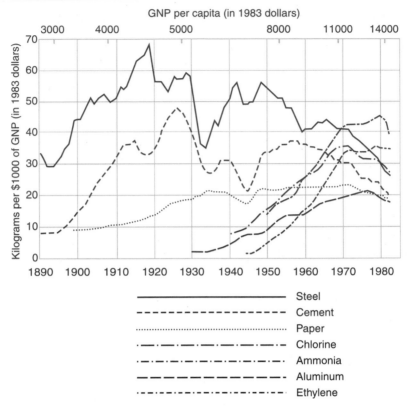

Figure 3.5
Trends in consumption of seven materials, 1890–1985. Source: Eric D. Larson, Marc H. Ross, and Robert H. Williams, "Beyond the Era of Materials," *Scientific American*, 254(6), 1986, pp. 34–41.

sophisticated applications, and production processes become more efficient. The material begins to replace conventional materials, and growth in demand increases rapidly. At a peak point, the demand levels off and thereafter more closely follows the fluctuations of the larger economy. Eventually the demand begins to taper off as newer materials are introduced that can compete more successfully.[13]

The history of steel consumption provides a good example. During the nineteenth century, steel became the primary material for the construction of the nation's infrastructure. Consumption grew dramatically and stimulated several technological innovations—the Bessemer process, the open hearth furnace, and the basic oxygen process, for example. The

Table 3.7
U.S. per capita consumption of materials, 1776 and 1996 (pounds)

	1776	1996
Aluminum	0	93
Cement	12	742
Clay	100	326
Coal	40	7581
Copper	1	23
Iron ore	20	603
Lead	2	13
Natural gas	0	8164
Petroleum	0	7520
Phosphate	1	340
Potash	1	48
Salt	4	404
Sand, gravel, stone	1000	19,061
Sulfur	1	111
Zinc	0.5	12

Source: U.S. Geological Survey, *Mineral Information Institute*, Washington, D.C., 1997, and National Mining Association, *Facts About Minerals, 1997–1998*, Washington, D.C., 1998.

growth in demand outpaced the growth of the economy up until the 1920s. During this period, the primary use of steel began to shift from construction to consumer products (automobiles, refrigerators, etc.). By the 1950s, the steel in consumer products was increasingly being replaced by aluminum and various plastics, and the demand for steel began to slacken. Increasing pressure from foreign suppliers forced further cutbacks, and production fell from 150 million tons in 1973 to 120 million tons in 1981. By the mid-1980s, these decreases resulted in the closing of a fair portion of the nation's productive capacity.

The intensity of the use of materials changed over the century as well. Table 3.7 presents a historical comparison of per capita use of minerals over 200 years. Not surprisingly, it shows a significant growth in the per capita use of fuels and construction materials. Using a ratio of material weight to gross domestic product (GDP) to compute intensity of use, a recent study found that timber was the most intensely used material in 1900, but it has fallen consistently since then. The use intensity of steel, copper, and lead also gradually declined over the past century, while the use intensity of plastics and aluminum increased substantially over this same period.[14]

Certain uses of industrial materials are inherently dissipative. Solvents used in cleaning metal parts volatilize into the atmosphere. Asbestos, once commonly used in automotive brake linings, wore off on roadways throughout the country. Detergents, flocculants, lubricants, soaps, dyes, paints, pigments, pharmacueticals, cosmetics, fertilizers, pesticides, and herbicides are all dissipative during normal use. Many uses of heavy metals such as arsenic, cadmium, chromium, and mercury are dissipative. It is estimated that 30 percent of all industrial materials are dissipated into the environment during production and use.[15] Recent studies of flows for specific materials note substantial losses due to dissipation. For instance, a U.S. Bureau of Mines study of tungsten (used largely as a reagent and in metal alloys) estimated that between 1910 and 1991, an estimated 204,000 metric tons of tungsten were lost to the environment and this represented roughly 58 percent of the total amount consumed in end-use products. A historical account of zinc production from 1850 to 1990 estimates that 63 percent of nearly 73 million tons of zinc produced during that period was lost by dissipation in wasted end-use products.[16]

The way in which a material is used can have a significant impact on the environment and human exposure. Lead offers a good example. Historically, the United States has been the largest producer and consumer of lead. Many of the conventional uses of lead in pigments, gasoline, chemicals, and ammunition were highly dissipative and this resulted in a wide dispersion of small amounts of lead into the environment. A U.S. Bureau of Mines report found that between 1968 and 1973, peak years for leaded gasoline, the dissipative uses of lead ran at over 400,000 tons per year, or 32 percent of the average annual reported consumption.[17] Figure 3.6 tracks the rise and fall of dissipative uses over a nearly 50-year period. Since the 1980s, government regulations have resulted in a phaseout of lead in gasoline and paint, which has produced a sharp drop in dissipation. By 1988, dissipative uses accounted for only about 7 percent of total consumption.

Today the largest single use of lead in the United States is in the manufacture of lead–acid batteries for transportation vehicles and power

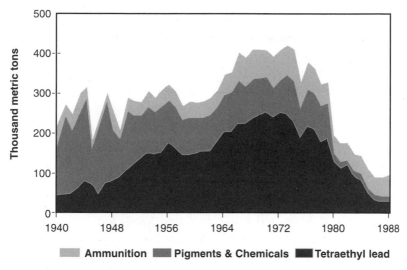

Figure 3.6
Dissipative uses of lead. Source: William Woodbury, Daniel Edelstein, and Stephen Jasinski, "Lead Materials Flow in the United States: 1940–1988," unpublished paper, U.S. Bureau of Mines, Washington, D.C., 1993.

storage units at utilities to provide "peaking power." Lead used in these batteries is relatively easily recycled and the commercial infrastructure for lead recycling is relatively sophisticated. Thus the flow of lead has moved from a relatively open system with significant dissipation to an increasingly closed system with high levels of recycling.[18] This does not mean that lead releases to the environment are insignificant. Lead appears in many construction and consumer products that are widely distributed and periodically disposed of as waste. Some lead continues to be released to the atmosphere during the combustion of coal and oil, and secondary smelters still release lead in flue gases that produce local impacts worthy of concern.[19]

Substitutions of Materials
The history of the consumption of industrial materials is a history of material substitutions. In structural applications, iron replaced wood; steel replaced iron; aluminum replaced steel; and most recently, plastic is replacing aluminum. In clothing fibers, rayon replaced cotton, which was replaced by nylon, which is now being replaced by polyester. In

these evolutionary processes, many materials display a kind of market cycle by which they move from invention to adoption to maturity to replacement.

Over the past century the U.S. economy shifted from one dependent on renewable materials to one based on nonrenewables. Figure 1.1 showed the relative material constituents of the economy over the century. At the beginning of the twentieth century, 75 percent of the economy's materials were derived from renewable resources; the remaining 25 percent was from minerals. Following World War II, two shifts occurred: the relative percentage of renewable to nonrenewable materials began to decrease, and the use of nonrenewable organic feedstocks (petrochemicals and plastics) began to grow substantially. Today, the nonrenewable organic feedstocks make up nearly one-quarter of the nation's materials, replacing significant shares of both renewable and mineral resources.[20]

The success of synthetic chemicals as substitutes for natural materials is due in part to improved performance properties and in part to lower production and labor costs. For instance, polymers outperform metals in various applications in terms of manufacturing versatility, ease of processing, lower density, and most important, cost of production. Plastics are attractive in manufacturing processes because the material weight is lower, less capital is required for small production runs, complex parts can be made in one operation, and products can be simpler to assemble. Plastic products may also outperform metals because plastics resist corrosion and denting, are softer and safer in use, and reduce the weight of products. The lower costs have come about as a result of rapid increases in the productivity of petroleum refining and petrochemical processing. Technological innovations and the increasing scale of production have led to lower labor costs in the production of chemical feedstocks and intermediates and in the manufacturing of finished goods. In many applications, including smelting, materials separation, purification, pulping, and parts cleaning, chemical energy can substitute for conventional labor energy. Lower labor costs (often, meaning fewer workers) mean lower product costs.[21]

If performance is fairly comparable, but a newer material is cheaper, there is a market incentive for substitution. In some cases this price differential is determined by the economic structure of integrated production processes. For instance, with synthetic plastics, the production of the feedstock materials can bear a disproportionate share of the cost of producing all of the downstream products. Today, plastics and synthetic fibers account for 12 and 4 percent, respectively, of the total value added in the U.S. chemical industry. Many of these polymers have been substituted for more conventional materials, such as wool, cotton, wood, and steel, because of their lower prices. However, the low prices were largely dependent upon the preexisting infrastructure for ethylene production paid for by the oil refining industry that was in the primary business of producing petroleum products, notably fuels.

In other cases the price differential occurs when the government provides subsidies. The federal government actively promoted material substitutions during World War II. Cut off from traditional sources of strategic materials, agents of the U.S. Geological Survey and the Bureau of Mines worked with private industry to encourage the substitution of lead for tin in solders and coatings, coated steel for copper in cartridge casings, molybdenum for tungsten as an alloy in steel hardening, and various materials for zinc in automobile parts.[22]

Finally, the consumer market is the primary testing ground for what industrial materials will prevail. Over the past century there was a steady sequence of material substitutions that led toward increasingly lighter, stronger, higher performance, and less costly materials. Many of these substitutes are nonrenewable materials that have come to replace more traditional renewable ones.

3.5   Disposal of Materials

All industrial materials become wastes. Indeed, one recent study of industrialized economies concludes that between one half and three quarters of all material inputs to these economies becomes wastes within a year. In 1997, 40.7 million tons of hazardous waste were generated by over 20,000 large-quantity hazardous waste generators in the United States. Of this amount, more than 2000 waste treatment facilities managed 37.7 million tons. The commercial products of industry also become waste either shortly after purchase or after many years of useful service. The nation generated 208 million tons of municipal solid waste in 1995. Of this, 56 million tons were recovered by recycling, 33.5 mil-

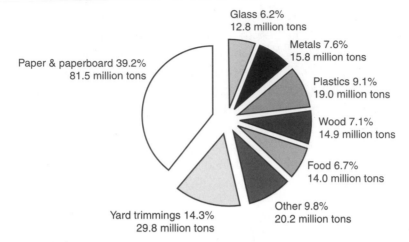

Glass 6.2%
12.8 million tons

Metals 7.6%
15.8 million tons

Paper & paperboard 39.2%
81.5 million tons

Plastics 9.1%
19.0 million tons

Wood 7.1%
14.9 million tons

Food 6.7%
14.0 million tons

Yard trimmings 14.3%
29.8 million tons

Other 9.8%
20.2 million tons

Figure 3.7
Materials composition of municipal waste. Source: U.S. Environmental Protection Agency, Office of Solid Waste and Emergency Response, *Characterization of Municipal Solid Waste in the United States, 1996 Update*, Washington, D.C., 1997.

lion tons were burned in incinerators, and 118 million tons were landfilled. The municipal waste stream is largely made up of discarded products, used packaging, and other postconsumer wastes. Figure 3.7 shows the materials composition of municipal solid waste in 1996. All of these wastes must be disposed of, and the conventional options include landfills and incineration.[23]

Since its earliest years the nation has relied largely on land disposal for managing solid wastes. Theoretically, landfills result in the slow decomposition of their contents and eventual assimilation of the constituents into the earth. Practically, landfills are like huge biochemical reactors. Water and solid waste are the inputs, and landfill gases and liquid leachates are the output. The most common constituents of municipal landfill gases are ammonia, carbon dioxide, carbon monoxide, hydrogen, hydrogen sulfide, methane, nitrogen, and oxygen. The amount of liquid leachate is proportional to the amount of liquid inputs (largely rainwater and water in the trash) and the leaching of chemicals from the solid waste can heavily pollute the leachate with acids, organic nutrients, and heavy metals.

Returning metals to the earth after use would appear to be relatively benign. The disposal of iron and aluminum is fairly nonproblematic. Because aluminum is corrosion resistant, it decomposes quite slowly, but iron rusts fairly rapidly and assimilates relatively easily. However, the nonferrous metals tend to be quite reactive during disposal. They can leach into nearby water bodies and some may become even more hazardous before they are assimilated.

Waste incineration has also been common since the early days of open burning at town dumps. As industrial boilers and ovens became more sophisticated, so did waste combustion technology. Today, in the United States, more than 150 large centralized waste-to-energy plants generate heat and electricity as they consume roughly 17 percent of the nation's municipal wastes. These facilities are similar to large chemical processing plants. Besides using waste as inputs, these facilities often require fuel for makeup heat, and they generate as outputs various stack gases, fugitive emissions, and bottom ash. Common incinerator gases include nitrogen, sulfur dioxide, carbon monoxide, oxygen, hydrogen chlorine, water vapor, particulates, and trace amounts of metals. In addition, quite toxic substances such as dioxins and furans have been found in incinerator stack gases. The fly ash and bottom ash are composed primarily of metals (lead, cadmium, zinc, etc.) and soluble chloride and sulfate salts. The composition of residuals released from incinerators is determined by the mixture of the injected wastes. A large proportion of organic materials in the waste stream may help fuel the combustion process, but it is also likely to add to the chemical complexity of the residual gases and ashes.

As wastes, polymers generate significant environmental impacts. The volume of plastics discarded each year is greater than the annual volume of all ceramics or metals discarded. Most plastics are not readily biodegradable, so they often persist in landfills. Incineration of chlorinated polymers is of particular public concern owing to the generation of toxic and potentially carcinogenic air emissions. Tires are especially difficult to incinerate. Nondegradable plastic released as litter, especially polystyrene foams and polyethylene films, is dissipated around the countryside as a nuisance and an eyesore.

Recycling and the reuse of materials has been practiced for centuries. Some materials, such as metals, recycle well, while others, such as plastics, are often degraded by the recycling process. The recycling of iron and aluminum is dependent on their principal use. The recycling of scrap from structural uses (buildings, pipelines, machinery) is normally quite high. The recycling of smaller items (nails, foils, wire, razor blades, mattress springs) is much harder to manage and less cost effective because of the dissipated use pattern. Recycling rates for beverage containers are strong where government deposit laws and recycling infrastructure create incentives for more centralized collection.

With the exception of recycling, most waste management strategies are largely dissipative. Landfilling typically involves the dumping of mixed waste streams that make all but the most persistent materials difficult to reclaim. Incineration converts solid and liquid wastes into gaseous wastes and concentrated ash. While metals in the ash may be reclaimable, the gases are permanently lost. In addition, incineration usually removes whatever energy value may remain in the waste materials.

3.6 The Materials Economy

As the nation grew, its materials flows increased and changed. The natural materials of the land gave way to increasingly sophisticated synthetic materials. Increasing consumption of energy not only made materials more available at higher levels of refinement but also subsidized the costs of many of the newer synthetic materials that increasingly replaced more conventional ones. The volume of material extraction, processing, and use has steadily increased, but so has the intensity of use. Materials have become lighter, stronger, more workable, and better performing. Relative costs have decreased and more people have been able to enjoy greater material benefits with more functional and convenient products and services. The great technological advances in transportation, communication, agriculture, housing, entertainment, and military equipment that marked the past century and a half have all depended on the ever-increasing sophistication of the materials base.

Within this materials economy, large volumes of industrial materials are processed into commodities by thousands of manufacturing firms, and these products are rapidly pumped through millions of consumption activities and converted into huge amounts of wastes. While most of the production by-products are released to the environment as pollution, an increasing share is collected as wastes and managed in a range of treatment and disposal technologies of varying sophistication. Yet eventually these wastes also enter the environment and there they settle into the environmental media in ways that alter and too often threaten the nearby natural systems.

The linear flow of materials in a materials balance model shows how raw materials consumed in the economic subsystem are converted into wastes. As high-quality materials are processed, used, and disposed of, they are transformed into wasted materials that are of lower quality, and because of this lower value, they are released to the environment as dissipated pollutants. The materials balance model assumes that the natural environment has an endless supply of raw materials for extraction and an unlimited capacity to assimilate wastes. This assumption poses some serious problems.

The rapid depletion of the best of the nation's material resources and the unremitting release of the great glut of material wastes upon the environment over the past 150 years was certain to have impacts. The immediate impacts on land and water were always clear; less clear were the more long-term effects on ecological systems and human health. A more subtle and less visible shadow follows the grand story of the nation's development of materials, and this history needs to be considered here next.

4

Industrial Materials and the Environment

Man has too long forgotten that the earth was given to him for usufruct alone, not for consumption, still less for profligate waste.
—George Perkins Marsh

In 1972, an international research forum called the Club of Rome issued a controversial report entitled *Limits to Growth*, which projected a world short on the physical capacity to sustain a burgeoning human population.[1] Concerned about the exponential growth of energy demand and depletion of the earth's raw materials, the authors used computer models to simulate global trends and warned that the earth would run out of resources within a hundred years. As alarming as this projection was, it would not have been considered news by many nineteenth-century observers. The loss of the forests to the sawmill; the rapidly diminishing supply and quality of various ores; the depletion of oilfields; and the ruined rivers, water supplies, and fishing grounds due to chemical fouling were well recognized and discussed from the end of the Civil War on.

The clearing of the eastern forests, the building of the urban infrastructure, and the extraction of minerals and fuels raised significant concern about material supplies among nineteenth-century professionals. World Wars I and II created huge demands for materials and the ensuing concern about supplies, more specifically access to supplies for national security, created additional waves of concern about the loss of such materials. While the opening up of global markets in materials alleviated some of this anxiety, the rise of the petrochemical industry with its great wealth of new materials offered seemingly unlimited substitutes for scarcer materials, and that tamed concerns about materials depletion—at least as long as the petroleum reserves held out. Concern over wastes also appeared early on, although the vastness of the American continent allowed most of these wastes to be hidden until well into the twentieth century. Still, wastes as an issue in mining, industrial production, and municipal refuse were of substantial concern within local regions, where the impacts were often significant.

Concern over the depletion of the nation's resources—its "natural capital"—was not discussed in terms of material dissipation because the concept was not well developed at the time. However, the depletion of material resources and the fouling of the environment were two sides of the problem of dissipation. High-quality materials supplies—largely minerals and forests—were extracted, processed, used, and disposed of in a manner that reduced the available stocks and dispersed them as wastes throughout the countryside and waterways, often jeopardizing the healthy functioning of the natural environment. Plenty of public attention was focused on both the falling materials reserves and the increasing materials wastes, and, on occasion the connection between the two was clearly drawn.

4.1   Understanding Dissipation

Dissipation is not a readily understood concept. The dissipation of materials is best explained by reference to thermodynamic principles. According to the second principle of thermodynamics, energy is dissipated during work. In accomplishing any task, a certain amount of energy is consumed and expended in the form of heat. Once the work is completed, it is said that the quality of the energy has been dissipated because its capacity to do further work has been diminished. The term *entropy* is used to denote this dissipated state of energy. Because there is a link between energy and materials, and materials embody certain amounts of physical, chemical, or nuclear energy, this thermodynamic principle can be applied to materials as well. Thus, any material transformation that involves work results in dissipation in the quality of the material and an increase in entropy. The quality of materials (capacity to do work) is increased when they are extracted, synthesized, and processed, but this requires energy inputs, usually from fossil fuels, and the result is always a loss in some amount of energy in the biosphere.[2] As most materials are used, their quality (capacity for further use) diminishes. Solvents become contaminated, tires become thinner, and textiles become worn. Yet it is disposal—the discharging of materials back into the environment in irrecoverable ways—that most leads to dissipation and increased entropy.

The dissipation of industrial materials presents two problems. First, the quality of materials is diminished by use and where the material is nonrenewable, the global stock of high-quality material is depleted. Second, the resulting lower quality material is distributed into the environment in the form of pollution and wastes. As these dissipated materials accumulate in specific environmental media or ecological niches of the environment, they threaten the vitality of natural processes.

Material is lost to the environment throughout a material's life cycle. Although mining and smelting are intended to collect and refine a targeted metal, some of the metal is lost in the overburden, tailings, and smelter wastes. Likewise, pumping and refining oil and gas involves losses of some of the product. In addition, these extractive processes disturb many nontarget materials that may not be commercially viable at the time, but are simply mixed into the wastes. Processing of materials also involves losses that can occur as unrecycled scrap, spills and leaks, fugitive emissions, or off-specification product. Processing usually involves intermediates and accessory materials that are released as waste as their purity or performance is diminished. In use, some materials, such as solvents and paints, are designed to volatilize, while many products simply erode from abrasion and wear. Disposal is the final source of dissipation because so many products are landfilled or incinerated after their useful life is over. Leachates and volatiles from landfills, and smoke and ash from combustion slowly disperse materials into the environment in forms that are nearly irretrievable. The sum total of the dissipated materials released to the environment can be staggering. Table 4.1 shows the worldwide emission of metals into the atmosphere.

In 1935, Arthur Tansley coined the term *ecosystem* to describe the interconnected webs of materials and organisms that are necessary to sustain viable habitats for plant and animal communities. Ecosystems are

Table 4.1
Woldwide atmospheric emissions of trace metals (thousand tons per year)

	Energy produc- tion	Metals refining	Manufac- turing	Commercial and trans- portation	Total anthro- pogenic	Total natural
Arsenic	2.2	12.4	2.0	2.3	19.0	12.0
Cadmium	0.8	5.4	0.6	0.8	7.6	1.4
Chromium	12.7	—	17.0	0.8	31.0	43.0
Copper	8.0	23.6	2.0	1.6	35.0	6.1
Lead	12.7	49.1	15.7	254.9	332.0	28.0
Manganese	12.1	3.2	14.7	8.3	38.0	12.0
Mercury	2.3	0.1	—	1.2	3.6	317.0
Nickel	42.0	4.8	4.5	0.4	52.0	2.5
Selenium	2.3	2.3	—	0.1	6.3	3.0
Tin	3.3	1.1	—	0.8	5.1	10.0
Zinc	72.5	72.5	33.4	9.2	132.0	45.0

Source: Adapted from J. O. Nriagu, "Global Metal Pollution—Poisoning the Bioshere," *Environment*, 32(7), 1990, pp. 7–32.

dependent on relatively balanced flows of materials and energy. Too much or too little material or energy accumulating in an ecosystem's stocks can jeopardize its biological capacities. As dissipated materials from human activities build up in ecosystems, they may achieve a "critical load," or a concentration over which they cannot be effectively assimilated. In such cases, the pollutants are said to threaten the assimilative capacity of the system, or the ability of the ecosystem to accommodate activities and receive waste products without becoming deteriorated. The assimilative capacity represents the upper bound on how much material an ecosystem can effectively accommodate in any given time. As the amount or rate of pollution and wastes increases, ecosystems reach their assimilative capacity, their structural and functional integrity is breached, and they become incapable of supporting the organisms that depend on them.[3]

All ecological systems can process and accommodate a certain amount of exogenous material through various assimilative procedures. In the atmosphere, pollutants may be dispersed great distances, particularly if they are volatile or easily adsorbed onto aerosol particulates. In the air, some reactive substances can be chemically transformed when exposed to light or water vapor and degraded to more stable compounds. Excessive pollutant loads in water may be widely distributed by currents and deposited into freshwater or seabed sediments. Chemicals in water can be assimilated by simple precipitation or distillation, acid–base reactions, hydrolysis, oxidation–reduction reactions, or various biochemical processes. Such biodegradation usually leads to a lower hazard, but there are exceptions. In soils, pollutants can be sorbed to solids and attenuated by the existing mineral strata, particularly where the medium is rich in sorptive clays, hydrous metal oxides, or humus. In such cases, the pollutant may undergo a chemical transformation that allows it to reside in an ecosystem without threat.

However, once dissipated materials reach a certain concentration, they compromise the effective functioning of an environmental medium. Atmospheric damage occurs where chemical reactions threaten the upper ozone layer or build up excessive carbon concentrations that trap solar ultraviolet radiation. Water quality is compromised when the biological oxygen demand (BOD) is overwhelmed, when substances concentrate as subsurface sediments or water/air surface films, or when contaminants leach into groundwater. Soil damage appears when chemicals displace oxygen in soil, or alter the porosity of the soil, leading to compaction or erosion. The acidic wastes of smelters or petroleum brines, for example, can inhibit plant growth, reducing soil cover and permitting erosion as unprotected soil is exposed to precipitation.

As environmental media are damaged, the healthy functioning of the associated ecosystems is threatened. An excess of ultraviolet radiation due to atmospheric damage can limit the plankton community on the ocean surface, leading to reduced nutrients for marine life. An increase in levels of BOD from the decomposition of organic pollutants in rivers and lakes can deplete the oxygen supply in water, leading to the suffocation of fish, which may further increase the BOD levels as the dead fish decompose. Compacted soils or soils made more acidic by precipitation rich in sulfuric acid may curtail microbial life, resulting in further compaction, reduced nutrient transfer potential, and major changes in plant communities.

So dissipation is a form of both environmental and economic destruction. It decreases the quality of materials stocks as well as the capacity of the ecological "sinks." It leads to disorder and devaluation. Dissipation is about more than merely economic losses, for each unit of value we lose is a threat that the environment gains. Today we can study and describe the processes of dissipation in tight scientific terms, but the concepts of loss and destruction have long been on the minds of those who cared for and wrote about the environment. Their knowledge of the specific mechanisms by which dissipation damages ecosystems was less analytical than ours today, but what they saw was enough for them to recognize the dangers.

4.2   The Stresses of Materials Depletion

The early settlers of the American continent found an environment rich in natural resources. Forests, water, wildlife, and fertile soil abounded and it did not take these explorers long to discover the rich ores of the bedrock. At first it appeared that as resources were depleted, either because of scarcity or pollution, all that was needed was a move further west. However, beyond the Mississippi River, resources became more fragmented. Water and wood were less available, but the soils were fertile and precious metals offered a compensating reward. The rapid settlement and development of the continent resulted in the excessive consumption and overexploitation of this ample bounty of resources, and the costs soon threatened to outweigh the benefits.

At the close of the nineteenth century, public concern over the wasting of the nation's resources fed a popular movement that is regarded as the predecessor of today's modern environmental movement. Drawing on a literary history from Transcendentalists such as Henry David Thoreau and Ralph Waldo Emerson, the conservation movement arose during the 1880s and 1890s among the urban middle class of East Coast cities. Professionals, writers, and public servants in these privileged settings had become enamored of the natural endowment of the American landscape and were concerned about its potential destruction.[4] This early conservation movement revealed a deep concern over depletion of the nation's natural material wealth. The writings of movement leaders like Gifford Pinchot, Robert Underwood Johnson, George Bird Grinnell, Charles Sprague Sargent, and others abounds with estimates of the natural resources of the country and various predictions of their destruction. These men were well aware of the literary works of Thomas Malthus, David Ricardo, John Stuart Mill, and others who constructed classical economics based largely on the relationship between price and scarcity. The conservationists were joined by naturalists, agronomists, geographers, geologists, and biologists who were also researching and writing about the overtaxation of the nation's material resources and the threat these activities posed.

In 1908 President Theodore Roosevelt called a special Governor's Conference on Conservation at the White House. The papers prepared for this pivotal meeting were filled with inventories and projections concerning forests, minerals, water supplies, agricultural lands, and wildlife.[5] Following the conference, President Roosevelt established the National Conservation Commission to study the condition of the nation's natural resources and asserted, "Conservation of our resources is the fundamental question before this nation ... our first and greatest task is to set our house in order and begin to live within our means."[6] The Commission's study was conducted under the direction of the chief forester of the U.S. Forest Service, Gifford Pinchot. Its report, which was published in three volumes in 1909, documented the rapid exploitation of forests, minerals, water, and land. It urged efficient use of materials, called for the development of material substitutes, promoted a broad international approach to conservation initiatives, and warned that the use of resources "carries with it a sacred obligation not to waste this precious heritage."[7] Strong as the study was on conservation of resources, it did not directly address the industrial uses of materials or the problems generated by those uses.

As the chief of the national Forest Service, Gifford Pinchot became the leading voice for conservation of the nation's resources. He sums up this view in his 1910 book, *The Fight for Conservation*, where he claims,

The five indispensably essential materials in our civilization are wood, water, coal, iron, and agricultural products.... We have timber for less than thirty years at the present rate of cutting. The figures indicate that our demands upon the forest have increased twice as fast as our population. We have anthracite coal for about fifty years, and bituminous coal for less than two hundred. Our supplies of iron ore, mineral oil, and natural gas are rapidly being depleted, and many of the great fields are already exhausted. Mineral resources such as these when once gone are gone forever.[8]

The assumption here, as elsewhere in these reports, is clearly based on a view of resources as fixed and exhaustible. The concern for materials scarcity is the same as that which drove early classical economists, but the solution is different. Rather than wait for the market to raise prices to account for growing scarcity, the conservationists believed in government intervention to preserve and manage resources important to the nation's economy.

Long after the conservation movement had run its course, concern over materials scarcity remained high. World War I revealed the important strategic value of materials. The nation entered the war with commanding resources. In 1913, 64 percent of the world's petroleum came from U.S. wells and the United States was the largest source of 13 of the 30 most important mineral commodities, including, coal, iron, lead, zinc, silver, tungsten, molybdenum, arsenic, and phosphate.[9] Yet the war also revealed the importance of foreign minerals such as potash, bauxite, nitrogen, and the organic chemicals needed for munitions, fertilizers, and pharmaceuticals. The war actually spurred the development of previously uneconomical materials such as chromite and manganese, which were needed to substitute for materials embargoed by foreign combatants. Further, the war opened up domestic markets for domestic suppliers who were suddenly without competitors. But the war was also a voracious consumer of materials and it ended with some of the nation's highest-quality ores, such as copper, diminished, and many predicting that the richest oil fields would soon run dry.[10] The director of the U.S. Bureau of Mines predicted in 1919 that "within the next two to five years the oil fields of this country will reach their maximum production, and from that time on we will face an ever increasing decline," and the director of the U.S. Geological Survey (USGS) warned of a "gasoline famine."[11]

One of the most influential advocates for materials management to emerge after the war was Charles Kenneth Leith, a University of Wisconsin economic geologist. In 1931, Leith observed that the nation had mined and consumed more minerals in two decades than in its entire previous history. For emphasis, he noted that "(a) single Lake Superior iron mine now produces every two weeks a volume of ore equivalent to the Great Pyramid of Egypt...."[12] Leith, Josiah Spurr, president of the

Mining and Metallurgical Society of America, and other notable geologists were concerned about the rapid depletion of high-grade domestic minerals, but they also saw the value in developing those resources. As a way to balance these conflicting objectives, Leith advocated a more rational approach to the management of the minerals in the global market. With Spurr and other prominent leaders of industry, Leith pressed for professional studies and federal government programs to open up access for all nations to the mineral resources of the globe. The historical significance of Leith and his colleagues' concerns could be well felt by the close of the 1930s. By then, Britain and the United States controlled nearly 75 percent of the world's minerals, and while this provided security for the English-speaking countries, it was a point of deep frustration for the so-called "have not" nations like Germany, Japan, and Italy, who soon responded with force.

World War II, like World War I, consumed vast amounts of industrial materials. Although dominance in materials had been critical to wartime success, the end of the war found the U.S. materials reserves depleted and drained. In a deeply somber article in 1945, Interior Secretary Harold Ickes reported that,

The prodigal harvest of minerals that we have reaped to win this war has bankrupted some of our most vital mineral resources.... We no longer deserve to be listed with Russia and the British Empire as one of the "Have" nations of the world. We should be listed with the "Have Nots," such as Germany and Japan.[13]

This grim picture was echoed by Elmer Pehrson, the chief economist at the U.S. Bureau of Mines, who claimed that the nation had consumed 80 percent of its lead deposits, 75 percent of its zinc deposits, and 60 percent of its petroleum and copper reserves. "Exhaustion is well advanced in a number of important subsidiary minerals," Pehrson noted, "so that we can no longer drift along with the easy going philosophy that the earth will provide."[14]

Not everyone was so pessimistic. William Wrather, director of the U.S. Geological Survey, noted parallels with the dire predictions for oil reserves in the 1920s and stated flatly, "Faith in the ability of American scientists, scientific methods, and in a bounteous nature leads me to predict that great mineral resources remain to be developed."[15] In 1947, an engineering consultant, J. Frederick Dewhurst, published a massive inventory of the nation's resources and needs with projections for the future.[16] This study noted that between 1925 and 1929, the American economy was the largest consumer in the world of every important metal, but it was also a huge producer. In his projections Dewhurst found that petroleum, copper, lead, and zinc, and, in the longer term, aluminum and iron, presented serious domestic scarcity problems for the future. But by considering the opportunities for foreign imports (which he did not study), he remained optimistic,

It is inevitable of course, as time goes on, that our supplies of mineral resources will come nearer to exhaustion and that our needs will be satisfied with less ease and at greater cost than before. Given a system that permits free access to the world's resources, however, there can be no question of a raw material supply adequate to support an expanding American economy for many decades to come.[17]

Even those more optimistic about the future recognized the exhaustible quality of the domestic stocks of natural resources. Some believed that rising market prices would promote the exploitation of lower grade materials and others believed that foreign imports could be counted on for the nation's future needs. While such reliance on foreign imports worried many who feared embargoes by hostile suppliers, others saw in foreign sourcing a means to integrate national economies and make such hostilities less likely. But even these internationalists recognized that the global supply of high-grade resources was limited; eventually the world's economies would have to make do with lower quality and less accessible materials. To contend with that future, the optimists looked to new technologies and the development of new, synthetic materials.

4.3   The Stresses of Materials Dispersion

Only slightly less well articulated than the national debate and concern over the supply of industrial materials has been the public attention focused on the issue of material wastes dispersed into environmental media. While materials scarcity has been a broad, national concern, concern over this other end of the materials cycle has largely developed at the local level. Wastes had no military, commercial, or trade implications, so federal agencies saw no direct responsibilities for managing them.

However, waste materials were a public issue. From the mid-nineteenth century on, the increasing volume of mining wastes, industrial discharges, and municipal refuse grew in public concern; the response, though, was viewed as a local, not a national, issue.[18]

In mining districts there is a long history of concern over the digging of mines, the movement of overburden, the creation of huge and looming slag piles, and the volatile gases generated by smelters.[19] For many minerals, the large amounts of solid mining and milling wastes associated with extraction of ores far exceed the quantities of material obtained. The processing of metals often results in large amounts of waterborne acidic wastes and sludges from pickling. Copper, gold, and lead mines are particularly polluting. Iron and aluminum are obtained almost exclusively from oxide ores and their compounds are relatively benign, but the scale of their extraction processes and the need to dispose of large amounts of mildly hazardous overburden generate the most significant environmental impacts of these materials.

The refining of metals generates large amounts of air pollutants from smelters, coke ovens, and furnaces. Nonferrous metals, such as copper, zinc, lead, and nickel, are typically found in nature as sulfides rather than oxides. These sulfides are first converted to oxides by a simple roasting process in which the sulfur is oxidized and driven off as sulfur dioxide. Thus, copper, zinc, lead, and nickel smelters are major sources of air pollution. The processing of these metals often also generates large amounts of by-products. Copper, lead, zinc, or nickel smelting typically yields tailings, slags, ashes, or flue dusts rich in antimony, arsenic, bismuth, cadmium, selenium, or silver. The concentration of these by-products in the waste stream and their market value usually determine whether they are recovered for use or discarded in the waste.

Early observers of nineteenth-century mining noted the pockmarked landscape of the Missouri lead mines, the gouged-out hillsides of the western placer mines, the mounds of gravel and mine tailings, the dieoff of vegetation near ore smelters, the clear-cutting of nearby forests for smelter fuel, and the polluted rivers flowing from mined lands. The respected mining journalist, J. Ross Browne bluntly wrote in 1868: "No country in the world can show such wasteful systems of mining as prevails in ours."[20] Yet throughout this period, as various states began to write laws on mining—several of which included the appointment of mine inspectors to address safety issues—there was little legal requirement to consider the effects of mining or its wastes on the environment. One notable exception was the California reaction to hydraulic-placer mining. This technique used high-pressure water streams that turned hillsides into slurries in order to sluice for precious metals. The large volumes of silt-filled water disrupted local river systems and in 1884, downstream farmers on the Yuba River sued the mining industry for relief. In a well-publicized finding, the U.S. circuit court "perpetually enjoined and restrained" the mining companies from further discharge of mining slurries.[21]

The early iron mines in New York state or the copper mines of Michigan tended to be underground shafts that left only a modest impact on the surface. First introduced in 1890 in Minnesota, open pit iron mining with steam shovels dramatically changed the environmental impact of mining. By 1910 large open pit mines for copper were developed throughout the West, with some of the largest pits 2 miles in diameter. Many of these open pit mines also produced substantial tailings and tailing ponds, while other mines generated overburden and slag piles that reached mountainous proportions. As the quality of the ore (the grade) diminished, the amount of waste increased. During the nineteenth century, only ore containing 25 percent copper was mined; by 1915, grades as low as 15 percent were being mined; and by the 1940s the grade had dropped to 2 percent. Each of these grades produced a sequentially larger volume of tailings and overburden.[22]

Local communities protested against these mines and their wastes. Civic leaders wrote articles and appealed to local governments. Journalists photographed mine wastes, published exposés, and wrote editorials. Farmers and downstream fishing interests complained about water runoff and nuisance suits were frequent. But local critics were often confronted by local business interests that benefitted from the mines. The conflict in values between those who benefitted and those who were harmed was played out in many small communities and local governments. It took the national influence of the conservation movement to bring about a gradual shift toward the reconsideration of mine wastes. A 1907 resolution of the American Mining Congress condemned unnecessary mining wastes and urged the conservation of ores "for the best interests of the whole people."[23] During the 1920s and 1930s, public attention focused on the destruction of forests and farm lands by strip mining of coal and the pollution of rivers and streams by the residues of these mines. West Virginia passed the first state law regulating strip mining in 1939 and by the mid-1940s the U.S. Congress was beginning to hold hearings on the potential need for federal regulations.

Industrial and manufacturing wastes were a second source of growing concern among local community and civic activists. Manufacturing resulted in large amounts of materials dissipated into the environment in the form of solid wastes, water effluents, and air emissions. Early efforts to manage the volume of wastes generated by industrial facilities relied on discharges to waterways or on land burial. Nineteenth-century factories, especially those in the textile, chemical, and iron and steel industries, were often located along waterways for easy transport of raw materials and products, power captured from water flows or steam, and cooling and quenching of heated processes. These waterways also provided the easiest and least expensive means of disposing of soluble or suspended wastes. Studies of the nation's rivers at the turn of the century suggest that by 1900, 40 percent of the pollution load was from factories.[24]

The earliest remedy for those concerned about industrial wastes was the common law tradition of nuisance suits. The aggrieved party needed to show personal damage resulting from the reckless behavior of the perpetrator. While such cases were difficult to win, enough victories over industrial waste discharges were achieved during the beginning of the twentieth century that industrial firms began to pay some attention to their waste streams.

Initially, toxic and hazardous waste materials were routinely mixed into more benign industrial wastes with the anticipation that such mixing would dilute and thereby lower the risks of the more dangerous wastes. For instance, phenolic and tar wastes from the by-product recovery ovens at municipal gas works were routinely mixed with other solid wastes, including municipal wastes, before disposal. As public concern and legal nuisance suits arose from industry neighbors and downstream property owners, firms began to dump wastes in landfills or open pits, often on their own property. Open disposal of wastes continued during the 1920s, even as engineering firms were developing and marketing new chemical and physical waste treatment technologies. While some corporate leaders argued for sound waste management practices, most firms maintained an ad hoc policy of delaying waste treatment investments wherever possible. Without laws to the contrary, firms had little motivation to manage their wastes otherwise.[25]

Some professionals during the 1930s argued that large volumes of industrial wastes implied industrial inefficiencies in the management of materials. E. B. Besselievre, a noted sanitary engineer, writing in an industry magazine, argued for better waste treatment by claiming, "what may at first seem to be an exorbitant expense in order to stop polluting a stream may prove to be a boon to a plant by the proof to the owner that he is throwing away a product that may prove of value."[26] Besselievre went on to suggest several treatment options, including the recovery of valuable by-products for reintroduction as raw materials.

The economic depression of the 1930s encouraged more consideration of by-product recovery and reuse. Recovery of silver from photographic film production sludges was a notable example, as was the recovery of propane at oil refineries, the conversion of waste propylene to isopropanol, and the production of various solvents from ethylene wastes. However, the markets for these early by-product materials were meager and therefore by-product recycling and production integration were not often mentioned as waste treatment options.[27]

Despite protestations by industry that hazardous wastes could be neutralized by mixing them with nonhazardous wastes and thereby safely discharging them, by the 1940s leading firms in the chemical industry were adopting primary waste separation practices for hazardous sludges and semisolid residues and reserving these wastes for special land disposal sites. For instance, Monsanto Chemical Company began sequestering its toxic wastes for land disposal in East St. Louis during the 1940s, while Hooker Chemical Company began isolated burial of toxic wastes at Love Canal, New York, during the same time.

Physical isolation by burial was the cheapest treatment technology, but it was not the only technology available. Acid neutralization was a proven technology. Incineration of chemical wastes had been developed and tested by various engineering firms, and Vulcan Iron Works had set up a rotary kiln incinerator during the 1940s with a small customer base made up primarily of chemical companies. Pumping wastes into abandoned oil and gas wells was also advocated and by the 1950s both Dow and Du Pont were using such deep well injection for their saline wastes. Still, up through the 1960s these technologies were more the exception than the rule.

Following World War II, the chemical industry grew rapidly and by 1980 it was the largest industrial source of hazardous wastes. While the large majority of these wastes were deposited in on-site landfills, the remaining amounts were dumped indiscriminately. During this period the leading firms in the industry reported depositing some 762 million tons of chemical wastes in over 3300 locations around the country.[28]

The third area where waste problems were identified involved domestic refuse. The problem of postconsumption wastes—municipal refuse, garbage, or "trash"—grew to crisis conditions in large urban centers during the 1890s. Such wastes had always been a factor of municipal life, but the scale of the problem increased with a vengeance as city populations swelled and increasing affluence led to a rising standard of living. For example, Boston authorities estimated that refuse collectors hauled off a record 350,000 loads of garbage and street debris from city streets in 1890, while some 225 street teams in Chicago gathered 2000 cubic yards of refuse daily.[29] In 1897 the American Public Health Association completed a decade-long study of municipal refuse that documented a rising wave of wastes in over 150 U.S. cities. Even at the time, U.S. households generated a greater waste stream than European counterparts. A 1904 survey of fourteen U.S. cities found an average rate of 860 pounds of mixed trash per capita per year compared with 450 pounds per capita in eight comparable English cities and 319 pounds per capita in seventy-seven German cities.[30]

Public concern over urban refuse emerged as a part of civic reform movements. Citizens groups, neighborhood associations, and newspapers campaigned for better municipal collection services, and gradually city boards of health were charged with responsibility for refuse collection. Boards of health in the larger cities tended to manage trash collection within municipal departments, but many other cities, skeptical of the influence of patronage, contracted with private services for waste collection. By 1915, half of the major cities had some form of municipally controlled collection systems, while another 39 percent provided this service under private contract.[31] However, carting away the wastes only moved the problem to another location. Much of the urban refuse was hauled out to open dumps on vacant land or deposited into lakes, rivers, harbors, or the open sea with no particular attention paid to sanitary protections. Such indiscriminate disposal bore its own liabilities. Residents abutting open dump sites protested aggressively, and public health physicians warned of the risks of infectious disease from exposure to such sites. By the close of the decade, waterways were so threatened by physical obstructions and refuse dumping that the U.S. Congress enacted what some today consider the first federal environmental protection law, the Refuse Act of 1899. This law, which at the time was justified as an appropriate federal initiative to protect interstate commerce, prohibited the obstruction of shipping channels by refuse thrown into surface waters.

Concern over the volume and health risks of open dumping of municipal wastes led to two frequently competing solutions: waste incineration and waste reduction through recycling. The first garbage furnace, modeled on European incinerators of the time, was built at Governors Island, New York, in 1885 and municipal officials were soon contracting with private companies for similar facilities across the country. These early combustion plants were often inadequately designed and poorly managed and as a result, of the 180 incinerators built in the United States between 1885 and 1908, 102 had been abandoned by 1909. Thus, waste incineration had many critics and few were more influential than New York City's Commissioner of Street Cleaning, George E. Waring, Jr.[32]

Waring, a military engineer noted for building the first separate sanitary municipal sewer systems, was appointed to his New York City post in the anti-Tammany Hall reforms of 1894. From this position Waring became a national champion in the drive for municipal "cleanliness." He made street cleaning professional and created an army of New York City waste collectors. In a well-read text, *Sewage and Land Drainage*, Waring claimed, "There is no surer index of the degree of civilization of a community than the manner in which it treats its organic wastes."[33] In New York's trash Waring saw a reclaimable resource. He argued that burning of municipal waste "means destruction and loss of matter which may be converted into a source of revenue."[34] Instead, Waring promoted the establishment of resource recovery centers that employed recent immigrants to scavenge city wastes and reclaim productive materials for recycling and reuse. During his tenure, New York City established a broad network of resource recovery centers that competed for financial returns with the private pushcart junk dealers and the "scow trimmers" who rode the waste barges floating off the New York coastline. Indeed, Waring's successor, John McGaw Woodbury, used the same argument for waste reclamation to criticize ocean disposal. In a 1903 professional discussion, Woodbury argued, "It is of the utmost importance that dumping at sea be stopped, not simply because it makes the beaches unsanitary and unsightly, but because it is a waste of valuable materials."[35]

Recycling and resource reclamation centers arose rapidly in the major metropolitan centers prior to World War I. By one estimate at the time, more than 70 percent of U.S. cities had programs to separate materials for reclamation. But reduction of wastes by recycling also had its critics. The costs of building resource recovery facilities and their operation proved to be higher than the costs of incineration, and the markets for the reclaimed materials generated by the recovery facilities were weak to nonexistent. "We have been seduced by the glowing promises of rich rewards which the reduction process has failed to give us," complained a New Orleans city councilor. By 1914, a survey of reduction plants across the country found that only twenty-two of the hundreds of municipal recovery centers built were still in operation. Following World War I, resource recovery fell out of favor as an uneconomical solution and most cities closed their facilities.[36]

The economy changed during the 1920s and 1930s and so did the waste stream. The growing affluence of the economy and the changing nature of its materials meant that the municipal waste stream grew much larger on a per capita basis and changed in composition to include disposable paper goods, plastics, toxic chemicals, and various synthetic materials. One estimate suggested that waste increased about five times as rapidly as population. In 1920, per capita generation of waste was about 2.75 pounds per day; by 1970, it had risen to 5 pounds per day; and by 1980 the number was closer to 8 pounds per day.[37]

The rapid growth in disposable products and packaging resulted in the creation of thousands of landfills on the outskirts of growing cities during the 1940s and 1950s. Local opposition to such landfills increased as suburbs developed around these open disposal pits, and waste management professionals were soon promoting "sanitary landfills" that involved nightly coverings with a layer of earth. Although this practice improved dump management, it did not reduce the volume of waste, and communities were increasingly finding that their landfills were creating prominent landscape forms. By the mid-1970s, combustion was back in vogue. Incineration was again touted as the best means to reduce the volume of waste and by 1980, some 60 new, privately managed "waste-to-energy," "resource recovery," and "mass burn" facilities were operating or under construction. Seven years later, an EPA survey identified 110 plants in operation, with another 220 planned. Once again, community and neighborhood resistance emerged, raising concerns about both public health and property values, and by the turn of the decade the large incinerator contractors were facing a dwindling market.[38]

As the nation developed, so did its wastes. Mining wastes were recognized as a source of stream and river pollution and smelters were accused of air pollution. Urban wastes were viewed as a nuisance, an aesthetic blight, and a hazard to public health. But up until the middle of the twentieth century, these wastes were considered primarily a problem of proper management, rather than a threat to the global ecosystem. The Club of Rome report dramatically brought to public attention the scale of anthropogenic wastes, and it was this report that first suggested that the global capacity to assimilate these wastes may be an even greater barrier to economic growth than diminishing supplies of high-quality, non-renewable resources.

4.4   Understanding the Reluctance to Conserve

Concern over the depletion of industrial materials arose after the Civil War and was revisited after each of the world wars. This was first of all a concern about materials for national security, but easily seen in these discussions was also a concern about commercial development. The loss of forests was recognized early in the nineteenth century. Concern over the scarcity of minerals, particularly copper and iron, marked the turn of the century. More recently, the worry has been about the limited supply of oil and natural gas and various strategic metals.

In 1972, the Club of Rome used the emerging analytical power of electronic computers and the new field of economic modeling to predict in *Limits to Growth* that within 50 to 100 years the world economies would collapse as a result of the exhaustion of natural resources and the buildup of ecologically destructive levels of pollution. In doing so, the study reinforced the finite image of "spaceship earth" enunciated a decade earlier by economist Kenneth Boulding. However, the study was roundly criticized by many economists, who found that the model unquestioningly assumed continuation of the exceptional exponential growth in population and economic activity that characterized the 1950s and 1960s and relied upon model inputs that significantly underestimated the earth's mineral reserve base.[39]

The problem of materials scarcity did not go away, but the responses split into two camps. One view, shared mostly by economists, held that as the supply of any resource declines, its price rises and the economy adjusts by substituting more plentiful resources for those that are scarce. Thus, iron replaced wood in structural applications as the supply of large, solid trees diminished. Similarly, plastics replaced metals as the supply of high-quality metals diminished and as metal prices rose. The most compelling case for the market effects came from two resource economists, Harold Barnett and Chandler Morse. Barnett and Morse used historical evidence from the United States to show that for many resources there had actually been an increase in usable materials relative to economic investments since the Civil War, because technological and productivity advances had continuously made lower grade materials commercially viable.[40]

The other view, held mostly by political scientists, argued that the total amount of natural resources is fixed and as the supply of some essential materials declines, so too will the economy. Materials substitution may well delay the final accounting, but once the resources are exhausted, the only possible result is a lower quality of material well-being. For instance, Robert Heilbroner argued that materials scarcity and environmental degradation derive directly from the industrial organization of the national economy, and that the rate of industrial growth must be dampened or the economy will collapse. "Ultimately there is a limit to the ability of the earth to support or tolerate the process of industrial activity, and there is reason to believe that we are now moving toward that limit very rapidly."[41] William Ophuls in *Ecology and the Politics of Scarcity* took this same theme and extended it to political theory. He argued that the United States has been politically stabilized by its continuous economic growth, which has depended on access to an abundant supply of material resources, and U.S. political history "is but the record of a more or less amicable squabble over the division of the spoils of a growing economy." But Ophuls saw this abundance running out and the nation unprepared for an economy of scarcity. He concluded by stating, "What is ultimately required by the crisis of ecological scarcity, is the invention of a new mode of civilization, for nothing less seems likely to meet the challenge."[42]

Both camps acknowledged the effects of technological advances. The economists argued that technologies emerge to alleviate the decline in available resources. Each new concern about scarcity has encouraged the development of technologies that make lower grade materials commercially viable and promote the discovery or development of new materials. The political scientists saw technologies as accelerating the depletion of natural resources. For example, Barry Commoner, a biologist by background, found, "The chief reason for the environmental crisis that has engulfed the United States in recent years is the sweeping transformation of productive technology since World War II." In particular Commoner fingered the chemical industry because its products and processes are so hazardous. He is optimistic that changes in technologies and materials can alleviate the crisis, although his is a cautious optimism. Because decisions about technologies and materials are primarily in the hands of corporate managers and investors, Commoner argues that the necessary technological changes will not occur without a change in the current system of social decision making.[43]

Indeed, there is little evidence of materials scarcity in the United States. There are a host of explanations for this and it is probably due to a combination of several factors, including the following: (1) as higher grade resources become more scarce, lower grade materials become more attractive and available, (2) rising prices have encouraged more mineral exploration, which has proven to be highly successful, (3) increasing costs for scarce materials have permitted new materials sources to enter the market, (4) technological improvements have increased the productivity of existing materials, and (5) global markets have expanded resource availability. The result has been an objective abundance of materials resources, which has chilled the call for resource conservation and quieted concerns over the diminished condition of the old domestic reserves.

This easy solution has not extended to the problem of wastes. Historically, wastes have been less of a factor in determining materials development because for a long time they were so easily discharged to the environment with little economic cost. As public concerns rose, more distant disposal sites were substituted for those in dense urban areas. But the volumes of waste grew dramatically, and spatial isolation could only delay the inevitable confrontation of residential development, ecological assimilation, and waste disposal. There is a finite amount of ecological space and capacity for sequestering wastes and over time the increasing demands threatened to outstrip supply.

Eventually, industrial wastes so polluted the water and air supplies, and the stream of municipal wastes became such a torrent that wastes became a major public issue. The materials industries responded by investing in waste treatment options such as managed landfills, retention ponds and lagoons, and hazardous waste incinerators, and local governments responded with sanitary landfills and municipal incinerators. But on the whole, waste treatment can only manage waste, not reduce it. The volume of industrial wastes has continued to grow. Material wastes may be a symptom of industrial inefficiencies, but they are not yet so significant as to impel a serious commitment to conservation of materials.

During the 1960s, local and state officials began more aggressive initiatives to control water effluents, smoke concentrations, and automobile exhausts. But by then the changing composition of industrial wastes was beginning to be noticed. Added to the metals and inorganic chemicals of the previous century were the new synthetic organic chemicals, and these created a new kind of concern. These concerns were recognized by the materials industries, but not because of the loss of valuable resources as raw materials or as wastes. Instead, the concern that arose was tied to public health—particularly the effects of exposure to toxic chemicals—and the focus of industrial attention was diverted from resource conservation to toxic pollution. Toxicity of wastes, not their volume, became the primary public concern. Of course, toxicity was not a new concern. Workers in production facilities were well aware of toxic chemicals as an occupational hazard. But public concern over toxic chemicals in the environment, slow in coming, finally emerged during the 1960s and once arisen, it would not recede even as the century closed.

5

Industrial Materials and Public Health

As crude a weapon as the cave man's club, the chemical barrage has been hurled against the fabric of life—a fabric on the one hand delicate and destructible, on the other miraculously tough and resilient, and capable of striking back in unexpected ways.

—Rachel Carson

The publication of Rachel Carson's *Silent Spring* in 1962 is often noted as the pivotal event in stimulating public recognition of the hazards of synthetic chemicals in the environment.[1] Important as this publication was, the public had been aware of chemical risks long before the 1960s. Not only the public, but public health scientists, industrial professionals, and government agents had ample evidence of many hazards associated with industrial chemicals from early in the twentieth century. Dramatic evidence of environmental damage and human health effects was revealed through tragic cases of chemical contamination, and specialists in public health, industrial hygiene, and sanitary engineering had access to many published studies on human and environmental exposure to hazardous materials. Knowledge about the hazards of industrial materials developed in tandem with the development of many of those materials. Thoughtful observers of the time were not ignorant of the potential dangers to human heath. Scientists, researchers, journalists, and government agents openly discussed and wrote about the effects of industrial materials on human health and safety.

These warnings did not go without responses. There were government studies, professional conferences, informative textbooks, corporate research projects, industrial guidelines, and even new professions established and populated by trained and talented people. The information was there and professionals responded, but, still, many dangerous chemicals were introduced to the market and their wastes showed up throughout the environment. This was not the result of naive opportunism or simple irresponsibility. There was an active struggle here between the economic drive to produce profitable products and a professional and civic caution that arose about the human and ecological effects of those products; and the result was a compromise at best.

5.1   Scientific Understanding of Chemical Hazards

Certain chemicals have long been recognized as poisons. Common knowledge arose from direct experience over many centuries and was transferred through generations in both oral and written form. Political, military, and medicinal uses of poisonous substances led to early experiments, often on prisoners and patients. Over time these experiments and direct observations built up a body of knowledge that today serves as a foundation in the scientific fields of toxicology and pharmacology.

Occupational exposures to industrial chemicals often generated the earliest indicators of a chemical's hazards. Mercury, used in the hat-making industry, was recognized as the cause of hatter's shakes as early as the eighteenth century. Lead was early identified as a chronic occupational hazard because of its easily recognized symptoms, such as wrist drop and lip discoloration. Other occupational hazards associated with metalworking were well recognized by the seventeenth century. One of the founding philosophers of toxicology, Aureolus Paracelsus, published a major work, *On the Miners' Sicknesses and Other Diseases of Miners*, in 1567. Bernardino Ramazzini published a more general classic, *A Discourse on the Diseases of Workers*, in 1700 and this text was still used well into the nineteenth century.

The modern science of toxicology arose in Europe in the latter part of the nineteenth century, and by the turn of the century a robust literature and body of knowledge on chemical hazards in the workplace was emerging.[2] The experimental tradition of toxicology developed along with the growth of organic chemistry. Because German scientists were leaders in organic chemistry, they were also leaders in toxicology and some of the field's early experimental work was carried out by chemists and physicians at the Pharmacological Institute of Berlin.[3]

Public concern over chemical hazards in the workplace developed quite slowly over the early period of industrialization.[4] The primary focus of both employer and employee was on acute and physical, but not chronic, chemical hazards. The immediate dangers of substances such as ammonia and chlorine that were prone to accidental explosions during processing were well recognized as chemical manufacturing developed. Within an hour of beginning operations, the first electrolytic cell Herbert Dow built for processing chlorine in 1895 exploded. Shortly after Dow Chemical began to produce bleach,

[e]xplosion after explosion jarred the plant as cells blew up. Yellow chlorine poured out onto the floors. Chlorine gas filled the buildings, but workers found that they could counteract the effects of the gas by inhaling the fumes of grain alcohol.[5]

Such safety risks were considered a private issue properly negotiated between employers and prospective employees under the prevailing doctrine of "freedom of contract." Injury at work could be brought before the courts under prevailing civil law, but the prospects of winning such cases were small. Professional concern about the health of workers tended to focus on infectious disease, as a part of the broader public health concern about influenza, typhoid, and diptheria.[6]

The earliest efforts to recognize chronic chemical hazards in the workplace grew more slowly in the United States than in Europe. However, by 1910, a number of events marked the emergence of substantial professional interest. This was the year that Alice Hamilton published her first studies on lead poisoning in the pottery and paintmaking industries, John B. Andrews published his findings on phosphorus poisoning in the match industry, and W. Gilman Thompson set up the nation's first occupational health clinic at Cornell University Medical College in Ithaca, New York. Much of the early work focused on simple documentation. Both government and academic specialists began creating inventories of substances known to cause occupational health problems. For instance, the U.S. Bureau of Labor in 1910 published a list of hazardous work settings along with their associated diseases.[7] Relying mostly on European research, the list focused on dusts and gases that could be inhaled, but also identified lead and cadmium as substances that could be ingested or leached through the skin. Recognizing that this inventory was preliminary at best, the Bureau requested that employers voluntarily record worker exposure data and develop statistics that linked specific exposures to specific responses. Over the next two decades enough companies initiated recordkeeping programs and reported their statistics that the Bureau was able to publish updates to the initial list in 1922 and 1933.[8]

The first national conference on occupational disease was held in Chicago in 1910 by the American Association for Labor Legislation, and the National Safety Council was established the next year. In 1915, the American Public Health Association established a special Section on Industrial Hygiene and the U.S. Public Health Service established a Division of Industrial Hygiene and Sanitation. Although this new division suffered for years with inadequate appropriations, it did formalize the distinction between sanitation issues (involving infectious diseases) and industrial hygiene (involving chemical exposure) and it served as an information clearinghouse and resource coordinator for state industrial hygiene programs, which were also emerging during this same period.

These early professional responses to industrial chemical exposure recognized and affirmed important principles of toxicology. First, they identified the three primary routes of human exposure as inhalation, ingestion, and dermal (skin) exposure. Respiratory intake was of particular interest as new research methods were developed to monitor air concentrations of chemicals. On-site investigations also revealed the importance of ingestion of dusts or residues on lunchtime foods and exposure from contaminated clothing or work surfaces. Second, these practitioners recognized the importance of the magnitude and duration of chemical exposure captured by the professional homily: "the dose makes the poison." Third, they drew the distinction between acute (short-term) symptoms that could be readily diagnosed and the chronic (longer-term) symptoms that might not appear for several years.[9] Alice Hamilton, one of the foremost experts on lead exposure, warned against focusing too simplistically on short bursts of intense exposure to lead and consistently stressed the importance of long-term exposure to even small amounts.[10]

During the 1930s groundbreaking research on the toxic effects of chemicals had begun at the Department of Pharmacology at the University of Chicago, particularly around the work of Eugene M. Geiling. Additional research on chemicals and human health effects was carried out at the National Institutes for Health and the U.S. Public Health Service.[11] World War II and the Manhattan Project brought about rapid developments in toxicology that included advances in quantitative biology, radiotracer technology, experimental pathology, and inhalation toxicology, and the establishment of important research centers at the University of Rochester in New York and Oak Ridge, Tennessee. In 1951 Adrian Albert's classic text, *Selective Toxicity*, was published in several editions, which offered a concise documentation of the site-specific actions of chemicals. During the 1950s, the American Society of Toxicology was founded; the journal *Toxicology and Applied Pharmacology* was begun; and Kenneth Dubois and Eugene Geiling published their comprehensive *Textbook on Toxicology*.

A link between cancer and industrial chemicals was suggested as early as the eighteenth century, but was firmly developed during the late nineteenth and early twentieth centuries. In 1775 Percivall Pott, an eminent English physician, identified high levels of cancer of the scrotum among patients occupied as chimney sweeps and concluded that coal ash soot was the causative factor, although his findings had little effect on public health policy during the next century. By the 1890s, various cutting oils were implicated as causative factors in the etiology of skin cancers; "parafin cancer" was linked to shale oil; and exposure to amine dyes was associated with bladder cancer. In 1895, abnormal levels of bladder cancer were identified among workers in the aniline dye industry. In 1911, Peyton Rous had found a sarcoma virus that induced cancer in chickens, and in 1915 a Japanese pathologist, Katsusaburo Yamagiwa, reported on research demonstrating that skin tumors could be induced in rabbits by applying coal tar to their skin. This work resulted in a rash of studies on chemicals and cancer. In 1932, E. Kennaway was able to identify the carcinogenic compounds in coal tars as a class of polycyclic hydrocarbons.[12] The National Cancer Institute published a survey of some 700 chemicals in 1941, of which 169 were found to be active in inducing tumors in animals.[13] Of these, various common industrial chemicals were highly suspected as human carcinogens, including benzidine, arsenic, lead and zinc chromate, hexavalent chromium, coal tars, and coke oven emissions.

The mechanisms by which chemicals cause tumor formation proved more difficult to establish. The McArdle Laboratory for Cancer Research was established in Wisconsin during the 1940s and Elizabeth and James Miller began important research on carcinogenicity there. In 1947, the Millers first demonstrated that azo dyes become covalently bonded to proteins in the liver, but not to the proteins of the resulting neoplasms.[14] Continuations of this work demonstrated that chemical carcinogenesis required the covalent interaction of some form of the chemical with the macromolecules of target organs. By the 1950s it became clearer that specific industrial chemicals were not often the direct carcinogen, but rather that metabolic processes created derivatives of the chemical that could be even more potent carcinogens than the parent compound.[15] Pioneering work in plant mutation assays during the 1960s promoted further research into the genetic mechanisms of carcinogenicity by Bruce Ames and the Millers, as did the development of analytical tools that could detect chemicals at increasingly lower levels.

An understanding of the relationship between industrial chemicals and mutagenic damage also developed as toxicology matured. In 1927, H. J. Müller found that X-rays were mutagenic in fruit flies. The first unequivocal evidence that chemicals could be mutagenic occurred with the studies of Charlotte Auerbach and J. M. Robson in 1942 that demonstrated that mustard gas is mutagenic in fruit flies, although this work was not reported until after wartime censorship was lifted. During the war, Frederick Oehlkers working in Germany found that urethane could cause chromosomal damage, and shortly thereafter I. A. Rapoport working in the Soviet Union demonstrated mutagenic effects from several industrial chemicals, including ethylene oxide, epichlorohydrin, diazomethane, diethyl sulfate, and glycidol.[16] The field of genetic toxicology was born in the 1950s and grew rapidly during the 1960s. One of the principal leaders in the field, Alexander Hollender, led the effort to found the Environmental Mutagen Society in 1969.

During the 1960s the occurrence of a large number of infants born with a rare limb malformation called "phocomelia" in Germany was linked to the maternal ingestion of the drug thalidomide during pregnancy. This stimulated a series of studies on the effects of chemicals on the development and delivery of the fetus. Jim Wilson published a landmark article in 1959 and a monograph, *Environment and Birth Defects*, in 1973 in which he laid out principles of teratology.[17]

The analytical capacity of toxicology and the environmental sciences grew substantially during the last half of the twentieth century. The development of carbon absorption instruments vastly improved sampling methods. Mass spectrometry, although developed earlier, was first applied to water samples in 1953. Infrared and ultraviolet spectrometry and paper and gas chromatography facilitated the identification of trace contaminants. Radioisotopic tracer methods came on-line during the 1960s, and advances in bioassays improved studies of health effects. In 1940 colorimetric methods made it possible to study concentrations of chemicals at 10 parts per million (ppm). By the 1960s paper chromatography permitted analyses at 1 ppm, and by the 1970s gas chromatography had reduced this sensitivity to a few parts per billion.

While toxicology has emerged as the central science in the identification of the hazards of industrial chemicals, the field has developed in a slow and painstaking way. The development of the science has often been delayed by simple lack of understanding (or at least consensus) on the mechanisms by which chemical exposures cause specific harms. Over the years, much of the most sophisticated work has focused on cancer and carcinogenicity. While this has resulted in a solid understanding of biological responses to certain types of toxins, other outcomes (e.g., genetic damage, endocrine and immune system effects) are less well understood. Indeed, the sum of all of this research is highly focused on a relatively few outcomes and a few chemical substances. The large majority of effects from chemical hazards remain unstudied and our understanding is often based on conjecture or colored by fear.

5.2 Chemical Hazards in the Environment

Waste discharges from the early production of many organic and inorganic chemicals have long been known to cause environmental damage. For instance, the production of caustic soda in England was marked by severe environmental damage from waste hydrochloric acid. British dominance in the alkalis was based on the Leblanc process, which required the application of sulfuric acid to salt to produce salt cake (sodium sulfate) with hydrochloric acid as a by-product. Although the hydrochloric acid could be treated with lime to produce bleaching powder, the bleach market was far smaller than the market for caustic soda, so large volumes of hydrochloric acid were released into local streams and volatilized into the air, where they killed vegetation and acidified the soil. By the late nineteenth century, conditions had so deteriorated that the British Parliament enacted the Alkali Act, which authorized inspectors to monitor pollution from alkali and bleach production facilities.[18]

In the United States, the early twentieth-century awareness of chemicals as a cause for concern in the environment was tied to the emerging fields of public health and sanitary engineering. Much of the initial work was quite local and linked to municipal reform movements that included control of industrial pollution as a local government responsibility.[19]

By the 1890s, air pollution was common in all of the nation's industrial cities. Physicians routinely complained that bad air led to various respiratory diseases, and respiratory diseases such as influenza and tuberculosis were life threatening. By the turn of the century, women's clubs, municipal engineers, city foresters, and civic improvement groups were banding together to fight for smoke control ordinances. The earliest efforts were confronted by a popular conception that industrial smoke was a necessary by-product of economic growth; in fact, some civic boosters argued that industrial smoke was symbolic of a city's prosperity.[20] But gradually a majority of citizens came to recognize the broader concerns about health and amenity. In 1907 the Russell Sage Foundation funded an extensive citizen survey of social conditions in Pittsburgh. The "Pittsburgh Survey" revealed that the city's residents saw smoke as a serious public nuisance and surveys initiated in Salt Lake City in 1926 and New York in 1937 further documented public concern over urban air quality. A 1943 air pollution incident in Los Angeles and another in Donora, Pennsylvania, in 1948 aroused concern over the health effects of industrial smog. However, it was the 1952 London soot-saturated fog, alleged to have resulted in 4000 deaths, that raised public concern to the level of alarm.

Water pollution was another common problem. Industrial discharges to rivers, lakes, and coastal areas had greatly increased during the nineteenth century and many urban waterways were turgid, slick, and odoriferous. That these waterways were also unhealthy was dramatically demonstrated by massive fish kills reported during the early part of the century.[21]

Professionals responsible for water quality during this period largely ignored the hazards of industrial discharges. For years the old engineering adage "running water purifies itself" and "the solution to pollution is dilution" had created popular assumptions and clouded the judgment of professionals who might otherwise have pressed for more analytical approaches in assessing the fate of industrial wastewaters. Indeed, most of the early work on water pollution was carried out by sanitary engineers and public health specialists whose concern about waterborne diseases focused their attention on organic and fecal wastes and the pathogens that they supported. Industrial wastes were neglected because they were not thought to carry the disease-causing pathogens that were the primary concern of most sanitary engineers. Some professionals even argued that the toxicity of industrial wastes acted as a kind of germicide to kill off dangerous germs and, according to one 1938 textbook, "Few (industrial) wastes are present in most streams in sufficient quantities to become poisonous."[22]

Not everyone was oblivious to the environmental effects of industrial wastes.[23] In 1912 the U.S. Public Health Service was established and the following year it began a series of research projects on stream pollution sponsored by what became the Center for Pollution Studies at Cincinnati. These studies were directed by Wade Frost, who envisioned them as "a systematic program to acquire accurate, practically useful knowledge of the pollution of all streams of the country."[24] For these studies Frost relied on Earle Phelps, who had previously served at the Massachusetts State Board of Health's Lawrence Experiment Station and had conducted the first research using dissolved oxygen as a measure of determining water quality. Phelps conducted a series of studies on the effects of various industrial wastes on stream water quality through the remainder of the decade.[25] Even though these stream pollution studies were eventually curtailed by budget cuts, Frost remained a strong advocate of government studies of industrial pollution, all the while believing that industrial waste disposal was "a duty of industry ..., and that the function of the government agency should be to stimulate and perhaps direct the efforts of industry toward this end rather than to take over the responsibility."[26]

Some industrial efforts to control environmental damage simply shifted industrial waste problems around. Most of the coke produced in the nation from 1850 to 1910 was made in the conventional beehive oven, which could also be used to produce municipal gas. In 1880 there were 12,000 such ovens in the country and by 1909 the number had increased to 104,000. As early as 1869, Pittsburgh had attempted to prohibit new beehive coke ovens that would add to the city's foul air. Because these ovens released volatile aromatics, sulfur-laden fumes, and coal dust, industry began to gradually replace them with by-product recovery ovens, and this transition was accelerated by increasing demands for steel during World War I. The by-product recovery ovens reduced air emissions but generated water effluents rich in phenols. During the 1920s this discharge became a severe problem in the Ohio River Valley, where chlorination of the river water for drinking purposes generated chlorinated phenolic compounds in the water supply.[27] In 1923, a Committee on Industrial Wastes of the American Water Works Association identified 248 public water supplies that had been contaminated by industrial wastes and in its annual report concluded that "most industrial wastes are detrimental to water supplies."[28]

In 1935 President Franklin Roosevelt established the National Resources Committee, which among its activities conducted a study of state health agencies concerning the extent of industrial wastewater pollution in each state. This study identified the industries generating the most damaging wastes. They included textiles, pulp and paper, food processing, coke and gas manufacturing, oil refining, chemicals, and coal mining. The study went on to identify the conditions that limited action on industrial wastewaters. They included the absence of industrial and municipal cooperation, the lack of uniform regulations, the cost of treatment, the failure of industry to reuse and recycle water, and the absence of "practical and economical" methods for treating the toxic wastes.[29]

After World War II, professional and governmental attention turned more directly to the health hazards of industrial wastes. The great advances in chemistry and industrial production during and immediately after the war greatly increased the volume and changed the character of industrial waste. A new recognition of industrial wastes was evident in changes in professional focus. In 1946, the American Chemical Society sponsored a symposium on "Industrial Wastes: A Chemical Engineering Approach to a National Problem" that reviewed industrial waste problems in several industries and included air as well as water contamination.[30] In 1945 Purdue University held its first conference on industrial waste, which thereafter became an annual event. The National Safety Council published its manual of industrial waste disposal in 1948 and the old *Sewage Works Journal* was expanded and renamed *Sewage and Industrial Wastes* in 1950.

During the first half of the century, community, health, and engineering leaders struggled to draw attention to the health hazards of industrial wastes. Throughout this history, the conflicting values between promoting industrial development and preserving public health were well recognized and frequently articulated. The dawning of the new environmental movement during the 1960s did not discover this conflict as much as provide it with new energy and broader public awareness. Biologists and ecologists documented the effects of toxic chemicals on reproducing birds. Mountain hikers and hunters pointed out the polluted air and damaged vegetation in backcountry wild lands. Boating enthusiasts and sport fishers drew the attention of news media to rivers and streams polluted with detergent phosphates. Even more dramatic was the accidental ignition of oil slicks on the Cuyahoga River in Cleveland and the blowout of an oil drilling platform off the California coast at Santa Barbara in 1969.

During the following two decades, public awareness of toxic chemical contamination led to public activism. National environmental organizations turned their attention to chemical pollutants in environmental media; trade unions agitated about carcinogens in the workplace; and consumers' groups mobilized by activist lawyer Ralph Nader organized people around pesticide use, nuclear energy plants, and dangerous food additives. Earth Day 1970 mobilized an estimated 20 million Americans to join in coordinated rallies across the country to express their concern that something be done to protect the environment.[31]

The new environmental consciousness of the 1970s spurred the emergence of thousands of grassroots community groups organized around polluted rivers and water supplies. In 1978, a small group of neighbors led by Lois Gibbs at Love Canal, New York, organized to demand attention to the buried industrial wastes that were polluting their backyards. This struggle soon escalated from confrontations with the state public health commissioner to a symposium at the White House that led President Carter to provide federal funds for the relocation of the Love Canal residents. Within a year, residents of hundreds of neighborhoods across the country were organizing in outrage over the leaking hazardous waste dump sites that were suddenly being discovered in their communities. In Shepardsville, Kentucky, residents found a rural dump containing nearly a thousand drums of hazardous waste; residents in Rancho Cordova, California, found their drinking water wells contaminated because an aerospace firm had pumped spent solvents into local gravel pits; and at Times Beach, Missouri, an entire community had to be relocated after a trucker sprayed PCB-contaminated oils as a road dust suppressant.[32]

National organizations such as the Citizens' Clearinghouse on Hazardous Waste, the National Toxics Campaign, and Clean Water Action emerged to organize these local groups into powerful national coalitions that confronted government and corporate leaders alike over the need for direct responses to the dangers of industrial wastes. The environmental advocacy organization, Greenpeace, launched national campaigns against waste incineration and the ocean disposal of hazardous wastes, and inner city groups mobilized to raise awareness and demand responses to the hazards of lead paint, asbestos insulation, and industrial waste treatment plants in their neighborhoods. The result was a highly agitated, grassroots "antitoxic chemical" movement and a broad public awareness of the toxic chemical contamination that could result from industrial wastes.[33]

5.3   Responding to the Hazards of Industrial Materials

During most of the nineteenth century, industries resisted state and local government efforts to hold them accountable for occupational diseases or environmental pollution. While it was commonly assumed that responsibility for the management of industrial chemicals lay primarily with the private industries that made or used the chemicals, there was plenty of criticism of how these industries carried out this responsibility. The plight of workers elicited the most public attention.

By the turn of the century, the increasing frequency of injury and death at work led to the enactment of workers' compensation laws. Compensating workers on a fixed-formula basis for their injuries at work was a German idea advocated by a broad coalition that included the National Association of Manufacturers, the National Civic Federation, and the American Federation of Labor. The state of Wisconsin passed the first workers' compensation law in 1911 and within 2 years twenty-one states had enacted workers' compensation provisions. These first laws covered only industrial accidents and avoided coverage for the occupational diseases arising from chemical exposure. Only after the U.S. Supreme Court upheld the constitutionality of these laws in 1917 did their scope begin to expand to cover chronic diseases and the effects of long-term chemical exposures.

Community victims of chemical pollution had fewer means for seeking redress, but after 1900, private individuals and communities began winning suits against polluting corporations. In 1915 the U.S. Supreme Court issued an injunction against a Tennessee smelter to prevent sulfur air pollution in Georgia, and smelters in Montana and Utah were continually in court to defend themselves against the claims of farmers and other nearby residents. Recognizing the potential for losses in courts, Anaconda, ASARCO, and other large mining companies began offering direct compensation to those with legitimate claims.

By the 1930s, the response of private industry had begun to shift from resistance to more proactive initiatives. Leading firms were investing in procedures and technologies that were designed to manage and control toxic chemical exposure. In 1932 the American Petroleum Institute (API) set up a Committee on Refinery Wastes and began investigating methods to control pollution from oil refineries. In 1935 the API published a manual, the *Disposal of Refinery Wastes*, that contained guidelines that were gradually adopted throughout the industry, reducing the damage from refinery by-products. Petroleum refineries and mine smelters experimented with bag filters, cooling chambers, and taller smokestacks to reduce air pollution. Likewise, during the 1930s, Du Pont, Dow, Union Carbide, and other chemical firms began to invest more heavily in protecting workers from exposure to toxic chemicals. This involved both simple engineering solutions and crude forms of protective clothing. Concentrations of chemicals in the air were to be reduced by improving air circulation with open windows, fans, and ventilators. Dermal exposure was to be controlled by providing gloves, goggles, or uniforms. The effects of chemical ingestion were to be minimized by providing lunchrooms separated from production areas. Dow established a separate medical department in 1946 and hired its first industrial hygienist 2 years later to monitor workplace exposures and recommend appropriate protections.

With better occupational health statistics and improved measurement techniques, after 1930 industrial hygiene specialists began to turn their attention to determining thresholds for acceptable levels of workplace exposure to industrial chemicals. This required an increase in scientific investigation of the toxic properties of industrial chemicals, which many still believed was an industrial responsibility. Under increasing government and professional pressure, the chemical industry responded by sponsoring and conducting its own research on toxic chemicals. Many of these studies were published in professional journals such as the *Journal of Industrial Hygiene and Toxicology*, *Industrial Medicine*, and the *Public Health Service Bulletin*, and were widely read by academic, government, and corporate professionals.

One of the most significant initiatives began in 1935 when Du Pont established Haskell Laboratory to conduct studies on the toxicity of chemicals. Du Pont established the laboratory following the death of several workers at its tetraethyl lead production facility and the appearance of bladder cancer among workers at its dyeworks.[34] Initial studies at Haskell Laboratory identified benzene as the cause of the cancers at the dyeworks, but could not establish an acceptable exposure threshold for ensuring safety. During the following years, researchers at the laboratory conducted studies to determine toxic doses for various chemicals, including carbon disulfide, chloroprene, duprene, and dioxane, and offered recommendations on preventive techniques to lower workplace risks.[35] W. F. von Oettingen, who served for many years as the chief toxicologist at Haskell Laboratory, wrote extensive scientific reviews on the toxic effects of the halogenated hydrocarbons, the aliphatic alcohols, and the aromatic hydrocarbons.[36]

In 1933 Dow established a biochemical research laboratory to study the potential effects of chemicals on humans and other organisms, with some of its earliest studies focused on benzene and lead.[37] Beginning in the 1940s Dow scientists began publishing studies on the toxic properties of some of its chemical products and intermediates, including benzene, toluene, ethylbenzene, styrene, and chlorinated solvents such as carbon tetrachloride, trichloroethane, and trichloroethylene.[38] Other chemical companies were also involved in toxicological studies. During the 1940s researchers at Hooker Electrochemical Company investigated the toxicity of several hydrocarbons manufactured by the company and concluded that six of these substances were toxic to humans when ingested and that dermal exposures were also toxic.[39]

These industrial initiatives were based on the emerging professional sentiment that the best way to address the toxicity of industrial materials was to control human and environmental exposure. Rather than examine and reconsider the use of toxic chemicals in industrial production, the focus centered on managing exposure. Since it was assumed that toxic materials were necessary for increasingly sophisticated production, reducing exposure rather than toxicity became the preferred strategy. In this commitment, industrial leaders were joined by government professionals. By the 1930s the U.S. Public Health Service was issuing recommended environmental exposure standards ("maximum contaminant levels") for a variety of chemicals, and the American Conference of Government Industrial Hygienists began publishing "threshold limit values" or "TLVs" for chemicals used in occupational settings.

During the first half of the twentieth century, many state and local governments were active in efforts to control chemical air pollutants. Oakland, California, prohibited certain smelters from the city, and Butte, Montana, struggled to close down facilities for the heap roasting of copper ores. Municipal smoke control ordinances were one of the earliest efforts to control industrial wastes released to the environment. During the early 1900s smoke abatement leagues were established in Cincinnati, Milwaukee, St. Louis, Chicago, and Pittsburgh, and these organizations led efforts to press for municipal action. By 1912 twenty-three of the nation's largest cities had smoke control ordinances that established smoke inspectors and smoke abatement programs largely focused on industrial sources. In 1941 Pittsburgh passed the strongest smoke control ordinance in the nation, and the smoke of "the smokiest city" was greatly reduced as industries switched from coal to natural gas for fuel. Indeed, it was this transition from one dominant fuel to another, rather than strong pollution control measures, that led to a gradual abatement of smoke during the middle of the century.[40] Additional efforts to control air contaminants focused on automobile emissions. Among the states, California led the way with an air pollution control act in 1947 to regulate the discharge of opaque smoke and during the 1950s with several laws restricting automotive exhausts. By 1960 eight states as well as seventeen counties and eighty-four municipalities had passed some form of air pollution control legislation.[41]

The first state legislation to control water pollution was an 1878 Massachusetts law giving the state board of health the authority to inspect and regulate river pollution caused by industrial wastes. By the 1890s many cities had ordinances for the control of water pollution, but almost all of these focused on sanitary discharges rather than industrial wastes. In 1905 Massachusetts became the first state to establish a central responsibility for occupational health within its state board of health. Factory inspectors were appointed and their annual reports provided the data for the establishment of other state programs. Among these was the influential Illinois Commission on Occupational Diseases that sponsored much of Alice Hamilton's early research. During the 1920s several states created state conservation commissions and some of these, such as the New York and Wisconsin commissions, provided technical assistance to private firms in order to reduce their waste effluents. None of these conservation commissions had enforcement powers, however, and the state public health departments that did have such authority were reluctant to move against industrial polluters for fear of political reprisals.[42]

Occupational exposure to toxic materials increasingly became a focus of trade union activity during the 1960s. The United Mine Workers fought a long battle with the coal mining industry to gain compensation for miners disabled by pneumoconiosis, commonly called "black lung disease." By 1970, the Oil, Chemical and Atomic Workers Union was campaigning for information on chemical exposure in its labor negotiations. In 1973 the union called a strike against the Shell Oil Company to force the company to agree to the worker's right to know about the chemicals they were exposed to, and after months of a highly publicized struggle, the company agreed. During this same period, asbestos workers brought a liability suit against the Johns-Manville Corporation, and after a long trial won substantial damages for a range of diseases they claimed had been caused by their exposure to asbestos products and covered up by the firm.[43]

Progress in responding to industrial pollution was not continuous or steady. Economic slumps and the waging of wars tended to slow and divert the progress. The economic panic of 1907, the Great Depression of 1929, and both world wars slowed and for a time reversed the efforts to combat pollution. During the first war, Secretary Franklin Lane was quoted as saying "war meant smoke ... people should stand it in contributing their 'patriotic bit'."[44] But gradually even with these setbacks, corporations and governments did develop responses to toxic chemical pollutants and exposures. Some of these, such as medical services and compensation for injured workers, were primarily changes in practice, but others such as smokestack filters and wastewater treatment systems, were based on new technologies. These practices and technologies tended to isolate the management of chemical exposure from the decisions about the chemicals being used. Indeed, within firms, the industrial hygienists, medical nurses, and waste management engineers were often well distanced, both administratively and physically, from those who made daily decisions about what chemicals to buy or make.

As early as the 1920s, government and corporate research scientists had begun serious studies on the effects of occupational exposure to industrial chemicals and these studies were published and read throughout the 1930s and 1940s. While many of the studies involved animal testing or analyses of acute workplace exposures, and the chronic effects on humans were yet to be well documented, the toxic nature of many of the common industrial chemicals was well known and generally accepted by midcentury.

Firms responded to these studies with varying strategies. In some cases they stubbornly resisted (and even hid) the research. Although Johns-Manville and other asbestos suppliers were well aware of their products' hazards, they concealed their own studies and resisted compensation claims for some 30 years. Likewise, the paint industry hid their studies of lead paint hazards and the polyvinyl chloride industry fought lawsuits by workers who alleged that exposure to vinyl chloride had led to their angiosarcoma of the liver. In other cases, firms initiated limited forms of worker protection (respirators, engineering controls) and waste management (surface impoundments and incinerators). While these initiatives added to production costs, most were minimal and easily absorbed in consumer prices. Although some of these initiatives were locally effective, the enormous growth of the economy and the rise in levels of production largely swamped the broader effects of the changes. Compared with the rapidly rising volumes of hazardous wastes and the increasing complexity of toxic chemical exposures, these responses were modest at best.

5.4   Assessing Toxicity Today

Our understanding of toxicity and the effects of toxic materials on the environment and the human body has grown significantly over the period of industrialization. While this understanding is far from complete, the knowledge we have today provides a reasonable ground for assessing the hazards of some of the more common industrial chemicals.

Toxicity may be expressed through various outcomes, including acute poisoning, neurotoxicity, teratogenicity, mutagenicity, or carcinogenicity. Some toxic chemicals have acute effects that appear as symptoms fairly rapidly following exposure. This may involve skin lesions, tissue necrosis, burns, headaches, dizziness, or death. Other substances cause chronic effects where the harm does not appear as recognizable symptoms for an extended time. Carcinogens are chronic toxins, with tumors or cell mutations appearing often years after a chemical exposure.

Many green plants generate toxic compounds in their leaves, fruits, and roots as a defense against predation, and several animals produce chemical toxins as weapons for defense or as chemical agents for securing prey. But nature is fairly cautious about toxins, using them in quite restricted and controlled ways. Toxic chemicals can be quite dangerous to living systems when they are spread about in reckless ways, that may result in unintended and uncontrolled exposures. A long period of evolution has determined which materials and how much of them can flow into and out of an organism's system without adverse effects. While some materials necessary for sustaining living organisms are toxic, they can be tolerated at low doses; a larger dose of those same materials may inflict substantial damage.[45]

Metals provide a good example. While some metals are critical for human metabolism and growth, most metals are acutely toxic at least at some dose and many metals have chronic toxic effects as well. Some metals are known to be carcinogens, while others demonstrate adverse reproductive effects. The toxicity of arsenic, cadmium, lead, and mercury are well recognized, but the salts of bismuth, copper, chromium, thallium, and zinc are also toxic enough to have found uses in agricultural pesticides.

Many common industrial materials are toxic. Some are quite toxic in their elemental form, such as chlorine and phosphorus. Other chemicals are more toxic in chemically bound forms; hydrogen cyanide and organometallics such as dimethyl mercury, and tetraethyl lead, are examples. Table 5.1 lists several frequently used industrial chemicals, along with their recognized toxic properties.

Toxic substances vary widely in their effects within an organism. Most toxic substances have a harmful effect only on certain, so-called "target organs." For a toxic substance to have a harmful effect, an organism must be exposed to it and it must be absorbed into the body through the lungs, the gastrointestinal tract, or the skin. For a toxic chemical to reach a target organ, it must cross a cellular boundary. The most common mechanism for cellular membrane transfer is passive diffusion, which involves the flow of toxins from high concentrations on one side of the membrane to low concentrations on the other side. The ease of membrane passage is facilitated by certain properties of a chemical such as molecular size (smaller molecules transfer more easily), and ionic state (non-ionized compounds transfer more easily). The solubility of a chemical often determines its success in reaching a target organ. Substances that are primarily water soluble may have difficulty penetrating blood capillaries, organ walls, or cellular membranes because of the size of water molecules. Substances that are fat soluble permeate such membranes more easily.

Following absorption, the toxin must be transported, typically by blood, to a target organ in a chemical condition in which it is bioavail-

Table 5.1
Toxic properties of selected industrial materials

Substance	Carcinogen	Reproductive hazard	Neurotoxin	Mutagen or teratogen
Acrylonitrile	Probable	Yes		
Arsenic	Known	Yes	Yes	
Benzene	Known	Yes	Yes	
Benzidine	Known			
Cadmium		Yes		
Chlorine		Yes		
Chloroform		Yes		
Chromium (+6 valence)	Known	Yes		
Ethylene oxide	Probable	Yes	Yes	Yes
Formaldehyde	Probable	Yes		
Lead	Lead acetate	Yes	Yes	Yes
Methyl mercury		Yes	Yes	Yes
Nickel	Probable	Yes		
Perchloroethylene	Probable			
Polychlorinated biphenyls	Probable	Yes		Yes
Polycyclic aromatic hydrocarbons	Probable	Yes		
Styrene		Yes		
Tetrachloroethylene				
Trichloroethylene		Yes		
Vinyl chloride	Known	Yes		

Source: Curtis D. Klaassen, ed., *Casarett and Doull's Toxicology: The Basic Science of Poisons*, 5th ed., New York: McGraw-Hill, 1996.

able, that is, accessible to that organ. Once on location, the effects of a toxic chemical—its mode of action—are determined by the extent of the time and concentration of the exposure, the kinetics of the interaction, and the sequence of events that follow the interaction. An absorbed exogenous chemical is often passed to the liver, where it may be metabolized (transformed) through oxidation, reduction, or hydrolysis and then conjugated (joined) with normal body constituents. This process, referred to as the biotransformation of a chemical, can lead to a more or less toxic metabolite that is more or less bioavailable (capable of reaching a target organ). Metabolites are usually but not always more polar and less fat soluble, and therefore less bioavailable than the parent chemical. An example of a biotransformation that generates a more toxic and more bioavailable metabolite is the oxidation of methanol to form formaldehyde.

Some substances that are absorbed into an organism are quite persistent and resist metabolic biodegradation. They may be stored for long periods without decomposition and released to cause harm long after the initial exposure. For instance, lead, which is readily stored in bone, causes cellular damage only when it is released to exposed soft tissues, and chlorinated compounds easily sequestered in adipose tissue cause harm only when liberated from the host fat molecules as they are metabolized for energy.

Some toxic substances are asphyxiants in that they deprive tissues of adequate oxygen or actively interfere with the delivery of oxygen. For instance, some inert gases simply replace oxygen in the blood supply. Cyanide inhibits enzymes such as cytochrome oxidase which are necessary for the effective utilization of oxygen within cells. Other industrial substances affect the blood as hematological toxins. Common industrial chemicals such as benzene, arsenic, and arsine gas, once absorbed into the blood supply, bond with the oxygen-combining sites on hemoglobin molecules and reduce their capacity to take up oxygen and deliver it to tissues. Lead behaves differently. Absorbed lead inhibits the enzymatic reactions that form the iron-rich heme that is required for the maturation of red blood cells.

Various toxic chemicals act as carcinogenic initiators by damaging the genetic makeup of a cell and, thereby, creating mutated cells. These cellular mutations can be caused by the direct action of chemical substances or more indirectly by the metabolites of those substances. Bis(chloromethyl) ether, for example, directly binds with the genetic structure of a cell, creating adducts that may then be processed into mutagens. Other chemicals, such as the polycyclic aromatic hydrocarbons and the β-naphthylamine used in dyes, need to be converted to electrolytic metabolites before they bind with genetic material and lead to cell mutations. Carcinogenic initiation is followed by a period of promotion in which various cellular factors allow the mutated cell to develop into a rapidly growing proliferation of cells that may (or may not) progress into a malignant tumor. Over the years, a group of industrial chemicals, including asbestos, vinyl chloride, benzidine, arsenic, and benzene have been proven to initiate carcinogenesis, while there is a second and third group of chemicals that have come to be suspected of such effects (given various experimental studies on animals) and are ranked as "probable" or "possible" carcinogens by the International Agency for Research on Cancer.

Most of those toxic chemicals hazardous to humans can be quite hazardous to nonhuman species, and their effects can be seen at the level of individuals, populations, communities, or ecosystems. Cellular and organ damage within individual plants or animals can be induced by toxic chemical interactions much as in humans. Toxic chemical pollutants in air, water, or soil can result in individual mortality or reproductive damage, population reductions, loss of diversity in communities, or the collapse of healthy ecosystems.

Toxic chemicals vary widely in their potency. Scientist are able to classify different chemicals according to their acute toxicity by relying on animal studies to quantify the probable lethal dose. Rating chemicals by chronic toxicity is more difficult because there is no clear indicator and so many factors affect the results. In the absence of clear-cut data, attention has focused on those substances that exhibit three concurrent factors: toxicity, environmental persistence, and bioaccumulative potential.

Environmental persistence refers to the ability of a chemical to withstand natural biodegradation processes. Some industrial materials, particularly structural materials such as aluminum, various alloys, and many plastics, are designed to resist natural degradation. Substances like polychlorinated biphenyls and lead are introduced into materials to resist biological degradation. However, long-lived toxic materials that resist decomposition can be problematic in the environment or in the human body. Persistence of toxic compounds means that they remain toxic even when they are disposed of and that they are likely to resist metabolic transformations once absorbed into an organism.

Even when the concentration of a persistent toxic chemical is below commonly accepted levels of concern, it may continue to be a problem if it is bioaccumulative. Very small amounts of toxic materials may be accumulated and concentrated within an organism by ingestion over time or through ingestions of successive trophic levels in a food chain until toxic concentrations are found within the higher members of the

Table 5.2
Persistent, bioaccumulative, and toxic industrial materials (excluding pesticides)

Polycyclic aromatic hydrocarbons

Benzo[*a*]pyrene	Benzo[*b*]fluoanthene
Benzo[*r,s,t*]pentaphene	Benzo[*a*]anthracene
Dimethylbenz[*a*]anthracene	Dibenzo[*a,h*]anthracene
Methylcholanthrene	Dibenzo[*c,g*]carbazole
Benzo[*k*]fluoranthene	Benzo[*j*]fluoranthene
Dibenzo[*a,e*]pyrene	Dibenzo[*a,h*]pyrene
Indeno[1,2,3-*cd*]pyrene	Dibenz[*a,h*]acridine
Dibenz[*a,j*]acridine	Benzo[*a,h,l*]perylene
Dibenzo[*a,e*]fluoranthene	Methylchrysene
Dibenzo[*a,l*]pyrene	Benzo[*a*]penanthrene
Nitropyrene	Benzo[*j,k*]fluorene

Cobalt and cobalt compounds
Mercury and mercury compounds
Vanadium and vanadium compounds
Polychlorinated biphenyls
Hexachlorobenzene
Octachlorobenzene
Tetrabromobisphenol A
Dioxin and dioxinlike compounds

Source: U.S. *Federal Register*, "Persistent, Bioaccumulative Toxic (PBT) Chemicals—Proposed Rule," 64(2), January 5, 1999, pp. 688–729.

food chain. As chemicals accumulate in the fat, bone, or muscle of living organisms, dilute concentrations may become transformed into more concentrated and therefore more potent forms.

Where chemicals exhibit these three characteristics—persistence, bioaccumulation, and toxicity—the potential for human and environmental exposure is high and it is likely that such chemicals will produce chronic effects in those organisms that are exposed to them. Table 5.2 lists chemical substances identified by the EPA as persistent, bioaccumulative, and toxic.

5.5   Understanding the Acceptance of Risk

Among the wide array of industrial materials in use today, a significant number may cause environmental damage and are disturbingly dangerous. While large gaps in scientific understanding of the environmental and health effects of these materials persist, the clear dangers of some of the most common materials are well recognized. The history of the development of the health and environmental sciences and the efforts to respond to the knowledge they generated demonstrate that this information has been available for decades. It cannot be argued that the current mix of hazardous materials results from a historical absence of knowledge about their potential for harm. It was not an absence of knowledge but a conflict over values that permitted dangerous materials to be developed, used, and disposed of with less than adequate protection.

Some have argued that professional divisions hampered the diffusion of environmental and health information to those who made decisions about materials development and use. Christopher Sellers has noted, "the sanitary engineers who developed water pollution as a specialty pursued questions that differed markedly from the toxicological ones of their industrial hygiene colleagues.... They gave little or no consideration to the diverse chemicals that comprised industrial pollution, much less to the substances' human toxicity."[46]

Others have argued that there were rigidities in the focus of professionals that kept them from seeing the hazards. Writing in 1954, Edward Cleary, the chief engineer for the Ohio River Valley Water Sanitation Commission, complained that professionals in "the field of sanitation relating to water supply and waste disposal have been peculiarly unresponsive to problems of toxicity" because they persisted in focusing on "bacteriological hazards."[47]

Indeed, those who were supposed to be focused on protecting human health and the environment from industrial wastes were cut off from useful knowledge about chemical hazards. Historian Craig Colten observes that during the 1950s, while chemical companies possessed "internal expertise on toxicology, it was seldom applied to chemical wastes except when public agencies pressured individual companies."[48] Yet, Colten goes on to argue that those who supervised waste disposal at landfills did know that the materials were hazardous and that they could leach into groundwater.[49] Joel Tarr, a historian of urban pollution, does not directly debate this conclusion, but argues that the knowledge was mediated by attitudinal and social barriers. In a footnote to a recent article he writes, "My personal belief is that a combination of "careless house-keeping,"..., a failure of communication, and a deliberate ignoring of possible harmful effects was involved."[50]

These explanations are too easy, and too neglectful of the political and economic forces at work in promoting and managing industrial materials. David Ozonoff, in a review of the public health decisions that permitted the production of tetraethyl lead to proceed, observes that "Enough was known about the dangers of lead to have instituted effective preventive measures at that time [1925].... The reasons that these measures were not put in place has much less to do with the state of the scientific art than it does with whose hands were on the levers of power.... The evidence *was* [his italics] ambiguous. But the possible health consequences of a wrong guess were clear, and the decision to risk those consequences tells more about the decision-making process than it does about the ambiguity of the health evidence."[51]

Decisions were made. Real values were at stake. It is clear that business entrepreneurs and corporate managers did go ahead with the production of materials that were known to be hazardous. It is not easy to establish that those who made such decisions were fully apprised of the available knowledge, but some government officials, some professionals, and some corporate staff did know that the production and use of these new industrial materials would eventually lead to environmental or human health risks. The risks were certainly undervalued and under-appreciated, and the costs were assumed to be less than the hoped-for benefits that the materials would provide. Within this context, there were efforts to address risks, but these were almost solely focused on managing exposures, not on reconsidering the production or use of toxic chemicals. Profits and the drive to amass wealth certainly affected these decisions, but so did technological development and the demands of an expanding market.

For an increasingly aware public, the results were unacceptable. Polluted air and waterways, thousands of leaking dump sites, wildlife in jeopardy, and rising levels of cancer and suspicious health problems among a population exposed to chemicals never before seen on earth were problems that called for more aggressive attention. Reliance on corporate self-restraint had failed. As time went on, it became more apparent that there was a role for the federal government in the management of industrial materials. There was nothing new here. Well before the close of the nineteenth century, the federal government had become a significant factor in the development and use of materials. What was different now was the demand for a federal policy on industrial materials that addressed environmental and public health issues as well as the supply and management of resources.

II
Reconsidering Industrial Materials Policy

6

The Federal Policy Response

The Congress recognizes that each person should enjoy a healthful environment and that each person has a responsibility to contribute to the preservation and enhancement of the environment.
—U.S. Congress, National Environmental Policy Act of 1969

For well over a hundred years, the federal government in the United States has struggled to develop policies for addressing the nation's materials. At times it has attempted to address materials with a full and comprehensive policy; more often the effort has focused on specific issues. During the twentieth century, federal interest shifted from the conservation of natural materials to materials scarcity, national security needs, environmental impacts, commercial interests, and the development of new materials. There have been a large number of federal initiatives and several federal agencies with policies directed at the extraction, use, and disposal of industrial materials. While there is no lack of federal attention to these materials, there has been little success in developing a comprehensive and coordinated approach to their management and use.

In its efforts to manage the nation's materials, the federal government has pursued five broad strategies: the conservation of natural materials resources, the protection of strategic resources, the development of a comprehensive policy, the regulation of materials as environmental pollutants, and the management of materials markets. Although these strategies have at times developed together, it is useful here to consider each of them separately.

Table 6.1
Primary federal mining and forestry legislation

1866	Mineral Land Act
1872	General Mining Act
1873	Timber Culture Act
1878	Timber and Stone Act
1891	Forest Reserve Act
1997	Forest Management (Organic) Act
1902	Reclamation Act
1905	U.S. Forest Service Established
1912	U.S. Bureau of Mines Established
1920	Mineral Leasing Act
1955	Multiple Use Mining Act
1960	Multiple Use–Sustainable Yield Act
1964	Wilderness Act
1970	Mining and Minerals Policy Act
1973	Threatened and Endangered Species Act
1976	Federal Land Policy and Management Act
1976	National Park System Mining Regulation Act
1977	Surface Mining and Regulation Act
1980	National Materials and Minerals Policy, Research and Development Act

6.1   Policies for the Management of Natural Materials Resources

As steward of a vast empire of public lands, the federal government has been a significant player in shaping materials markets through its control of large amounts of mineral and forest resources. Table 6.1 lists some of the most significant federal legislation concerning mines and forests. The period from the 1880s to the 1930s was dominated by a deep concern about the management of the country's natural resources, including land, water, forests, and minerals, but particularly those resources that existed on federal lands.

The existence of the federal lands has a rather remarkable history resulting from the activism of the leaders of the conservation movement. While much of their work developed in the East around urban parks and wildlife reserves, the most dramatic attention focused on the federal lands of the West. Spurred by the advocacy of George Perkins Marsh, Charles Sprague Sargent, Berhard Fernow, John Muir, and Gifford Pinchot, the movement leaders were committed to preserving the magnificent lands of the West from the kind of rabid development and destructive exploitation that they believed had damaged so much of the East. Most of these western lands had been acquired from Mexico by the Treaty of Hildago Guadalupe and were still under federal ownership, although following the Homestead Act of 1862, they were rapidly being given away to eager settlers.

In 1873 Congress enacted the Timber Culture Act, which made it easier for farmers to privatize public lands by settling on them, clearing them—and planting trees. So the first strategy of the conservationists was to protect the trees. George Bird Grinnell, Theodore Roosevelt, and the Boone and Crockett Club lobbied hard for some kind of federal forest protection. Finally, in 1891 a section was introduced into a law (now called the Forest Reserve Act of 1891) intended to reform the Timber Culture Act by authorizing the president "to set apart trust reserves where, to preserve timber, he shall deem it advisable" and to administer these reserves under the secretary of the interior.[1] President Harrison immediately created six reservations and, with additions over the next 2 years, fifteen reserves totaling 15 million acres were set aside. Another thirteen reserves involving 21.4 million acres were set aside by Grover Cleveland in 1897 with a sudden and surprising executive order; and during his presidency Theodore Roosevelt added or enlarged thirty-two reserves totaling 75 million acres.[2] In particular, Grover Cleveland's stealthy reserve policies angered the forest and mining industries, and substantial pressure arose to limit the presidential power to reserve public lands. Indeed, the very purpose of these reserves remained in controversy until the passage of the Forest Management Act (also called the Organic Act) of 1897 provided a negotiated resolution by declaring that the reserves were for the conservation of water, minerals, and forests, and that they could be wisely used under federal supervision.

In 1905 Congress created the U.S. Forest Service within the Department of Agriculture and transferred the forest reserves to the new division. President Roosevelt appointed Gifford Pinchot, who was a strong advocate of a utilitarian, multipurpose approach to federal land management, as the first chief forester of the new agency. Pinchot proved to be a commanding presence and his philosophy of multiple use became the dominant approach to the management of the forest reserves. In 1907 Pichot pressed Congress to rename the forest reserves as "national forests" to suggest their broader mission.

Mining has long been a notorious industry and early federal efforts to control mining were weak and ineffective. Hard-rock mining literally exploded its way across the West during the middle of the nineteenth century. Much of this mining was conducted on federal lands with no effort by the government to extract revenue or control procedures. Indeed, because much of the land had only recently been purchased from Mexico, there were no federal or state laws in effect. In reporting on the conditions of the California gold rush in 1849, General Persifor Smith, then in command of the Pacific Division of the Army, remarked:

I do not conceive that it would be desirable to have the mines worked for the benefit of the public treasury. To do that would require an army of officers and inferior agents, all with high salaries, and with opportunities and temptations for corruption too strong for ordinary human nature. The whole population would be put in opposition to the government array, and violent collisions would lead even to bloodshed.... The advantage that the whole country will derive directly from the opening of the mines, and the indirect advantage to the treasury of augmented commerce, will, in my opinion, more than compensate for any outlay it has made or may make.[3]

In a modest effort to control and rationalize mining on federal lands, Congress passed the Mineral Land Act of 1866, which established a process for claiming mineral patents that was modeled after the informal process by which the prospectors of the 1850s determined mineral rights. These provisions were further augmented by the landmark General Mining Act of 1872. Under these laws, a mining patent could be obtained by making a "valid" mineral discovery, paying a token sum ($2.50 per acre for a surface mine and $5.00 per acre for an underground mine), and investing $100 per year in mining activities for 5 years. In 1873, a similar mining law was established to provide easy and inexpensive access to federal lands for the mining of coal. The discovery of oil in Pennsylvania in 1859 and its subsequent development did not elicit as much federal attention because for many years most of the claims were made on non-federal lands.[4]

In managing the national forests, Gifford Pinchot was careful to court mining interests and that attitude was perpetuated by those who succeeded him. In return, the mining interests have supported most Forest Service policies, often to keep pressure for a revision of the mining laws from building in the public or Congress. During the 1920s and 1930s, the mining laws were widely criticized because they generated so little federal revenue and because they were often misused by land-grabbing speculators. In 1920 Congress passed the Mineral Leasing Act to allow oil companies access to federal lands, but instead of a patent system, Congress opted for a leasing system. These leases were to be competitively bid and to generate royalties on the extracted oil. Over the years this mineral leasing concept has been extended to the mining of coal, natural gas, phosphate, and sodium on federal lands, while the patent system has remained the cornerstone of hard-rock mineral development.

By the 1950s the abuses of the mining patent system were again under heavy criticism, not only by conservationists, but also by the Forest Service and the forestry industry. This time Congress responded with the Multiple Use Mining Act of 1955, which closed loopholes and prohibited any activity other than mining on patented claims. Finally, in the 1976 Federal Land Management Policy Act, Congress formalized the requirements for recording unpatented claims and established a process by which inactive claims could be invalidated.

Throughout this history the federal government has had a significant opportunity to guide the development and management of many of the nation's resources, largely because a sizable proportion of those resources are found on lands owned and controlled by the government. Federal ownership has led to federal management, but the process has been rife with struggle among various pressures for rapid exploitation and a desire for stewardship. However, the federal government has not confined its management role to its own materials. Concern over the depletion of the nation's resource stocks has drawn federal agencies into managing materials in the private market as well, and this has been equally contentious.

6.2   Policies for Addressing Scarcity of Materials Resources

Over the years, many public and private leaders worried about the dissipation of the nation's privately held materials resources. The federal government offered early and valuable responses. In 1879 Congress authorized the establishment of the U.S. Geological Survey (USGS) and in 1880 the new agency began its first inventories of mineral resources. In 1910 Congress established the U.S. Bureau of Mines to promote minerals and mining, and in 1923 the minerals information collection activities of the USGS were transferred to the Bureau of Mines.

Following World War I, a long debate emerged about the most appropriate response to the dwindling materials resources of the country and the recognition of their strategic value. Those who represented domestic materials industries argued for materials self-sufficiency, believing that domestic materials industries, even those built from the remaining lower grade ores and fuels, were a hedge against future military aggressors. Others, such as Wisconsin professor Charles Leith, and Josiah Spurr, the director of the Mining and Metallurgical Society of America, argued that international free markets in materials reduced the motivation for nationalistic aggression and encouraged peaceful international relations.[5]

The advocacy for free trade in materials following the war was frustrated by the efforts of various nations to close their own doors. Canada imposed an embargo on timber and pulpwood headed for U.S. sawmills and pulp mills; Britain cut its price for tin smelting to force the closure of the only U.S. tin smelter; and British and French interests joined forces to block U.S. access to Middle East oil.[6] To promote the idea that free trade was necessary for sound international relations, Leith and Spurr formed a blue-ribbon committee of experts in 1921 to develop a comprehensive minerals policy for the nation. Their report, published by the Mining and Metallurgical Society, promoted mineral extraction, calling mineral resources "wasting assets fixed geographically by nature" and concluded that the "international exchange of minerals cannot be avoided if all parts of the world are to be supplied with needed materials."[7] Arguing that the best way to conserve the nation's weakened mineral reserves was to tap richer sources in other countries, the report called for a free market for materials unencumbered by national duties, tariffs, and embargoes, and government stockpiles of metals whose supplies were critical to American industry and armed forces. Although the society's report failed to develop a comprehensive national minerals strategy, it did draw industry and government attention to the implications of resource depletion and the need for free international markets.

During this period the military also began a thorough reassessment of the materials necessary for national defense. In 1919 the former chairman of the War Industries Board, Bernard M. Baruch, recommended to President Woodrow Wilson that adequate supplies of critical materials should be set aside in case of national emergencies, and the following year the assistant secretary of war prepared a list of strategic materials defined as those "raw materials essential to the prosecution of war, which cannot be procured in sufficient quantities from domestic sources, and for which no domestic substitute has been found."[8] In 1932 this list was further expanded to include "critical materials" that were important for national defense but were less difficult to procure in adequate supply. Recognition of the value of these materials led the military to agree with Leith's and Spurr's recommendations for national stockpiles of strategic materials.

It was the prospects of a new war that finally committed the country to a national warehouse. In 1934, a special Planning Committee on Mineral Policy (which included Professor Leith) recommended stockpiles of seven strategic materials. Although Congress passed the Strategic War Materials Act in 1939 authorizing stockpiling, President Roosevelt resisted the initiative. Only after Germany and Japan began their military expansions did the federal government move to establish a materials inventory capable of withstanding a protracted war. In 1940 the federal administration established several corporations for acquiring strategic and critical minerals, and the next year the targets were readjusted and additional non-mineral materials were added to the list of purchasable commodities.[9]

Efforts to build adequate strategic stockpiles during the war were routinely criticized because the stockpiles grew too slowly and were poorly managed. Following the war, the stockpiles were established on a permanent basis by the Strategic and Critical Materials Stockpiling Act of 1946, but only after long national debates. In part, the stockpiles served as strategic reserves for possible future military needs, but they also turned out to provide the federal government with a strong leverage in foreign policy.[10]

The Korean War, coming at a time of a rapidly growing economy, put a further strain on the nation's depleted material reserves. As a result, in 1950 Congress enacted the Defense Production Act, which provided private industries with federal loans and guarantees, exclusive purchasing contracts, open purchase offers, and rapid amortization of new facilities for tax purposes in order to encourage the production and importation of materials critical to the economy and military needs. However, there was a growing public interest in developing a more comprehensive approach to materials. The minerals mined and the wood harvested from federal lands generated a bounty of privately produced materials enhanced by large government subsidies. Unquestionably there were significant public benefits. Yet, over and again these policies had arisen in an ad hoc and sector-specific manner. For some it seemed quite obvious that the nation would benefit from a more coordinated approach to materials development and use. No less than a presidential commission established by President Truman in 1949 to recommend reorganization of the executive branch of government would conclude:

To meet the needs of the future and to promote more orderly development and exploitation of the Nation's resources, as well as to guard the heritage of the people, the unification of the responsibilities and services of the Government dealing with such matters seems clearly called for.[11]

6.3   Efforts Toward a National Materials Policy

Since the mid-twentieth century, there has been active federal interest in setting a more proactive national policy for materials (see Table 6.2). There have been two major federal commission studies and three federal laws calling for a national materials policy. The first detailed argument for a national materials policy was presented in 1952 by the President's Materials Policy Commission.

The Paley Commission and the Mid-Century Conference
At the close of the Korean War, the country was faced not only with a sober recognition of its depleted domestic resource base, but also with the threat of Cold War isolation from many important mineral deposits

Table 6.2
Primary federal materials policy legislation

1939	Strategic War Materials Act (53 Stat. 811). Established the National Defense Materials Stockpile.
1946	Strategic and Critical Materials Stockpiling Act (60 Stat. 596). Appropriated funds for acquisition of critical materials.
1950	Defense Production Act (64 Stat. 798). Provided for loans and long-term supply contracts for defense-related materials. National Science Foundation established (P.L. 81-507)
1970	National Naturals Policy Act/Resource Conservation Act (P.L. 91-512). Established National Commission on Materials Policy and required several studies.
1970	Mining and Minerals Policy Act (P.L. 91-631). Promoted minerals research, reclamation of minerals, and more protective mining disposal practices.
1976	Resource Conservation and Recovery Act (P.L. 94-580). Established "cradle-to-grave" responsibility for solid and hazardous wastes.
1979	Strategic and Critical Stockpiling Revision Act (P.L. 96-41). Revised and updated policies for national defense materials stockpiling.
1980	National Materials and Minerals Policy, Research and Development Act (P.L. 96-479). Mandated an executive-level study of national materials demand, supply, and need and the development of a national plan.
1984	National Critical Materials Act (P.L. 98-373). Established the National Critical Materials Council to promote research and development on critical and advanced materials.

throughout the world. To consider these dual threats, President Truman established the President's Materials Policy Commission in 1951 and appointed as chair William S. Paley, then president of the Columbia Broadcasting System. The Commission, which became known as the "Paley Commission," was charged by the president with making "an objective inquiry into ... assuring an adequate supply of production materials for our long range needs and to make recommendations that will assist me in formulating a comprehensive policy on such materials."[12]

A year later the Paley Commission delivered a far-sighted report called *Resources for Freedom*, which carefully converted the materials scarcity issue into one of costs:

The threat of the Materials Problem is not that we will suddenly wake up to find the last barrel of oil exhausted or the last ton of coal gone, and that economic activity has suddenly collapsed. The real and deeply serious threat is that we shall have to devote constantly increasing efforts to acquire each pound of materials from natural resources which are dwindling both in quality and quantity, thus finding ourselves running faster and faster in order to stay standing still. In short the essence of the Materials Problem is costs.[13]

Rejecting a materials self-sufficiency approach and courting the opposition of the domestic minerals industry, the Paley Commission recommended lowering domestic trade barriers and encouraging American firms to invest in the development of foreign materials. In doing so, the report recommended that materials be purchased on the global market, observing the "least cost principle" and respecting the need for national security. Recognizing the intense resource consumption caused by the Korean War and the needs of the rapidly expanding national economy after the war, the Commission's report advocated "a national materials policy for the United States" to avoid the economic dislocations resulting from increasingly scarce national resources. Noting the absence of a national materials policy, the Commission suggested that the President's National Security Resources Board should assume the task, noting that:

[t]he overall objective of the national Materials Policy for the United States should be to insure an adequate and dependable flow of materials at the lowest cost consistent with the national security and with the welfare of friendly nations.[14]

Understanding that the federal government might need additional encouragement to adopt and implement the Commission's recommendations, Paley established a nonprofit organization called Resources for the Future to follow up on the report. Other advocates began to push for a presidential conference similar to the 1908 White House conference. Paley joined this effort and offered his new organization as the sponsor. Initially financed by the Ford Foundation, the organization began to outline a research program and to organize a founding conference called the "Mid-Century Conference on Resources for the Future." Political uneasiness kept the White House from sponsoring the conference, but President Eisenhower did provide the keynote address at the opening session. The 1600 scientists, economists, and business leaders attending the conference broke into eight working groups to consider the protection and development of the materials resources of the country. In the closing sessions the participants further urged the development of a broad and comprehensive national materials policy.[15]

The concerns raised about material supplies in the Paley Commission report were softened in the years that followed by a robust economy that showed little evidence of scarce materials. The Paley Commission principle that price and substitution would adjust for diminishing material supplies appeared vindicated, and the robust economy of the next decade saw the interest in a national materials policy wane. Resources for the Future did keep the idea alive. In 1962 the organization published a national inventory called *Resources in America's Future*, but by then the idea of a national materials policy raised Cold War fears of a centrally planned economy.

Materials Policy and the Environment

During the late 1960s the new environmental consciousness and the rising costs of materials led to a renewed interest in materials, this time focused on materials conservation and environmental quality. In 1970, Congress passed the Mining and Minerals Policy Act, which declared

It is the continuing policy of the Federal Government in the national interest to foster and encourage private enterprise in (1) the development of economically sound and stable domestic mining, minerals, metal and metal reclamation industries, (2) the orderly and economic development of domestic mineral resources, reserves, and reclamation of metals and minerals to help assure satisfaction of industrial, security and environmental needs, (3) mining, mineral, and metallurgical research, including the use and recycling of scrap to promote the wise and efficient use of our natural and reclaimable resources, and (4) the study of and development of methods of the disposal, control and reclamation of mineral waste products, and the reclamation of mined lands....[16]

Congressional debates over resource management in 1969 had raised the issue of a comprehensive materials policy and with the passage of the National Materials Policy Act in 1970, Congress established the National Commission on Materials Policy,

... to enhance environmental quality and conserve materials by developing a national materials policy to utilize present resources and technology more efficiently and to anticipate the future materials requirements of the Nation and the World, and to make recommendations on the supply, use, recovery and disposal of materials.[17]

The Commission's report, which was released in 1973, added to the Paley Commission focus on materials for production a parallel focus on the need to protect the environment. The report provided 108 detailed recommendations heavily weighted toward conservation of materials, accelerated recycling of wastes, increased attention to waste management, and more efficient materials use. In striking a balance between the two often competing values of goods production and environmental protection, the Commission recommended that "environmental costs be taken into account in the computation of costs and benefits of any action to extract, transport, process, use or dispose of any material."[18] The Commission recommended a high-level federal agency—a new Department of Natural Resources—to achieve coordinated materials and energy policies.

In preparing its full report, the National Commission on Materials Policy contracted with the National Academy of Sciences for two studies: one considering resource conservation and depletion and the other focused on the relationship between national materials policy and various national and international environmental protection policies. The second report affirmed the need for a national materials policy to create a balance between materials use and environmental protection, but went further by urging the establishment of a population policy, and by calling for an amendment to the U.S. Constitution that declared the "right of an individual citizen to a safe, healthful, productive, and aesthetically and culturally pleasing environment shall not be abridged." This report, under the direction of Nathaniel Wollman of the University of New Mexico, went well beyond the more placid recommendations of the contracting commission and urged eliminating the use of certain dangerous materials, reexamining the value of various novel and easily obsolete products, and guiding economic growth so as to discourage those economic sectors that stress the environment and to encourage those sectors that do not.[19]

The report of the National Commission on Materials Policy was released in the wake of the Club of Rome's widely read report, *Limits to Growth*. The result was a broad array of conferences and research reports. The National Academy of Sciences conducted studies on mineral and materials development, and the U.S. General Accounting Office prepared reports on materials research needs.[20]

In 1974, the Senate Committee on Public Works held hearings on resource conservation and recovery that resulted in the drafting of a new bill on waste management and materials recycling and conservation.[21] Although the resulting legislation was titled the Resource Conservation and Recovery Act of 1976, (RCRA) the broader conservation vision was lost in the final version. The largest section of the law covered the management of solid and hazardous wastes through a cradle-to-grave record-keeping system. It also created a permit system for all waste transfer, storage, and disposal facilities on or off industrial properties. Little guidance and even less budget were provided for conserving materials or recycling resources. The legislation did set up a Resource Conservation Committee and called for, yet again, another study on resource policies. This study, chaired by EPA Administrator Douglas Costle, considered ten specific policy areas ranging from tax subsidies on virgin materials to beverage container deposits, product regulations, and a national litter tax, but offered relatively weak recommendations, most of which were ignored.[22]

As the government began to implement RCRA, the hazardous waste portion of the act became its central focus and materials conservation and recovery were largely neglected. By the close of the decade, the tragedy at Love Canal had emerged as the icon of environmental attention and public health issues rose to national prominence. Thereafter, hazardous waste management became a central focus of national materials concern.

6.4   Policies for Regulating Materials as Environmental Pollutants

While the federal government has never adopted a comprehensive national policy on materials, many of its other policies, particularly those regulating protection of the environment and public health, have had substantial effects on disposal of materials and indirect effects on their use. From very limited beginnings in the 1940s, Congress began to shape a federal environmental protection responsibility that over the decades would grow into a broad collection of regulations and dramatically affect the way in which industrial materials are extracted, processed, marketed, used, and disposed of. Table 6.3 identifies the primary federal regulatory environmental and health and safety laws that affect industrial materials.

Table 6.3
Primary federal environmental and health and safety laws that affect materials

1899	Refuse Act
1924	Oil Pollution Control Act
1948	Water Pollution Control Act
1955	Air Pollution Control Act
1960	Federal Hazardous Substances Act (P.L. 86-613)
1965	Solid Waste Disposal Act
1969	(Coal) Mine Safety and Health Act
1970	Occupational Safety and Health Act (P.L. 91-596)
1970	Clean Air Act (P.L. 91-604)
1972	Clean Water Act (P.L. 92-500)
1972	Consumer Product Safety Act (P.L. 92-573)
1974	Safe Drinking Water Act (P.L. 93-523)
1976	Toxic Substances Control Act (P.L. 94-469)
1976	Resource Conservation and Recovery Act (P.L. 94-580)
1977	Surface Mining and Reclamation Act
1980	Comprehensive Environmental Response, Compensation and Liability Act (P.L. 96-510)
1984	Hazardous and Solid Waste Amendments (P.L. 98-616)
1990	Clean Air Act Amendments (P.L. 101-549)
1990	Pollution Prevention Act (P.L. 101-508)
1990	Oil Pollution Control Act (P.L. 101-340)

Throughout the early years of the twentieth century, the federal government responded to the problems of industrial materials with various service and research programs rather than regulatory initiatives. In addition to the earlier U.S. Public Health Service, in 1930 Congress established the National Institutes of Health, while the U.S. Food and Drug Administration was set up in 1938. The Public Health Service began working on national drinking water standards during the early 1920s, but agreement was hard to reach and maximum permissible concentrations for lead, copper, and zinc were only released in 1925. It was not until the 1940s that the Public Health Service began developing "maximum allowable concentrations" (MACs) for a variety of chemical substances used in the workplace. While these MACs were not legally binding and they only represented goals that firms might try to achieve, they did stimulate more research on industrial chemicals and they did raise awareness of the potential hazards of these chemicals.[23]

The administration of Franklin Roosevelt was marked by more aggressive federal interventions, and in regard to industrial pollution, this meant a massive funding program for urban sewers, sewage treatment plants, and drinking water purification facilities. By 1938, the federal government had provided direct funding or grants-in-aid for 1165 of the 1310 new municipal sewage treatment plants built in that decade.[24]

During the 1950s, the U.S. Food and Drug Administration formalized an experimental program for testing drugs and began an aggressive research program. Congressional amendments to the Food, Drug and Cosmetics Act in 1958 contained the famous Delaney clause that prohibited the use of any known or suspected carcinogen in prepared foods. The stringency of this clause stimulated further development of quantitative toxicology and initiated efforts to establish risk assessment as a public policy-setting tool.

Historically, the federal government avoided environmental protection or health and safety issues because most government leaders believed that this responsibility should be chiefly shouldered by private industry, the states, and professional organizations. Long before federal interests were drawn into this area, many leading states were establishing government programs to assess environmental and occupational health and safety conditions, provide advice, train professionals, and in cases where it was appropriate, adopt regulations. By the mid-1960s this perspective was changing. This transition occurred because rising standards of living created new expectations, increasing levels of environmental education created a more aware public, new national agendas (exploration of space and alleviation of poverty) increased the acceptance of federal government services, and mounting evidence suggested that the states' efforts were largely inadequate.[25]

Prior to 1970 Congress enacted several laws aimed at supporting or prodding state efforts. These included the Water Pollution Control Act of 1948, the Air Pollution Control Act of 1955, the Hazardous Substance Labeling Act of 1960 , the Clean Air Act of 1963, the Water

Quality Act of 1965, and the Motor Vehicle Pollution Control Act of 1967. These early efforts to increase environmental protection by providing financial aid and technical assistance to the states proved ineffective as the public became increasingly incensed by dramatic examples of environmental pollution and hazardous work conditions. But it was the confluence of the growing ecology movement of the late 1960s and the activism of an ambitious Congress that created the bold, more interventionist federal regulatory laws governing use and disposal of industrial materials during the first years of the 1970s.[26]

Confronted with a swelling popular movement for environmental protection, President Richard Nixon created the Environmental Protection Agency in 1970 to coordinate and administer federal pollution control programs. That same year Congress enacted a new Clean Air Act and authorized the new EPA to set uniform, federal ambient air quality standards for industrial pollutants (as well as vehicle exhausts). In addition, the law required the EPA to work with the states in issuing permits for the discharge of primary and secondary pollutants from stationary sources such as industrial smokestacks and air vents. In 1972 Congress passed the Clean Water Act, declaring "the objective of this Act is to restore and maintain the chemical, physical, and biological integrity of the Nation's waters."[27] This law updated earlier water pollution control laws and required the EPA to set uniform, national water discharge standards for both toxic and nontoxic pollutants, and in conjunction with the states, to establish permit systems for the discharge of municipal sewers and industrial pollutants into all surface waters. The Safe Drinking Water Act of 1974 required the setting of uniform standards for drinking water quality and required programs to protect groundwater from industrial discharges.

Finally, as indicated, in 1976 Congress enacted the Resource Conservation and Recovery Act to close the circle of media-specific environmental regulations. The Surface Mining and Reclamation Act of 1977; the Comprehensive Remediation, Compensation, and Liability Act of 1980; and the Oil Pollution Control Act of 1990 rounded out these laws by setting the conditions and responsibility for the cleanup and reclamation of land or marine resources damaged or contaminated by the improper disposal of industrial wastes.[28]

Early on during this period, two laws were passed specific to occupational safety and health. The Mine Safety and Health Act was passed in 1969 following a tragic mine disaster in West Virginia that killed 78 miners. The more general Occupational Safety and Health Act was passed in 1970. These laws established three new federal agencies—the Mine Safety and Health Agency, the Occupational Safety and Health Agency (OSHA), and the National Institute on Occupational Health and Safety. These agencies were to conduct research on the workplace dangers of industrial hazards (including chemicals) and to set occupational health and safety standards. In its first few years, the Occupational Safety and Health Agency moved aggressively to adopt hundreds of national consensus standards for workplace health and safety.

Other laws enacted during this period focused more directly on industrial chemicals in products and as products. The Consumer Product Safety Act of 1972 built upon the 1960 Federal Hazardous Substances Act to establish a Consumer Product Safety Commission with authority to test products; to require specific labeling, design, packaging, or composition of products for sale to the public; and to prohibit the sale of substances deemed to be too hazardous for consumer use. The Toxic Substances Control Act of 1976 (TSCA) focused directly on industrial chemicals. It opened by noting that:

(1) human beings and the environment are being exposed each year to a large number of chemical substances and mixtures; (2) among the many chemical substances and mixtures which are constantly being developed and produced, there are some whose manufacture, processing, distribution in commerce, use, or disposal may present an unreasonable risk of injury to health or the environment.[29]

The law goes on to state that it is intended to gather health effects information on existing chemicals, promote testing where information is inadequate, and regulate the way in which new chemicals enter the market. For a new chemical, the law required that the EPA make a short health effects review of the substance prior to market entry to determine whether the substance required further testing, special labeling, special conditions of use, or an outright prohibition on use.

Until the mid-1980s, pollution control and waste treatment remained the dominant strategies for managing toxic and hazardous wastes. Pollution control technologies mandated by the federal environmental regulatory laws scrubbed, screened, and filtered industrial wastes at the point of environmental release and contributed substantial volumes of toxic sludges, filter cakes, and combustion ash to the growing national hazardous waste stream. In 1984 Congress passed amendments to the Resource Conservation and Recovery Act that prohibited the disposal of liquid hazardous wastes in landfills, and the limits of waste treatment and disposal capacity soon became a national crisis. Something needed to be done, and more of the same regulatory tightening did not appear promising.

Instead, environmental advocates, government officials, and some progressive business managers began to consider ways to reduce, not simply control and manage, waste materials. By focusing on the inefficiencies of industrial production processes, it appeared possible to reduce the generation of pollution at its source (sometimes called "source reduction"). This thinking was given a significant boost in 1986 with the publication of two well-read reports. The EPA prepared a special *Report to Congress* on the minimization of hazardous waste ("waste minimization"), and the U.S. Congressional Office of Technology Assessment published a critical report identifying the opportunity for significant waste reduction and cost savings to firms that implemented programs designed to prevent (rather than control) pollution ("pollution prevention"). Waste minimization and pollution prevention soon became the code terms for a raft of government and industry programs designed to reduce the generation of hazardous wastes. In 1990, Congress enacted the Pollution Prevention Act, which established the reduction of pollution at the source as the priority strategy in the nation's hazardous waste management policy.[30]

Over the past several decades, the federal government has made substantial efforts at protecting the environment from the toxic and hazardous effects of industrial chemicals. Many of these policies have been regulatory and geared to managing and controlling wastes rather than reducing or altering the toxicity of those wastes. With the exception of the Toxic Substances Control Act, almost none of these laws and policies are directly focused on the development of industrial materials in such a way that the materials themselves are made safer or less hazardous to the environment. Overall, the government's wide-ranging regulatory approach has never been well integrated or unified, and redundancy and gaps are often noted. However, the combined effort is extensive and its impact on industry has been significant and controversial.

6.5   Policies for Managing Industrial Materials Markets

Even as the environmental agenda swept materials issues into the broad area of waste management, the idea of a national materials policy did not die. The rising price of oil and other critical materials in 1973 led Congress to establish a National Commission on Supplies and Shortages in 1974. However, the Commission's report, *Government and the Nation's Resources*, provided a strong argument against government interference in materials markets.[31] Noting that resource exhaustion was not imminent and that rising prices were a temporary phenomenon, the report recommended against import restrictions, government-sponsored research on materials conservation or substitutions, and any national "coordinated materials policy." The report, with its *laissez-faire* approach, was received with little enthusiasm by anyone except the hard-pressed mining industry.

The change in administration in 1976 opened a new opportunity for materials policy. The initiative arose from the concerns of western congressional representatives over constraints on nonfuel minerals production. The House Committee on Science and Technology reviewed bills in 1977 and 1978 on minerals policies and in the closing days of President Carter's administration, Congress passed the National Materials and Minerals Policy Research and Development Act. The new law directed the president to continue federal materials research and development and to prepare a national materials plan within 2 years.

The plan was prepared under President Ronald Reagan's new administration by the Cabinet Council on Natural Resources and Environment and delivered to Congress in 1982. It was roundly criticized for its limited objectives, its lack of emphasis on materials other than minerals, and its recommendations for opening up more public lands for mining interests. The House Committee on Science and Technology noted that "it is no accident that the Cabinet Council on Natural Resources and Environment, headed by the Secretary of the Interior, placed primary emphasis on minerals, mining and related public land policies with almost no attention to basic materials processing, conservation, substitution, or new materials development."[32]

Investing in Materials Research
The government-sponsored scientific teamwork that was so important to the development of synthetic rubber and the Manhattan Project during World War II provided ample evidence of the importance of government support of materials research and the strategic importance of maintaining established teams of scientists that could further develop those materials for commercial products. In 1945, President Truman called upon Vannevar Bush, then president of the Massachusetts Institute of Technology, to lead a commission that could chart a course for how best the government might support research and technology during peacetime. The Bush report, titled *Science: The Endless Frontier*, drew a linear connection between government-financed basic research, through industry-financed applied research, to the development of commercial innovation and recommended that the federal government concern itself first and foremost with the support of basic scientific research. In 1946 Congress established the Office of Naval Research and 4 years later, the National Science Foundation, and both agencies quickly committed themselves to supporting research in the materials sciences. Between 1947 and 1957, federal support for materials research rose from $1 million to $10 million and by 1957 made up 70 percent of all university research in the field.[33]

Beginning in 1956, the Atomic Energy Commission and the Air Force each made recommendations for federal sponsorship of national materials research centers that could keep the country a leader in new materials research. An interagency Coordinating Committee on Materials Research and Development was set up by President Eisenhower in 1958 and it recommended the establishment of interdisciplinary materials research laboratories. Recognizing the value of these recommendations, Herbert York of the military's Advanced Research Projects Agency provided funding in 1958 for a new joint federal–university program in interdisciplinary laboratories (IDLs) for research into advanced materials. During the 1960s, the Atomic Energy Commission and the National Aeronautics and Space Administration expanded this program by funding additional university labs. By 1972, when all of these labs were transferred to a central administration at the National Science Foundation, they were renamed the materials research laboratories (MRLs).

During the mid-1980s, when a frustrated Congress abandoned its efforts to promote a comprehensive federal materials policy, it chose to focus more narrowly on development of new materials. Concern over Japan's significant advances in automobiles and electronic products led to a growing demand for government support for materials research and development. The passage of the National Critical Materials Act of 1984 sought to focus government resources on the commercial and military needs for the more advanced, synthetic materials required by manufacturing industries. This act established the National Critical Materials Council in the Executive Office of the President and charged it with overseeing policies related to both "critical" and "advanced" materials. Although this council functioned within the Executive Office, it was largely ineffective at coordination. By the close of the decade the council was reorganized into the Committee on Materials (COMAT) within the President's Office of Science and Technology. COMAT itself became a strong advocate for funding for advanced materials research, and during this period appropriations increased for materials research sponsored by the National Science Foundation and the National Institute of Science and Technology.

These various initiatives promoted by the White House and the National Science Foundation were important in encouraging research on new materials. However, there was nothing comprehensive or integrative in their mission. While they offered a significant boost to materials research and development and to the production of advanced materials, the broader policy connections to materials conservation, environmental protection, and future supply issues were left unaddressed.[34]

6.6   The Mixed Record of Federal Policy Initiatives

Throughout its history the federal government has been an active participant in assisting private industry in the development, management, and disposal of industrial materials. Government activities have ranged from subsidies and investments to regulations and prohibitions. Many of these policies have had a significant impact on stocks and flows of industrial materials.

Federal resource management has had a substantial effect on the extraction, use, and dissipation of natural resources, particularly the forests, waters, and minerals of the western United States. The departments of interior, agriculture, and energy have sponsored, assisted, and rationalized the private conversion of these materials for commercial purposes. At times the government has also been quite aggressive in market interventions. Wartime procurement, stockpiling of critical materials, tax incentives, and protective tariffs have helped to shape the demand and competitiveness of materials of specific importance to the national economy and its security. More recently, the federal government has been a significant sponsor for research and development of advanced materials that promote new technologies and products.

Finally, both federal and state governments have invested heavily in policies that regulate the production and use of toxic and hazardous industrial materials in order to protect human health and the environment. The costs of these regulations have been high for both the regulated and the regulators. There is much controversy and a host of studies about the effects of this panoply of regulatory laws. Nevertheless, even without conclusive evidence of the full benefits of these laws, the immense amount of attention and investment that has gone into writing, implementing, and enforcing federal laws concerning the environmental effects of industrial materials remains impressive.

The full range of the government policies on industrial materials suggests a staggering amount of public investment in materials development and management. The policies are typically justified by their sponsors as benefits to the general public as consumers and citizens, but there are many private interests that also benefit along the way from extraction to end use, and little analysis exists to demonstrate just how effectively the benefits actually get to the intended customer. Whatever benefits these policies have brought, they have not been broad or sophisticated enough to account for the full life cycle of a material or the future of the system by which the nation extracts, processes, uses, and finally disposes of the natural resources with which it was once so well endowed.

By the close of the century, the federal government could show a long and proud history of individual initiatives to promote materials development, materials stockpiling, and environmental regulation, but far less attention to materials conservation, materials recycling, or materials detoxification; and there is no such thing as an integrated national policy. Nearly a half century after the Paley Commission called for a comprehensive national materials policy, and 25 years after the National Commission on Materials Policy did the same, the federal agenda remains a complex assortment of individual commitments and as lacking in overall vision as it was a century ago.

7

The Performance of Industrial Materials Policies

First, and most important, evaluation provides reliable and valid information about policy performance, that is, the extent to which needs, values, and opportunities have been realized through public action.
—William N. Dunn

The early materials policies of the nation were based on the assumption that people were scarce and the resources of nature were abundant. In large part, industrialization was driven by the need to improve the productivity of people because labor was in short supply. Factories arose and machines proliferated in order to multiply the daily output of a limited number of workers. Synthetic chemicals replaced natural materials because they required less labor to produce and often required less labor to use. Initially, industrial waste quantities were small, and the receiving media of the environment appeared endless. But as the population of the country increased, the demand for products increased and so did the volume of wastes. Today, people are abundant and natural resources in the form of raw materials or places to put wastes are increasingly scarce.[1]

Thus, throughout the nation's history, government policies have focused primarily on the development and use of industrial materials and much less on their conservation. Today the nation has an ample supply of inexpensive and functionally effective materials, and this comfortable abundance has meant that policies to conserve and formally manage the supply of materials have been modest at best. The aggressive overexploitation of the nineteenth century has been curtailed, but the voracious demands of a rapidly growing domestic economy and the enormous costs of two world wars have seriously drained the nation's once-rich natural reserves. In contrast, the policies put forward to manage wastes have been late and limited. Every day produces another mountain of domestic trash and commercial discards, and the effluents and emissions of industrial production, while better managed today, still contribute thousands of tons of pollution to overburdened air and water resources. Only in the past 30 years have federal policies emerged to protect the public from dangerous exposures to toxic materials. These policies suffer from inconsistent enforcement and inadequate scientific knowledge, and the consequences are difficult to assess.

A detailed assessment of the effectiveness of current policies on material conservation and human health and the environment is warranted. Only a brief review can be offered here, but even an overview provides some useful lessons. This chapter covers two areas of materials policies: those policies addressing material dissipation and those focused on toxicity. Each discussion focuses on three factors: the information available to provide an assessment, the degree to which the policies can be shown to have been influential, and an analysis of the factors that have limited their effectiveness.

7.1   Information on the Flow of Industrial Materials

Assessing the performance of policies requires data. Ideally, an assessment of national policies on industrial materials would be based on national trend data that track materials as they move from reserves to feedstocks to commodities to wastes. Unfortunately, there are no such data, at least no comprehensive data. There are some national data on the production of materials, almost no data on the use of materials, and a substantial amount of quite heterogeneous data on the release of materials as wastes or pollutants. Table 7.1 lists several of the most significant sources of federal data. Most data come from government or trade association surveys. The sources are self-reports and most likely this results in underreporting. The data are assembled by public agencies with widely divergent missions, differing definitions and metrics, and little coordination in data collection processes. All of this makes analysis difficult. For those who try to measure materials flows, it is necessary to establish some common metrics (units of analyses, such as weight, mass, energy

Table 7.1
Sources of government information on national materials stocks and flows

Subject	Source[a]	Period of Issue
Resources Data		
Mineral reserves	USGS	Annual
Water resources data (by state)	USGS	Annual
Petroleum supply annual	DOE	Annual
Natural gas annual	DOE	Annual
Materials Processing Data		
Mineral industry surveys	USGS	Annual
Mineral commodity surveys	USGS	Annual
Mineral yearbook	USGS	Annual
Metal industry indicators	USGS	Monthly
TSCA Chemical inventory update	EPA	(4 years)
Materials in Commerce Data		
Census of manufacturers	B. of Census	Quin-quennial
Current industrial reports (by product)	B. of Census	Varies
Current business reports (inventories and sales)	B. of Census	Annual
Manufacturing energy consumption survey	DOE	(3 years)
U.S. General imports (selected commodities)	Int. Trade Ad.	Monthly
Data on Material Wastes		
National air pollution emissions trends	EPA	Annual
Characterization of municipal solid waste in the United States	EPA	Annual
Preliminary biennial RCRA hazardous waste report	EPA	Biennial
Toxics Release Inventory	EPA	Annual
Ambient Environment Data		
National air quality and emission trends reports	EPA	Annual
National water quality inventory	EPA	Biennial
National public water system supervision program	EPA	Annual
National water conditions	USGS	Monthly
National oceanagraphic and atmospheric status and trends	NOAA	Periodic

[a] USGS, U.S. Geological Survey; DOE, U.S. Department of Energy; EPA, U.S. Environmental Protection Agency; B. of Census, U.S. Bureau of Census; Int. Trade Ad., International Trade Administration, NOAA, National Oceanographic and Atmospheric Administration.

content, toxicity, dollar value, and information content) and, by default, weight is the most commonly accepted metric.[2]

There are annual data on the production of materials. The U.S. Geological Survey prepares an annual report on the production and consumption of mineral commodities, including data on mineral imports and exports and market prices. These data, which are first released as annual mineral commodity summaries and later finalized into an annual yearbook, are generated from corporate surveys and personal inquiries conducted by USGS commodity specialists. Data on mining and material commodity imports and exports were compiled and summarized annually by the U.S. Bureau of Mines and the U.S. International Trade Commission until 1994, when both offices were closed by congressional budget cuts. Those cuts also closed the congressional Office of Technology Assessment which, until its closing, produced excellent analyses of specific materials issues.

The U.S. Bureau of the Census compiles data on the manufacturers of products and publishes monthly and annual statistics on materials production and trade. There are no annual national data on how materials are used in commerce, the transportation and distribution of products, or the use of products.

Some twenty federal agencies are involved in collecting environmental data, although no unified data system exists and there is no periodic report that would provide a comprehensive picture of the state of the environment. A significant amount of data is collected by the EPA on materials as wastes or environmental pollutants. The best of the historical trend data track pollutant emission by environmental media. There are substantial data on water discharges, but the best historical data track air emissions and are focused on six criteria pollutants. Most of the data on national air and water emissions are based on engineering calculations and quantitative models rather than direct monitoring. A more reliable data set focuses on toxic chemicals released to the environment. Under the federal Toxics Release Inventory (TRI), data on toxic chemical releases are collected from the annual reports of thousands of individual facilities, and these data are considered fairly reliable. Still, the toxics release data cover just 643 chemicals released from only a fraction of the nation's industrial facilities, and are available only since 1987.

Data on ambient air and water quality collected at various field stations around the country are limited because of the fragmented and inconsistent way in which they are collected. National data on sediment and soil uptake are fairly weak, as are national data on chemicals sequestered in the tissue of plants, wildlife, and humans.

Data are also collected on municipal solid wastes and hazardous wastes and on recycling of municipal wastes. Data on the generation of municipal solid wastes are based on engineering calculations rather than monitoring of actual garbage. Because of this, the national data tend to miss regional variations and fluctuations that are due to seasonal or economic activity. Data on hazardous wastes are collected on a biennial basis by the states, but because there are many exemptions under the federal definition of hazardous waste and because reporting procedures vary significantly by state, it is difficult to use the state reports to develop a national picture. Data on materials recycling programs are limited by the absence of common definitions and metrics. For instance, there is no common agreement on what constitutes recycled materials, and domestic discards set out at curbside for recycling are usually counted as recycled even when temporarily saturated secondary markets mean that those materials are landfilled or incinerated.

7.2 Assessing Policies on the Stocks and Flows of Industrial Materials

Policies promoting the development and conservation of the nation's materials are distributed among several federal agencies, including the departments of commerce, agriculture, energy, defense, and interior. Research on materials policies is conducted by the National Academy of Sciences, particularly the National Research Council and the National Academy of Engineering. For decades the U.S. Geological Survey and the U.S. Bureau of Mines have been the federal government's principal agencies for monitoring and promoting the development of the nation's minerals. As indicated earlier, over the past half-century there has been a long sequence of national advisory bodies on materials. During the past decade these have included the National Critical Materials Council and the Committee on Materials in the Executive Office of the President, and the National Strategic Materials and Minerals Program Advisory Committee and the Committee on Mining and Mineral Resources Research in the Department of the Interior.

Material Supplies

Federal policies promoting materials supplies have largely focused on the development and the maintenance of supplies of materials needed for a growing economy and the strategic reserves necessary for wartime. The development of materials has been guided by government policies that include tariffs, direct subsidies, taxation, government procurement, land use controls for public lands, and various environmental protection regulations, while wartime reserves are managed through the strategic stockpiles.

The flow of cheap European materials into the United States during the nineteenth century constrained the development of the domestic materials industries. The federal government turned to protective tariffs to try to compensate for price differences. The first federal tariff was levied in 1816 on foreign iron imports to protect the domestic iron industry. Additional tariffs were imposed to protect the mining and inorganic chemical industries. For instance, tariffs were levied on inorganic chemicals during the 1890s and tariffs were used during the 1920s to protect the Lake Superior copper industry from South American imports and to protect the manganese industry from Russian imports. The Fordney–McCumber Tariff Act of 1922, which placed a duty on a wide range of organic chemical imports, is credited with providing the cornerstone for the domestic organic chemical industry. There was no guiding federal policy on tariffs and these protections were often applied for political purposes and maintained for years after their objectives had been achieved. In 1934, Congress passed the Reciprocal Trade Agreements Act, which provided for a more systematic approach to tariffs. Since that time, the number of protective tariffs on materials has been declining, although tariffs currently continue on some lead and zinc ores. With these exceptions, today most mineral ores enter the United States with no tariffs, while refined metals and alloys pay only a modest duty.

At times the federal government has provided direct subsidies for development of minerals necessary for national security. The uranium and titanium industries were largely developed through federal sponsorship, and in 1950 Congress passed the Defense Production Act which, among its provisions, authorized joint federal–private industry programs that supported hundreds of mineral exploration projects. Between 1958 and 1974 the government assisted many mining firms by underwriting the costs of exploration with low-interest loans that were forgiven if no commercial production resulted.[3]

The extraction and mobilization of hard-rock minerals has been greatly subsidized by free and open access to Forest Service and Bureau of Land Management lands, which today make up roughly 71,400 square miles, or nearly 20 percent of the nation's total land area.[4] Contemporary mining policy remains grounded in the General Mining Law of 1872, under which the federal government offers open access to hard-rock mineral exploration on federal lands. By simply filing a claim, anyone can acquire title to a mineral resource. The maintenance of a claim requires some evidence of mining, although the government collects no royalties on minerals extracted from these claims. In essence, the minerals are free: their cost is determined by the cost of locating them and then mining and refining them. As of 1996 there were an estimated 2.9 million mining claims throughout the thirteen western states. A U.S. General Accounting Office study found that of 240 randomly selected claims, 237 had never been mined and most were used for other purposes.[5]

Policies regarding oil, coal, natural gas, sodium, and phosphate rock differ from the hard-rock minerals. Under the Mineral Leasing Act, the federal government retains title to all reserves of these materials on the public domain and requires leases for their exploration and extraction. Since 1953 these mineral leasing policies have been extended to cover ores in the seabed and submerged lands as well.

The organic chemicals industry—and particularly the petrochemicals and petroleum-based plastics—also receive public subsidies indirectly through government energy subsidies that promote extraction of oil and natural gas. It is estimated that energy subsidies in the United States in 1989 amounted to about $36 billion a year, of which about $22 billion went to fossil fuels. Because petrochemicals consume about 6 to 8 percent of the energy content of the fossil fuels consumed each year, this would mean that each year something over $1 billion in public subsidies is provided to the feedstocks of the petrochemical and plastics industries.[6]

The Internal Revenue Code gives special tax advantages to mining. While revenues derived from mining are subject to federal and state taxes, mineral producers can deduct from their taxes an amount up to 70 percent of the costs of exploration and development in the year that the expenses are incurred and then depreciate the remaining 30 percent over a 5-year period. In addition, operating losses and unused investment tax credits can be carried back 3 years or forward 15 years. Since 1913, percentage depletion allowances have permitted mineral producers to deduct a percentage of taxable net income as a business expense. Allowable annual percentages range from 5 percent for sand and gravel up to 22 percent for lead, zinc, and sulfur. Unlike ordinary tax depreciation for capital investments, the percentage depletion can aggregate over time to many times the value of a mine. Treating mine exploration and development as a current expense allows for a large immediate tax deduction instead of a gradual cost depletion as the value of the mine diminishes. In addition, the taxation of royalty income at capital gains rates benefits mine leasing arrangements and lowers the supply price of minerals.[7]

The provisions of the General Mining Law and the extensive tax incentives have promoted virgin mineral extraction and discouraged mineral conservation and metals recycling. For well over a century these lucrative subsidies have found critics. Somewhere between $2 and $4 billion worth of minerals are extracted from federal lands each year. Since the passage of the General Mining Law, it is estimated that the federal government has given away minerals worth more than $245 billion. One recent study concluded that 15 federal subsidies averaging $2.6 billion annually benefit resource extractive industries that favor the use of virgin materials in industrial processes. Because minerals are essentially free, there is no incentive to conserve them, other than the capacity of the market to absorb them at a price that would cover their extraction and processing costs. Not only does this distort the market, it deprives the government of revenues. Government budget analysts estimate that the percentage depletion allowance results in a loss to the federal treasury of nearly $270 million each year, with another $27 million lost each year owing to the current expensing of exploration and development costs.[8]

Government procurement, particularly military contracts, has had a significant impact on the commercial viability of various materials. The military often acts as the major purchasing arm of the federal government, developing and buying materials and technologies on a scale that dwarfs most private enterprises. Sometimes the military will simply buy what is available on the commercial market; more often it will contract for new materials that meet specific military objectives. Through subsidized research and development and a guaranteed market, the Department of Defense can promote new materials or technologies and significantly alter the commercial market. It is hard to find a better case than the military support for the development and commercial success of the solid-state electronic transistor. Developments in silicon and germanium crystallography, copper and lead solders, and dopant diffusion technology were all sponsored by the Army Signal Corps, which then became a major customer of transistor-based communication equipment.[9]

Additional government procurement occurs under the Strategic and Critical Materials Stockpiling Act. The inventories in these stockpiles are not small. There are more than ninety specific materials currently in the U.S. strategic stockpiles and at times the inventory of some of the metals in the stockpiles has exceeded the annual world production of those metals. Table 7.2 lists several of the larger stocks of critical and strategic materials. The size of these stocks gives the stockpiles a potentially dominant role in setting materials prices and because of this, the American Mining Congress has consistently argued for their elimination. Manipulation of the strategic stockpiles has played an important role in federal budget considerations, foreign policy, domestic economic stimulation, and materials price stability.[10]

Although legally bound to national security issues, the stockpile program has been used by various federal administrations to achieve economic objectives. At times, changes in stockpile purchases and sales have had a broad impact on world materials prices, and since the 1950s the stockpiles have played a major role in promoting metals exploration, development, and pricing. Continued U.S. government purchases of many basic metals following World War II eased the industrial transition back to a domestic economy. In 1958 a large percentage of the stockpiles were

Table 7.2
Selected materials inventories in national defense stockpile, 1999

Aluminum oxide	82,000 tons
Antimony	15,000 tons
Asbestos	32,000 tons
Bauxite	9,447,000 tons
Cadmium	2,665,000 pounds
Chromium	2,486,000 tons
Cobalt	28,156,000 pounds
Lead	277,000 tons
Manganese	2,143,000 tons
Mercury	9,778,000 pounds
Silver	26,203,000 ounces (troy)
Titanium	35,000 tons
Tin	72,000 tons
Tungsten	5,418,000 pounds
Tungsten ore and concentrate	73,440,000 pounds
Zinc	198,000 tons

Source: U.S. Department of Defense, *Strategic and Critical Materials: Report to Congress*, Washington, D.C., September 1999.
Note: Quantities include all compounds in each class.

sold off and this tended to depress metal prices. The replenishment of the stockpiles during the 1960s had an even greater market effect. Sales of tin from the stockpiles have been used to influence world tin prices, and in 1965 President Johnson used the threat of releases from the stockpile to force a rollback of copper and aluminum prices. The domestic chromium industry, which nearly collapsed during the 1950s, was resuscitated with stockpile purchases, and the domestic lead and zinc industries have been supported at various times with stockpile purchases.

These various policies support an abundant supply of materials. The supplies necessary for wartime demands are well provided for by the strategic stockpiles. The remaining federal policies are largely directed at economic objectives. In general, they tend to support and "lock in" the current pattern of materials development, and this inhibits the search for more environmentally benign materials. For instance, the current tax subsidies bias the economy toward extractive industries and against the development of secondary materials industries (including recycling and reuse). They also decrease tax receipts that would otherwise be collected as government revenue.

Up until the late nineteenth century there really were no policies on the conservation of materials. Early oil drilling and coal mining were grossly wasteful, with much of the valuable product flowing into rivers or mixed into tailings. The conservation movement brought the management of resources and resource extraction to national attention. Policies were established for forest management on government-owned lands, but little was directed at minerals management or the management of forests or minerals on private lands. Since the 1950s there has been increasing attention to the conservation of nonrenewable materials such as metals and fossil fuels. But rather than try to preserve some materials for the future, the focus has been on managing extraction rates to meet present and future demands. Therefore, mineral conservation has been defined in terms of maintaining an extraction rate that is below the rate at which technologies can create new reserves from the reserve base.

Conservation practices differ among materials. Through careful resource management and the adjustment of supply to demand, the oil and natural gas industries have been fairly stable, allowing for a planned approach to corporate investments and resource extraction. Such a consistent market permits an orderly approach to development of materials and encourages a private interest in resource conservation. This has not been the case with coal or various metals, where prices fluctuate rapidly and supply and demand are seldom matched for long periods of time. The volatility of these markets makes long-term planning and well-managed extraction rates difficult, and the result has been little effort at conservation of materials.

Wastes

Policies regarding the management of wastes present a different picture. While the management of nonhazardous domestic and commercial wastes has largely been viewed as a local and municipal responsibility, the wastes from mines and petroleum wells have been dealt with by state authorities. Since the 1970s, hazardous waste policies have been set by the EPA, with the state environmental protection agencies responsible for program implementation and enforcement.

Under the Resource Conservation and Recovery Act, the transport of hazardous wastes must be carried out by licensed waste haulers who track the movement of wastes from origin to final disposal through a sophisticated transport manifest system. State reports on hazardous waste generation are collected and compiled by the EPA every other year. Because there are many exemptions and complexities in reporting wastes under RCRA, the quantities in the biennial reports tend to underrepresent the total amount of hazardous waste generated in the United States. Still, reports from 1993 show that 258 million tons of hazardous wastes (including industrial wastewaters) were generated by over 24,000 RCRA permitted facilities. This represented an increase of nearly a thousand generators, but a decrease of approximately 47 million tons compared with the 1991 biennial reports.[11] Of these hazardous wastes, one industrial sector, the chemical industry, generates between 80 and 90 percent of the total. Although this is a large volume, nearly 90 percent of these wastes are treated and disposed of by the chemical-producing companies, often in facilities on their own sites.

A substantial amount of this hazardous waste is mixed with other materials for solidification or is chemically treated in large artificial ponds called "surface impoundments," where it may remain for months before it is eventually released. About one-quarter of the waste is dumped at privately owned landfills; another 5 percent is incinerated in on-site, hazardous waste incinerators; and the remainder is injected into deep wells.

The absence of a reliable data collection system makes it difficult to track the volume of hazardous wastes over time. A patchwork of evidence suggests that hazardous waste generation increased quite steadily from the 1940s up through the 1970s and since then there has been a reduced rate of growth. If true, it would be tempting to attribute this slower rate of growth to the imposition of environmental regulations, particularly those that increased the costs of treatment and disposal and liability for environmental damage. However, a more careful look at the data suggests that environmental regulations actually increase the amount of hazardous waste by diverting materials from conventional releases to air and water into the hazardous waste stream.[12] Thus, it would be more accurate to credit environmental regulations with both a positive and a negative effect on the volume of hazardous wastes.

Liability for the thousands of inactive or abandoned hazardous waste disposal sites is managed through the 1980 Comprehensive Environmental Response, Compensation and Liability Act (CERCLA) under which the EPA identifies potentially responsible parties and enforces cleanup activities. Although the EPA has identified some 27,000 sites, progress in achieving cleanups has been slow. While the agency has taken more than three thousand actions to remove hazardous wastes from these sites, as of May 2000, the Superfund site cleanup program had completed construction work at just over 650 of 1400 priority sites.

Mining wastes are covered under RCRA as well, but are classified as "special wastes" that are not subject to the stricter hazardous waste regulations. This is no small issue. While the total annual production of industrial hazardous wastes hovers around 260 million metric tons (40 million tons if wastewaters are not included), the annual generation of mine wastes is close to 2 billion metric tons. Of this, nearly half is simply dislocated overburden, but the remainder is composed of beneficiation tailings and of this, an estimated 1 percent (roughly 1 million metric tons) consists of potentially hazardous materials. The special waste classification leaves these wastes in a kind of confusing limbo that does little to address their true costs. One significant cost that is apparent involves the cleanup of thousands of abandoned mines across the country. The federal government has accepted liability for the cleanup of these mines, a liability estimated at somewhere between $33 and $72 billion.[13]

Responsibility for most domestic household wastes and the wastes from retail and service activities is usually relegated to local and municipal governments. For many years statistics on these solid wastes showed a continuing increase in municipal garbage. Municipal waste grew from 87 million tons in 1960 to 209 million tons in 1996. Per capita wastes increased from 2.7 pounds per person to 4.3 pounds per person per year over this same period. Until 1994, the volume of municipal waste has increased year-by-year, but since then the volume has remained fairly consistent. Most observers see this leveling off, not as an indication of decreasing rates of discarded materials, but rather as the result of aggressive local and municipal programs to promote waste recycling and composting.[14]

Indeed, the costs of municipal waste management provide a significant subsidy to the continuous flow of industrial materials. States and local communities have spent billions of dollars on the construction and operation of waste treatment and disposal facilities, almost none of which is recouped from materials or product suppliers. The enormous cost of managing waste materials—an amount estimated at $43.5 billion per year—is largely paid through municipal property taxes and community fees. The direct costs to the environment are almost totally hidden except when a leaking landfill or failed waste treatment plant is finally documented. Roughly a fifth of the federal CERCLA cleanup sites are former municipal waste disposal facilities.[15]

The Performance of Materials Stocks and Flow Policies

The national policies that guide material supplies are based on a nineteenth-century assumption that expanding supply and falling prices are the prerequisites to economic growth. The natural abundance of the continent supplied the national economy well into the twentieth century. As wars and the expanding consumer markets reduced the domestic supply of materials, the nation embraced a free and open global materials market, and increasingly imports came to replace dwindling domestic resources. Uncomfortable with significant import dependence, the federal government hedged its military needs by creating national stockpiles. However, there were no similar policies to address the continued resource depletion caused by the ever-growing demand for industrial materials. Those who spoke of materials conservation meant expanding reserves faster than demand. While some continued to worry about future materials scarcity, there were few policies that addressed conservation by slowing demand or preserving reserves for the future.

Even though the higher grade ores have been depleted, today it does not appear that the nation will run out of materials. Most economists have been sympathetic to the view put forward by the Paley Commission that the economy will not run out of usable materials because the rising price of scarce materials will drive demand toward the use of previously uneconomical materials or will encourage substitutes. Technological advances in the extractive industries do raise productivity and permit the substitution of more plentiful lower grade ores for the scarcer higher grade ores, but the costs are higher. For instance, many gold deposits are now so depleted that they can be economically mined only through a cyanide leaching process that has resulted in the poisoning of scores of western streams and lakes. Still, the relentless pursuit of the lowest-priced minerals has resulted in a sporadic and opportunistic approach to resource extraction that leaves significant amounts of currently uneconomical ores in overburden and slag piles. As prices rise, these ores become profitable, but this ensures higher prices and more dissipated mineral supplies for the future. This may satisfy an economy with ever-increasing wealth that can afford the rising prices, but it does not address the very real physical condition of depleting concentrations of high-quality resources.

If there are limits to the flow of industrial materials in the United States, they appear more at the disposal than the production end of the materials cycle. A significant slowdown in the construction of municipal landfills and local neighborhood resistance to the siting of waste treatment plants—especially incinerators—has raised the specter of a finite capacity for materials disposal, especially in densely settled metropolitan areas.[16] Still, wastes find space. While an occasional barge or train loaded with municipal wastes cannot find a dump, overall the physical capacity for managing wastes appears ample, although increasingly expensive. Even hazardous wastes appear to find a home, although this often involves exporting them to less-industrialized countries.

It is not the physical capacity to manage the waste stream, but the increasing ecological costs that suggest the true limits to the earth's assimilative capacity. The acidification of forest lands and lakes by sulfur-laden emissions, the accumulation of carbon in the upper atmosphere, the loss of species because of damaged habitat, and the contamination of groundwater by leaking dumps and agricultural runoff suggest that the environmental media have been overloaded and their assimilation has been compromised. During the past several years damage from weather-related destruction has suddenly leaped upward in terms of scale and costs, and there is a growing body of evidence that links these natural disasters to stratospheric pollution and carbon buildup in the upper atmosphere. Thus, it is not the actual flow of materials that puts the environment at risk today, as much as the secondary effects of the flow on the sensitive balances of ecosystems. The earth's resources may be more threatened by the echoes of the cascade of wasted materials than by the flood itself.

7.3   Information on the Hazards of Industrial Materials

Even with decades of toxicological studies of industrial chemicals, basic data on the potential effects of these chemicals on human health and the environment are too often lacking. Confidence in the safe use or release of industrial chemicals requires sound scientific studies of their effects on health and the environment. Unfortunately, such information is absent more often than it is present.

Several federal agencies generate data on chemical health hazards. The Centers for Disease Control, the National Center for Toxicological Research, the National Institute for Occupational Safety and Health, the Agency for Toxic Substances and Disease Registry, and the National Institute for Environmental Health Sciences provide federal focus and resources for research into the environmental and human health effects of toxic chemicals. These agencies provide data for public health policies, but data for occupational health policies are particularly difficult to acquire. Data on workplace accidents and acute physical injuries and deaths are available, but data on chronic illnesses or deaths caused by chemical exposures, particularly long-term, lower level exposures, are very difficult to acquire or trust. The federal Occupational Safety and Health Act requires that larger workplaces keep an annual record of employee injuries and illnesses that result in lost work time. Each year a sample of these data is collected by the Bureau of Labor Statistics, but the bureau only releases the data as industrywide aggregate figures. These data may be an adequate record for covering acute injuries requiring a worker to leave the work site, but for illnesses caused by chemical exposures, and particularly for those illnesses that occur later in life, this information source is inadequate.

The Toxic Substances Control Act gives the EPA the authority to require chemical manufacturers and importers to conduct tests and submit health effects information on industrial chemicals. Much of these data are posted on an on-line electronic database called TSCATS (Toxic Substances Control Act Test Submissions). Information is collected on several hundred priority chemicals and includes production, use and disposal data, records of adverse environmental or health effects, unpublished health and safety data, and notification of previously unknown risks. Every 4 years the agency requires firms to submit production, use, and disposal data on inorganic chemicals under its "inventory update rule." In addition, TSCA authorizes the EPA to require further health and environmental tests where the agency finds that a chemical may pose an unreasonable risk to human health or the environment, or where a chemical might enter the environment in substantial quantities. Finally, when a firm seeks to place a new chemical into commerce, it must submit a "premarket notification" to the EPA that includes available data on health and environmental effects and a description of reasonably ascertainable test data.

Even with these broad authorities, the EPA has found collection of chemical information difficult and slow. For instance, the EPA reports that less than 10 percent of the new chemicals reviewed each year under the premarket notifications contain adequate test data on health effects. In 1984, the National Academy of Sciences conducted a study of toxicity testing and concluded that minimal toxicity data were available on only 22 percent of 259 randomly selected chemicals produced or imported in large volumes. In 1997 the Environmental Defense Fund released a study based on a sample of 100 industrial chemicals which claimed that a basic set of data on health and environmental effects was available for no more than 29 percent of the largest-volume chemicals produced or imported into the country.[17]

More recently, the EPA has prepared its own estimates. The agency also focused on high production-volume chemicals These are defined as the 2863 chemicals imported or produced in the country at more than 1 million pounds per year and excludes polymers and inorganic chemicals. The adequacy of health and environmental test data was determined by the presence of the six common tests that are required under the international Organization for Economic Cooperation and Development's Screening Information Data Set (OECD SIDS). These include acute toxicity, chronic toxicity, developmental and reproductive toxicity, mutagenicity, ecotoxicity, and environmental fate. While these six tests do not provide a complete picture of the potential hazards of a chemical, they are considered a minimum set for screening chemicals for further study.[18]

In its 1998 analysis, the agency found that 43 percent of the high production-volume chemicals had no SIDS test data available at all and only 7 percent had a full set of the SIDS test data. Of the 203 high production-volume chemicals that also appear on the Toxics Release Inventory, 54 percent had full SIDS test data available, but 20 percent lacked two or more tests. Of those chemicals not on the Toxics Release Inventory, 46 percent had no test data available. Not all of these chemicals are equally likely to result in human exposure. Of the high production-volume chemicals on the Toxics Release Inventory with environmental releases greater than 1 million pounds per year, 26 percent lacked full SIDS tests. There are also gaps in the test data for chemicals of concern in the workplace. Of the high production-volume chemicals that are a priority for the Occupational Safety and Health Administration, 47 percent lacked evidence of the four basic SIDS tests for human health endpoints. With these findings, the EPA issued a challenge in 1998 to all 830 chemical companies manufacturing or importing high production-volume chemicals to complete the basic SIDS tests on their chemicals.

By the middle of 1999, over 60 percent of the targeted firms had committed themselves to generating the basic SIDS screening data. Such information about chemical toxicity and environmental fate will not be easy to assemble and it is costly to acquire. The EPA estimates that a complete set of SIDS tests costs about $205,000 per chemical and it will cost somewhere between $400 and $700 million to fill all of the missing data gaps for the high production-volume chemicals. Even with full SIDS tests completed for all chemicals, the EPA recognizes that more substantial tests will be required on chemicals with higher exposure potential (chemicals common in large environmental releases, workplace exposures, and consumer products).

It is too early to predict the outcome of this ambitious program. A complete set of screening data on the largest-volume chemicals would provide a better basis for assessing the hazards of industrial materials. But the program only covers chemicals with high production volumes and it does not cover polymers and inorganic chemicals. Moreover, the

SIDS screening tests provide only the beginning data and do not cover critical concerns such as carcinogenicity, neurotoxicity, and endocrine disruption. While this effort is laudable, it is hardly enough.

7.4 Assessing Public Health and Environmental Protection Policies

Federal policies concerning industrial materials and public health and environmental issues are implemented by the Environmental Protection Agency, the Occupational Safety and Health Administration, the Consumer Product Safety Commission, the National Oceanic and Atmospheric Administration, the Fish and Wildlife Service, the Forest Service, and the Food and Drug Administration. The protection of public health and the environment is shared among federal, state, and local governments. While the federal mix includes incentive-based and information-based policies, the dominant policies are regulatory (so-called "command-and-control") policies, requiring the control of pollutants and establishing limits on human and ecological exposure to toxic substances. The regulations specify scientifically based exposure standards that set thresholds for ambient conditions, discharges by facilities, and workplace exposures. Although often informed by scientific studies, most of these standards are the results of compromises negotiated among parties with quite divergent interests. Some standards have been adopted from professional associations such as the American National Standards Institute or the American Conference of Government Industrial Hygienists, while others are derived from formal regulatory negotiations that involve scores of competing protagonists. Once established, these standards form the basis for environmental permits that most states negotiate with individual facilities. In general, the states are responsible for monitoring facilities and ensuring compliance with permits.

The federal Clean Air Act sets standards for ambient concentrations of pollutants and emissions from both industrial and vehicular sources. The National Ambient Air Quality Standards include primary standards designed to protect human health and secondary standards to safeguard public welfare. Until 1990, only six criteria pollutants and eight hazardous air pollutants were targeted (the 1990 Clean Air Act amendments extended the list of hazardous air pollutants to 189). The standards were to be set so as to "provide an adequate margin of safety" without considering the economic costs of attainment. This has proven to be impossible because for many of these airborne pollutants, there is no scientifically proven safe level of exposure, and the costs of achieving zero exposure by controlling air emissions are starkly prohibitive.

The Clean Water Act requires the establishment of surface water quality standards according to the intended uses of the water (fishing, swimming, or agricultural, and industrial uses) and sets technology-based effluent discharge standards. Under the National Pollutant Discharge Elimination System, facility discharge permits have been issued to some 48,000 industrial facilities and 15,000 municipal treatment works.[19] Initially, 125 chemical substances were identified for regulation as priority pollutants in surface waters under the act, although additional, nonconventional pollutants ranging from ammonia to phosphorus have been added over the years. For each of these substances the EPA has established ambient water quality criteria. To date, the EPA has set standards for 81 contaminants in drinking water supplies under the Safe Drinking Water Act.

Industrial chemicals are regulated under the Toxic Substances Control Act. Between 1976 and 1990, the EPA reviewed over 10,000 new chemicals and today it receives about 2000 premarket notifications per year. While TSCA provides authority to phase out the use of chemicals found to present an unreasonable risk, only four substances (PCBs, CFCs used in aerosols, friable asbestos in schools, and dioxin-contaminated wastes) have actually been phased out under TSCA.[20]

The performance of the environmental regulatory laws has been widely studied.[21] Despite somewhat conflicting evidence and mixed results, there are some notable conclusions. While the population of the country has increased by 28 percent since 1970 and the economy (gross domestic product) has grown by 99 percent, the amount of measured pollution emitted to the air and discharged to surface water has generally decreased. Five of the six priority air pollutants have been reduced. Evidence also indicates that the concentrations of some of the most significant pollutants in rivers and streams have decreased. The majority of USGS stream water quality monitoring stations recorded a reduction in four of the six pollutants tracked since 1974, with the sharpest reductions involving phosphorus contamination. Land disposal of untreated hazardous wastes has been greatly reduced, and industrial releases of the nation's priority toxic chemicals have consistently fallen.

Other environmental indicators are not so reassuring. The United States is the largest generator of carbon dioxide emissions in the world, accounting for 24 percent of the global contribution. While overall air quality appears to be improving, nearly all urban areas exceed safe levels for atmospheric ozone at least part of the year, and a significant number of areas exceed safe levels for carbon monoxide and sulfur dioxide as well. Levels of atmospheric mercury continue to rise. Of the surface waters in lakes assessed by the states in 1994, nearly half could not support their intended uses. The overall quality of drinking water has remained unchanged over the past two decades. The amount of toxic chemical wastes generated at industrial facilities and sent off-site for treatment has increased substantially over the past 10 years.

The environmental release of some pollutants has been dramatically decreased. This is true for lead, phosphorus, PCBs, and CFCs, which have all seen substantial reductions. While emissions data may be affected by various factors, including growth in economic activity, ambient data provide a better indicator of material conditions in the environment. Air quality trends reveal that there has been a healthy reduction in most of the criteria air pollutants that have been the focus of environmental regulatory activity since 1970. Between 1970 and 1995, air emissions of volatile organic compounds have decreased by 25 percent, sulfur dioxide by 41 percent, carbon monoxide by 28 percent, dust particulates by 79 percent, and lead by 98 percent, while nitrogen oxide emissions have risen slightly.[22]

Progress in reducing pollution to surface waters is more difficult to gauge. The primary indicators of progress in surface water quality are the biennial reports of the states on the percent of waters capable of fully supporting their intended use. The 1994 reports indicate that just 57 percent of rivers and streams and less than half of the lakes, ponds, and reservoirs in the country are capable of fully supporting their intended use.[23]

A better picture of materials of concern as sources of pollution can be discerned from the ten years of data collected under the Toxics Release Inventory. Figure 7.1 presents a graph of the declining release over a 10-year period. Altogether the covered industrial facilities in 1997 released

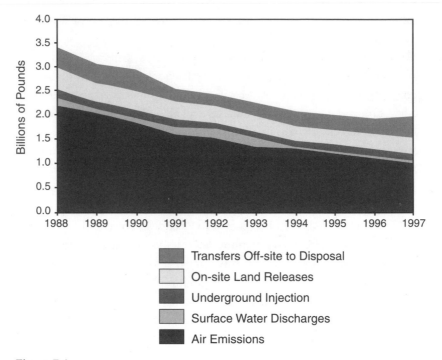

Figure 7.1
On-site and off-site releases of toxic chemicals, 1988–1997. Source: U.S. Environmental Protection Agency, Office of Pollution Prevention and Toxics, *1997 Toxics Release Inventory*, Washington, D.C., April, 1999.

to the environment a total of 2.6 billion pounds of toxic chemicals. Another 3.4 billion pounds were transferred to recycling, treatment, or disposal facilities, making a total of just under 6 billion pounds of toxic chemicals released and transferred off-site.[24] A review of the trends in the data suggests that over the years the releases to the environment have been decreasing while transfers to recycling facilities and energy recovery facilities have remained fairly constant. Trends in the TRI data need to be considered carefully because over the decade of reporting, additional industrial sectors have been added to the reporting universe and some chemicals have been added and some subtracted from the list. Even considering these changes, the trends suggest that reporting firms are improving their pollution control capabilities and sending more of their captured wastes to treatment facilities.

The success of environmental regulatory laws depends on effective enforcement and compliance with regulations, and this has been less well studied. Under the federal laws, a significant proportion of the enforcement responsibility is shouldered by state agencies. Because of this, compliance and enforcement vary significantly across the country. Wealthier states with more environmentally sensitive constituencies tend to enact more stringent regulations and enforce those regulations more assiduously. States dominated by specific industry types (chemicals, mining, specific types of manufacturing) tend to be more tolerant of noncompliance by those sectors than other states with more heterogeneous industrial mixes. In general, states do better at regulating air pollution from industries than from power-generating utilities and are better at managing water pollution from manufacturing than from mining facilities. Indeed, a recent nationwide analysis of industrial wastewater discharge permits found that almost one-quarter of all those facilities that are required to comply were operating without a current permit.[25] Even with good state performance, compliance is a continuing problem. Early studies of compliance with air and water regulations across all the states found that a significant proportion of permitted facilities were out of compliance.[26]

The causal connection between policies and results is often difficult or impossible to establish because so little data exist and there are so many confounding variables that may alter the explanation. The EPA attributes the reduction in air emissions to pollution control technologies at industrial and electric utilities and the cleaner industrial and automotive fuels required under state and federal air emission permits. If this is correct, it suggests that pollution controls are effective, but it also suggests that as industrial production and automobile use increase, the total amount of pollution may eventually reverse these declines. One study completed during the 1980s tried to relate total pollutant emissions by industrial sector to investments in pollution control equipment and found there was little correlation.[27] Indeed, some measures of air quality were improving at an impressive rate before the passage of the Clean Air Act in 1970, suggesting that factors other than the legislation may be contributing to air quality improvements.

The success of government policies in reducing the effects of occupational health and safety hazards is even more difficult to assess.[28] The Occupational Safety and Health Administration has been disappointingly slow in issuing chemical exposure standards. Between 1972 and 1984, the agency issued only twenty-four specific chemical standards (including a group of fourteen carcinogens) and nine "emergency temporary standards," and since then has avoided specific chemical standards altogether in favor of more generic process standards. To date the agency has established relatively few "permissible exposure limits." In fairness, industrial interests have challenged all but two of OSHA's final health standards and six of the nine emergency standards in federal courts, which has greatly discouraged progress on additional standards. The cotton dust and benzene standards were appealed up through final review by the U.S. Supreme Court, a procedure that added several years to the standard-setting process. Enforcement has also been a major issue. Since the early 1980s regional inspection activities have been seriously understaffed and the number of facility inspections, other than those requested by concerned employees, has been modest.

The results are no more impressive. Bureau of Labor Statistics data suggest a general reduction in industrial accidents since the enactment of the Occupational Safety and Health Act, but this decline may be explained by nonregulatory factors. There is little consensus on occupational illness trends. While the U.S. Public Health Service estimates that there are 390,000 new cases of occupational disease annually, the National Institute of Occupational Safety and Health estimates that 100,000 deaths annually are attributable to occupational disease.[29] During the 1970s, several studies tried to estimate the proportion of cancer deaths that could be attributed to environmental and occupational exposure to carcinogens.[30] The estimates ranged from 1 to 20 percent, but the efforts became mired in fierce controversy and no conclusions are well accepted today. Because the health outcomes of occupational exposures are so difficult to assess, some studies rely on chemical exposure as an indicator of a policy's effectiveness. Some of these studies have suggested that during the first two decades of federal occupational health standards, exposure to cotton dust, vinyl chloride, and lead decreased.[31] A more recent study based on company data for a handful of chemical substances suggests that worker exposure to regulated substances is still decreasing.[32]

The data required to assess long-term trends in public health and environmental protection and to link current materials management policies to those trends are woefully inadequate. Air emissions and toxic chemical releases are the only two data sources that can be used to systematically document historical trends. Both of these data sets indicate a steady decline in the amount of targeted chemicals released to the air. The TRI data indicate a decrease in aggregate releases of reported chemicals into the environmental media and an increase in the amount of these chemicals managed through waste recycling, treatment, and disposal facilities. The most direct policy impact has been on those substances highly regulated by a phaseout requirement and those regulations that have changed the dominant fuel uses of industry, power generation facilities, and transportation. Only when the significant growth in the nation's population, economy, and automobile use is considered is there enough evidence to speculate that regulatory policies have slowed and in some cases reversed the rate of movement of industrial materials into the environment. Altogether, the data do not suggest a very high level of effectiveness for the nation's public health and environmental protection laws.

The Limited Effectiveness of Environmental Policy

It is particularly difficult to evaluate policies intended to prevent harm because the most significant effects may be those that never occur. Still, a policy approach based on the control of hazards is fraught with limits. The slow pace of regulatory standard setting, the low level of enforcement and compliance, and the less than dramatic outcomes in terms of environmental quality or health effects need some explanation.

First, there are technical limits. In a recent review, Sheldon Krimsky of Tufts University observes that "like the science that informs it, the process of regulation has taken a reductionist approach; seeking chemical by chemical solutions; focusing on too few etiological outcomes; neglecting additive, cummulative, and synergistic effects; and allowing a balkanization of regulatory authority." With more than 50,000 chemicals identified as having some toxic characteristics, no strategy that deals with one chemical at a time can hope to do more than nibble away at the necessary testing, evaluation, and regulation writing that would be necessary to cover most of the chemicals likely to negatively affect human health or ecological functioning. While numerous chemicals have been reviewed and regulated by the EPA and OSHA, the statutes that require the substantiation of proof of harm or unreasonable risk for each chemical before action occurs curtail the overall effectiveness of the agencies' efforts. Even when there is adequate information, it is difficult to evaluate the risks associated with chemical exposure. Where quantitative risk assessment procedures are used, there are always difficulties in extrapolating animal test data to humans, in determining the effects of low doses, and in estimating the impacts on populations that vary widely in degree of exposure and susceptibility.[33]

Environmental and occupational health regulations rely on control technologies such as wastewater treatment, flue stack filters and scrubbers, respirators and ventilators, incinerators, and landfills. None of these technologies are 100 percent effective. Indeed, the development of technologies that are nearly totally effective is prohibitively expensive because controlling the last few percentages of pollutants is much more complicated than controlling the greater part of them. Even with the best equipment and practice, some pollution is still released and some exposure will occur and for some toxic compounds, scientific studies show that small amounts are of as much concern as large ones. Pollution control equipment that releases small amounts of contaminants may over time produce quite dangerous conditions as the volume of these small releases is multiplied by increasing levels of production.

The true limits of a materials management system that relies on the control of pollution and exposure lie in the finite capacity of the planet to absorb wastes. A control strategy simply moves pollutants around. Once a material becomes a waste, a control technology cannot eliminate the waste. Air emissions from a coal-fired utility are typically rich in sulfuric acid, arsenic, and enough other toxic chemicals to be classified as hazardous emissions. An electrostatic scrubber that is used for flue gas control will capture a portion of that emission, which is collected as a residue. Initially, these residues were released into wastewaters, thereby turning an air pollutant into a water pollutant. Today, much of these residues are landfilled, turning an air pollutant into a land pollutant. On a finite and contained planet, such control strategies are better at displacing than reducing pollution.

There are also political limits to a materials management system based on controlling hazards. The distribution of political power and economic resources in the United States makes aggressive hazard control strategies infeasible. The privileged position of many business interests creates an upper bound on many regulatory policies that are deemed to interfere too far in private industrial practices. Regulatory agency directors and their budgets are determined by political processes, which can result in retaliation if aggressive enforcement crosses partisan commitments or well-placed private interests. The political restraints and budget reductions that characterized federal environmental and occupational health policy during the 1980s were a direct result of business sentiment that government regulations had gone too far. Since that time, there have been few new federal environmental laws and no new occupational health laws enacted by Congress, and the executive administrations of both political parties have touted regulatory reform and regulatory streamlining as central objectives.

Finally, there are institutional limits. Regulatory agencies are often constrained by decisions of the past. Many policy observers today note the need for integrated and harmonious government programs to work with industries in multimedia, preventive approaches and to work on environmental protection in coordinated and holistic ways. Yet the federal and state laws that govern environmental and occupational health protection were enacted media by media (Clean Air Act, Clean Water Act, etc.) and problem by problem (worker health, chemicals, pesticides, products, etc.), and the administrative agencies at both the federal and state levels are divided into discrete units in order to implement these specific laws. The result is a balkanized array of regulatory divisions, with unique authorities and independent budgets that are quite resistant to coordination or integration.

7.5   The Costs of Industrial Materials Management Policies

Controlling pollution and managing human exposure costs money—lots of money. Today, an estimated 3 percent of the nation's gross domestic product (GDP) is spent on environmental protection. Government regulations, particularly environmental protection regulations, tend to drive up the costs of materials. Regulations require firms to seek permits, invest in new equipment and pollution control technologies, hire additional personnel, and monitor compliance. Regulatory activities also cost the federal government money, mostly in terms of personnel, research and development, and grants-in-aid, primarily to the states. In 1999 the total EPA budget stood at about $7.6 billion. In contrast, the OSHA budget was little more than $353 million. While the combined budgets for the two represent a hefty sum, it is just over 0.4 percent of all federal expenditures. Indeed, it is substantially less in relative terms than the EPA and OSHA budgets of 1977, when the EPA budget alone was 1.7 percent of the total federal budget.[34] While federal expenditures on environmental protection have been declining relatively, state expenditures have been increasing. State expenditures in 1992 were nearly $4 billion. In most states, local government spending exceeds state expenditures, but this is primarily for municipal garbage collection and landfill management.

Pollution control is a substantial expenditure for private industry as well. Industry spends some $121 billion per year on pollution abatement and control. In 1994 industry spent $17 billion on capital investments for pollution control, while the chemical industry alone spent $1.9 billion. These expenditures have also decreased relatively over time. In 1975 nearly 6 percent of total private capital expenditures went to pollution control; by 1990, the amount had dropped to 3 percent. Yet, across the entire economy, total spending for environmental protection (capital investment and operations) has been on a steady upward curve. In constant dollar terms, total spending increased from $26.5 billion in 1972 to $114 billion in 1992 and that year it represented roughly 2.3 percent of the gross national product.[35]

Environmental damage from industrial pollutants must also be considered a cost. For instance, some studies have estimated that acid rain causes more than $4 billion in damage to crops, forests, and buildings each year, while other research has estimated that the cost of ozone damage to agriculture and forest lands exceeds $5.4 billion annually.[36]

In 1997, the EPA issued a series of reports on its 6-year study of the costs and benefits of the 1970 Clean Air Act up through 1990. In conducting its study, the agency compared the actual costs and benefits of the regulations with the costs and benefits projected if the law had not been passed. The direct costs to the private and public sectors over the 20 years were calculated to be $523 billion in 1990 dollars. Estimating the benefits required modeling the effects of the law on air emissions and then translating those effects into dollars saved in reduced adverse health effects, improved visibility, and reduced crop damage. The emissions models demonstrated that as a result of the law, there was a 40 percent reduction in sulfur dioxide emissions from electrical utilities, a 75 percent reduction in suspended particulate emissions from industrial facilities and utilities and a 50 percent reduction in carbon monoxide, a 30 percent reduction in nitrogen oxide, a 45 percent reduction in volatile organic compounds, and a near 100 percent reduction in lead in the emissions from motor vehicles. Using these percentages as a base, the agency estimated that the economic benefits of the law over the 20 years ranged between $5.6 and $49.4 trillion in 1990 dollars.[37]

Analyses of federal water pollution control policies are less developed. One study completed during the 1980s concluded that the private costs of controlling water effluents lie between $25 and $30 billion annually (in 1984 dollars), while the benefits (not including ecological or human health and amenity benefits) lie between $6 and $28 billion.[38] Even if these numbers are off by a substantial amount, it appears that benefits are not exceeding costs.

The technologies that have been developed and used to control and manage the hazards of industrial materials are both sophisticated and costly. The professional attention required to address potential exposures and releases is a large overhead cost that increases the price of many materials. Workers' compensation premiums, litigation, and the potential for litigation add further to these costs. Generally, it is far cheaper to prevent the generation of pollutants at the source than to collect, treat, and manage them once they are considered as wastes. Likewise, it is cheaper to avoid the use of a hazardous substance than to invest in costly exposure control at the workplace. The old health adage that an "ounce of prevention is better than a pound of cure" holds true for pollution as well. Unfortunately, a serious commitment to pollution prevention arose decades after the hazard control approach was firmly established in law and practice, and even substantial cost savings are often ignored in favor of the status quo.[39] A substantial lowering of the costs of pollution from the dissipation of industrial materials will require a serious reworking of the systems by which we manage these materials.

7.6   Reassessing Industrial Materials Policies

The policies of the nation have aided and supported the development of a highly functional array of industrial materials, with little attention to the conservation of resources or the disposal of wastes, and only modest attention to the hazards of those materials or the processes that produce them. Beyond the production of materials, our national policies are largely underdeveloped and limited in effectiveness.

The data for managing the vast empire of materials that we have created are inadequate. There are large data gaps in the tracking of materials through the economy. In most respects we have no way of measuring or evaluating whether the materials policies that we have established are functional or effective. The millions of dollars of public funds that go into subsidies, research, regulations, and enforcement are spent without a means of measuring their performance. We have stockpiled a huge amount of industrial materials as a resource for wartime, but we have only the slimmest of information on the vast amount of industrial material that is sequestered in the production processes or product infrastructure of our peacetime economy.

We waste a huge amount of our resources. There are virtually no economic incentives for preserving resources for the future and no national programs for conserving materials. We increasingly rely on nonrenewable materials, but show little interest in planning for their depletion. Our extraction and processing of materials is inefficient and enormously wasteful. In our households and businesses we convert products to waste with astonishing rapidity. Indeed, because we import a substantial amount of our industrial materials, we waste a substantial amount of materials from other countries as well.

The scientific knowledge we have about materials is dangerously myopic. The materials sciences provide a substantial base of knowledge for the development of new materials and production processes, but little knowledge for managing and conserving current materials. Estimates of reserves remain speculative; theories about enhancing productivity are sketchy, and little work has been conducted on ways to recycle and reuse materials already in commerce. Knowledge about the fate of large amounts of industrial materials released to environmental media, particularly into the upper atmosphere, groundwater, and lake and ocean sediments, remains incomplete. The sciences of toxicology and pharmacology are still developing basic models, and many of the mechanisms by which the human body copes with toxic chemicals remain a mystery.

The regulatory system of exposure control is costly and clumsy. The high costs of pollution control are only partially embedded in the market costs of materials. Additional costs are borne directly by the public as government expenditures. Other costs—quite significant ones—are passed onto the environment and are therefore not generally viewed in terms of dollars. These costs do not show up on ledgers and are not represented by monetary exchanges associated with the development or use of materials. Instead they represent a kind of social overhead—the complex costs of living each day. It has become increasingly evident that these costs are not borne by consumers of the products that cause pollution, but are inflicted on the environment and nearby residents. Hazardous waste dump sites, waste treatment facilities, and industries with substantial records of pollution are often located near low-income communities and communities largely made up of recent immigrants and people of color. The residents of these communities present a strong case that they bear a disproportionate share of the costs of environmental pollution and exposure to hazardous chemicals.[40]

These limits are not merely discomforting wrinkles on the current approach to materials management. The absence of adequate information, the neglect of conservation, the focus on hazard control, and the high cost of regulation add to the environmental and public health risks that we have spent decades trying to manage. They reveal more starkly the limits of the current materials management system. We have invested heavily in addressing the effects of the materials in our economy while mostly ignoring the materials themselves. Considering the high costs and limited effectiveness of our current approach, it is worth considering a fundamental shift in approach.

It is on this basis that it makes sense to take a long look at the way in which materials are made and used and whether there are safer systems for making new materials and using existing ones. The disappointing assessment of our current approach to managing materials after they are made and often after they are released as pollutants or wastes begs for a consideration at much earlier stages in their life cycle. What is needed is a more comprehensive consideration of the use of materials and a commitment to the development of safer and more sustainable materials that ensures that the full costs of consumer goods are born by those who benefit from them.

8

Reconsidering Materials Policies

Successful problem solving requires finding the right solution to the right problem. We fail more often because we solve the wrong problem than because we get the wrong solution to the right problem.
—Russell L. Ackoff

The current approach to development of industrial materials will continue to yield a plethora of new and exciting materials in the future. The competitive forces of the market, where they are uninhibited by constraints, will reduce this rich array of materials to those that are truly useful and valuable. However, if this course proceeds unaltered, it will produce an economy that is as full of highly hazardous and inefficiently used materials as the economy that exists today. Tighter government regulations on exposure could lead to more protective outcomes for the environment and public, but this will be increasingly costly and the effects will be limited. The mix of materials will improve in the long term only if conscious efforts are made to develop and manage materials with higher regard for the environment and the life that it supports.

Any approach to the development and use of industrial materials needs to take into account the role materials play in maintaining and improving the quality of human life. Whatever strategies that might be adopted in the future, they will not be popular, or easily accepted, if they reduce the material well-being of those affected. Still, this commitment should not be overstated. The material well-being of people living in industrialized countries like the United States was not unacceptable 50 years ago when the economy was based on roughly half the total volume of materials in use today and the amount of toxic chemicals was less than a quarter of the volume used today. Material well-being has as much to do with how much benefit is derived from materials and how safe they are as it does with the pure volume of raw materials processed.

8.1 The Nature of a Sustainable Society

The concept of sustainable development arose during the 1970s among those promoting economic development in the industrializing countries of the world. It developed as a response to the limits-to-growth debate and was first formalized internationally in the World Conservation Strategy of 1980. The early metaphor of sustainability came from work in forestry and agriculture. To increase the productive capacity of agricultural products in poorer countries without drawing on expensive inputs of fertilizers and pesticides, it was necessary to find a balance between the amount of a crop that could be harvested each year and the capacity of the land to support continued harvests in each succeeding year. The appropriate balance was called the "sustainable yield," and it defined the amount of crop that could be harvested without compromising the capacity of future harvests to yield an equal crop. The general elements of this concept are quite simple: The amount of activity should meet the needs of the present while reserving the resources necessary to support continuation of the activity in the future.[1]

In 1987, the World Commission on Environment and Development adopted the concept of a sustainable balance as the guiding principle for managing environmentally compatible economic development. The Commission, which had been established by the United Nations, found that, "(t)here are thresholds which cannot be crossed without endangering the basic integrity of the system. Today, we are close to many of these thresholds; we must be ever mindful of the risk of endangering the survival of life on earth." As a response the Commission proposed that further economic development be sustainable, meaning "development that meets the needs of the present without compromising the ability of future generations to meet their own needs." The concept involved raising the economic standards of the world's poor while limiting the negative environmental effects of the technological development and social organization of all societies. Thus there was an economic, an environmental, and a social imperative in the concept of sustainable development. In terms of industrial production, the Commission recognized the possibility of increasing the efficiency of energy and material use per unit of economic output, but noted that growth in population and rising levels of income would intensify the global environmental problems linked to resource use and could overwhelm the benefits of increased resource efficiencies.[2]

Since the World Commission report, the concept of sustainable development has received a hearty reception, with endorsements by the 1992 United Nations Conference on Environment and Development, the International Chamber of Commerce, and many national governments. In the United States, President William Clinton established a President's Commission on Sustainable Development that issued a number of reports in 1995 and 1996, but with little follow-up or effect.[3] Throughout this country scores of local communities have organized popular sustainable development initiatives that have produced plans, goals, and policies. The energy and rhetoric of these projects is full of hope, but as visionary as many of these local efforts have been, they have been short on pragmatic programs, particularly where they have addressed the issue of local industrial production. Many of these plans set goals for real and relative reductions in the consumption of materials and energy, particularly in the residential and transportation sectors, but actual programs to reduce energy and materials use in industry have been more cautious, reflecting a concern that industry must remain vigorous and competitive.

Converting local economies to following the principle of sustainable development requires a forthright response to industrial production because that is where the majority of choices are made on resource use. Communities need to include sustainable forms of industrial production. To meet the principle of sustainability, industrial managers need to redirect their business strategies, transform their technologies, redesign their products, and improve their environmental performance. An important contribution to these strategies will be a reconsideration of the development, use, and disposal of industrial materials.

The handling of industrial materials is important throughout the discussion of sustainable development at national and international levels as well. The economies of some of the world's poorest countries depend on their extractive industries. Cobalt mining is critical to the economies of Zaire and Zambia; bauxite mining supports Guinea and Jamaica; and Malaysia relies on its tin mines. Such industries are necessary to meet current needs, yet many of these mining operations are among the world's most polluting and environmentally damaging. Of even more significance is the profligate consumption of consumer products in the highly industrialized countries and the significant amount of waste that results. The depletion of the reserves needed by future generations and the severe taxation of ecological systems by the current rates of pollution and waste disposal are driven by this excessive rate of consumption. In other words, among other objectives, sustainable development requires a transformation of the current approach to the global materials economy.

8.2   Seeking a More Sustainable Materials System

The current materials management system in the United States is not sustainable. It meets the current needs of some people, but certainly not all people, and it makes little effort to ensure the ability of future generations to meet their own needs. The analysis presented in earlier chapters suggests that the current system:

- depletes the global stocks of high-quality natural resources;
- dissipates those resources into forms that are irrecoverable;
- uses and wastes large amounts of energy in production, use, and disposal of materials;
- pollutes the ecological systems of the planet with dissipated residues;
- converts relatively safe chemical compounds into quite toxic compounds; and
- exposes large numbers of people to chemical compounds that are inadequately studied.

From an ecological perspective, the basic model upon which this materials management system is based is flawed. The materials balance model is an inherently linear model that promotes the flow of materials from raw materials to wastes. It assumes that human society is an isolated (but open) system somehow separate and protected from the global environment. The model focuses attention on the human use of materials and by default diverts attention from the environmental consequences of this use. Materials come from "out there" and are discarded back "out there." The model is based on an assumption that the natural world is an unlimited bank with an endless supply of raw materials and a bottomless pit with an infinite capacity to accept wastes. High-quality materials are not conserved for future generations and the capacity of the environment to support future demands is compromised.[4]

In contrast to the isolated system of the materials balance model, many ecologists view human economies as subsystems embedded in the broader natural ecosystem. Rather than seeing a single-pass linear construction, they propose a cyclical model that tracks the movement of materials from extraction through use and disposal and back into production and reuse. This concept is modeled on the continuous flow of nutrients and wastes in healthy ecosystems. Sometimes called a "circular flow economy" or a "life-cycle model," this approach follows the flow of materials through the environment and through human economies in a continuous set of cycles. Materials are mobilized and used in both natural and human compartments and are held in stocks in both compartments as well.[5]

A material's life cycle involves a reoccurring set of cycles. Rather than the "cradle-to-grave" flow in a materials balance model, this is said to be a "cradle-to-cradle" flow. This materials flow model (figure 8.1) focuses attention on the molecules of materials, urging us to "follow the molecules" as they are extracted, transformed by production, manipulated in use, disposed in the earth, and recycled back for extraction. More than a model, it opens a way of thinking—"life-cycle thinking"—about the ways in which raw materials, products, and wastes are fundamentally similar. Viewed from a life-cycle perspective, the ecological or human health significance of a material is little different if it is experienced as a feedstock, a product, or a waste.[6]

In this model, the concept of waste is reinterpreted. The materials leaving each step of a life cycle are transformed, but they are wasted only to the degree that they are dissipated. The utility of products is conserved by enduring use or continued reuse. Residuals that are not consumed by being recycled back into a process are passed along in concentrated form so that they can be used in another process, or are carefully sequestered in well-managed stocks until they can be used at some future time.[7]

The capital of human economies is matched and complemented by the capital stock of natural ecological systems, which has previously been referred to as "natural capital." But natural capital involves more than the stocks of renewable plant and animal life, replenishable natural fluids, and nonrenewable mineral resources. A full accounting of natural capital must also include the global dynamics that provide the "natural services" upon which all life depends.[8] These services include a wide range of ecological processes and functions that maintain the hydrological cycle, the composition of the atmosphere, the recycling of biological nutrients, the regeneration of soils, the assimilation of wastes, the pollination of plants, and the maintenance of species. These resources and services are viewed as global assets and their contribution to the human economy is enormous. One study conducted at the University of Maryland estimated that in 1994 the economic value of the earth's natural services was about $33 trillion, which was twice the size of the global gross national product for that year.[9] It is these services that are neglected by the myopia of the materials balance model. By considering the entire materials cycle, these natural services are given a well-deserved place of respect as contributors to the well-being that is generated in human economies.

A life-cycle approach reshapes the concept of recycling as well. Throughout the planet, materials are constantly cycling within natural and physical biogeochemical systems, such as the hydrological cycle and the carbon–oxygen cycle. Materials are also cycling in economic systems. Some of this cycling is recycling and involves the continuous use and reuse of materials in a closed system. Precious metals offer a good example because their price and ornamental value promote continued use and recycling. A far larger volume of materials are drawn from the natural, planetary systems, used, and returned to those natural systems as wastes. It could be said that these materials are being recycled back into natural systems, from which they might some day be reclaimed by the same or other open economic systems. Wood products provide an example of this form of cycling. Most wood products that are discarded and buried in landfills eventually are degraded into nutrients that may one day become resources for a new generation of trees that could be harvested for new wood products.

Conceptually, this materials flow model could be constructed as a set of three nested cycles. The global environmental system is a closed cycling system. Most industrial materials move about in a second cycle

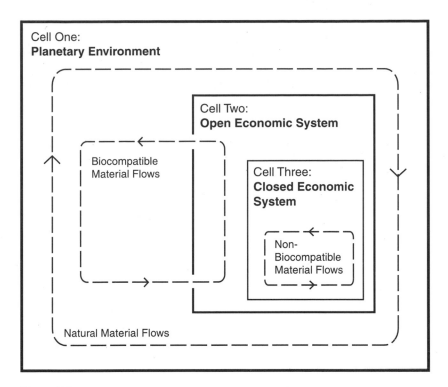

Figure 8.1
Materials flow model.

where they are exchanged between the environmental and economic systems in one or several passes. But some materials that are either hoarded or intensely recycled cycle within the economic subsystem as if it were a closed system. Figure 8.1 shows a materials flow model composed of these three nested cycling loops.[10]

Ideally, a sustainable materials system could be designed to make use of these three cycling loops. Loop 1, which involves natural capital and its cycles (the environmental services), needs to be preserved and protected. Left on their own, the ecosystems of the planet will continue to provide "free" and vital services to human life as long as they are not too brutally compromised. In loop 2, materials can continue to be exchanged between the economic subsystem and the natural systems of the earth as long as they do no great harm. Those materials that are reasonably compatible with ecological processes could be cycled back and forth between the economic and natural compartments of the second loop by extracting them from the environment and discarding them in a compatible environmental medium. Loop 3 incorporates a closed recycling system with materials cycling about only within economic systems. In these closed systems, materials that would not meet the criterion of "reasonably compatible with ecological processes" could still be used, but would need to be continuously recycled within the confines of this third loop.[11]

If we were to convert to this materials flow model today, it might look quite like the current materials management system except that the volume and properties of materials within the various compartments would be dramatically changed. If persistence and toxicity were among the factors used to define the criterion "reasonably compatible with ecological processes," then unlike today, it would mean that some of the most dangerous materials would be cycling only inside the third loop. This would be in direct contradiction to current conditions, where the materials most likely to be recycled (e.g., iron, copper, gold, and silver) are among the least hazardous, and the most dangerous materials (e.g., benzene, styrene, and halogenated hydrocarbons) are the least likely to be recycled. Indeed, the volume of materials forced by these criteria to remain as stocks inside the third loop would be tremendous. Because human life must be lived largely within loop 3, there would be a significant incentive to do something about the scale and danger of the materials it contains.

There are two obvious strategies for reducing the potential dangers and the overwhelming volumes of the materials locked up in the third loop: detoxification and dematerialization. Detoxification of the materials in the economic system would permit continuous recycling without concern for human health effects and allow an increasingly larger proportion of the materials to be cycled back into the natural systems of the second loop. Dematerialization would reduce the volume of materials cycling in the second loop and remaining locked in the third loop by increasing the rate of recycling, intensifying the use of materials, and replacing materials with services.

Viewing the materials used in human economies as borrowed from and returned to natural stocks recasts the issue of materials conservation.

A good economics principle suggests living on the dividends that can be drawn from capital, without drawing down the capital stock, so that "the resource stock should be held constant over time."[12] This notion of protecting natural capital is a concept sympathetic to the conservation movement's earlier promotion of the idea of resource conservation. The concept of sustainability requires that the stock of natural capital not be lost or degraded in order to protect resources for future generations. The renewable income can be used to meet current needs, but not the principle, unless new principle is generated.[13]

A lending library provides a useful model. Books are borrowed from the library, their contents educate or entertain their readers, and the books are returned when their function is completed. Once returned, the books rest on the library shelves until they are again borrowed and used. A value has been derived from the activity, but the capital stock is conserved, or at least most of it has been conserved. The use of books, like the use of materials, always creates some degradation. In keeping with thermodynamic principles, any transaction reduces the quality of a resource. Many libraries try to protect books from damage and loss, and measure their success by the number of readers per book. Today, libraries are experimenting with different information media (e.g., films, videos, compact disks, the Internet) to supplement books. Just as libraries become more sustainable as they generate ways to transfer knowledge and entertainment without moving books, reducing material flows becomes a means to conserve natural capital in the larger economy.

This is what is meant by dematerialization. Reducing the flow of materials without decreasing the quality of life conserves both human-made and natural capital. Materials still flow within the loops of the system, but the rate is slowing rather than accelerating, while the value derived from those materials (the "income") is constant or increasing.

The library analogy can be extended further by considering the production of books. Some books are produced on chlorine-bleached paper from pesticide-laden tree plantations, using petroleum-based inks containing heavy metals, and the pages are bound by adhesives rich in formaldehyde. While the book resting on the shelf is not particularly hazardous, its life cycle requires many dangerous chemicals. These chemicals expose workers and pollute the environment during production and contribute trace contaminants to the product that make it more haz-ardous when it is eventually discarded. None of these chemicals are nec-essary to the educational or entertainment value of the book's text, yet they clearly affect the natural capital from which the book was produced. As the library seeks books made from less hazardous materials, extends the useful life of books, and converts to information transfer systems that are less reliant on books, the toxic legacy of book production is reduced.

This is what is meant by detoxification. Reducing the use of highly toxic materials by substituting chemicals, redesigning products, reform-ing production processes to use less-hazardous substances, or recycling those toxic substances that cannot be replaced generates a safer mate-rials system that is less threatening to the natural systems with which it interacts.

Replacing the materials balance model with a materials flow model requires a new way of thinking about materials and a substantial trans-formation of our current institutional arrangements for managing them. Some of these shifts will involve changes in materials development and industrial production. Other shifts will require changes in how materials are used and how products are consumed. Changes will also be needed in how materials are treated and disposed of after use. Policies, both public and private, can be helpful in guiding these changes, but these policies must promote a more sustainable materials system. At a mini-mum this will require new solutions to the two core problems this book has focused on thus far: dissipation and toxicity. An effective response to these problems will require dematerializing and detoxifying the current materials system. At this point, it is worth considering these strategies in more detail.

8.3 Dematerialization

Dematerialization involves reducing the dissipation of materials by de-creasing the amount of material required to satisfy social needs. Less material used means less natural capital drawn upon, less resource deple-tion, and less material released as wastes. Dematerializing requires decreasing the amount of materials used and increasing the productive-ness of those materials that are used.

Dissipation occurs because the economy places a high value on virgin materials initially extracted from the ground or newly synthesized, a lower value on materials that have been previously used, and no value on the wastes and by-products of production or consumption activities. Over the years the continued dissipation of industrial materials has been too easily dismissed. Concerns about the rate of depletion of the nation's natural resources have largely been quelled by assuming that rising prices would continually make lower grade and substitute materials attractive and by creating large national stockpiles for emergency use. The tremendous flow of wastes has largely been accommodated by relying on the assimilative capacities of the nation's once ample "environmental services." Dire warnings aside, the country has comfortably denied and avoided the prospects of diminishing supplies of both resources and services. The result has been a continuously dissipative materials economy with an increasing rate of materials throughput and an astonishing lack of commitment to materials conservation or management of materials flow.

Dematerialization is a response to the problems of accelerating materials dissipation. It is a term that has grown in use since the mid-1980s. Several definitions have been proposed. Some have focused on waste: "From an environmental viewpoint ... (de)materialization should ... be defined as the change in the amount of waste generated per unit of industrial products."[14] Others have focused on a reduction in the weight of materials used in industrial end products. Still others have focused on reducing the embedded energy in finished products. Finally, some have focused on quantity: "More broadly, dematerialization refers to the absolute or relative reduction in the quantity of materials required to serve economic functions."[15]

Why should we care about dematerialization now? Over the past two centuries, public and professional concern about the potentially serious consequences of widespread dissipation of materials has built a foundation of knowledge and a commitment that can no longer be dismissed. There can be little doubt that the totality of the anthropogenic materials flows, including the movement of minerals, fuels, water, soil, and plant materials, has produced noticeable ecological consequences. Many of these are system-altering effects that will take years to correct, even if we change our ways today. And today we can see only the most obvious effects. It is reasonable to expect that time will reveal other ecological changes of equal or greater concern. Faced with these profound but early symptoms, and recognizing our uncertainty about future consequences, it would be prudent to curtail the current materials flows by dematerialization as a purely precautionary measure.

Rather than a necessary result of a materials cycle, waste should be viewed as a symptom of inefficient materials use. This is an important principle for those who promote the idea of industrial ecology. They, too, are proponents of the cyclical, closed-loop model and see it as a blueprint for returning industrial wastes to usable feedstocks. Indeed, industrial ecology is predicated on the idea that industrial systems have characteristics similar to basic field ecology. In this analogy, flows of industrial materials and energy could be modeled on the metabolic pathways that convert the wastes from one organism into nutrients for another organism. Although the concept of an industrial system modeled on an ecological system has been advocated since the 1970s, the concept has been energized by a group of industrial engineers connected with the National Academy of Engineering. The idea of industrial ecology, which was popularized in a 1989 article in *Scientific American*, has encouraged a wide range of useful studies of material flows, promoted a closed-loop perspective on materials management, and helped to place dematerialization in a broader conceptual framework.[16]

A closed-loop system could include all materials, products as well as wastes. Products could be returned to their producers following the conclusion of their useful life, rather than ending up as valueless discards. A returned product offers the opportunity for repair and resale, or the stripping off of still-functional components for inclusion in future products, or the deconstruction and separation of the constituent materials for sale to other producers as feedstocks. Each of these steps would result in fewer material inputs to the economic system and a greater amount of social use per unit of material.

Dematerialization can be carried out in at least three different ways: closing the loop on material flows, intensifying the use of each unit of material, and substituting services for products (materials). Recycling provides a well-recognized form of closing material loops. For instance, the aluminum industry maintains extensive aluminum can recycling processes that generate new cans from old ones. A much simpler strategy for closing material loops involves the direct reuse of products. Durable products can be passed from one user to another or sent into secondary markets for resale. Recycling materials and reusing products reduces material demands by increasing the number of users per unit of material.

The intensity of materials use can be improved by designing products that require less material to perform their functions. The advances in electronic components are often cited as prototype examples of how product size has been reduced. The weight of some products has also been greatly reduced. Tin plating on tin cans is much thinner, airplane fuselages are much lighter, metal structural beams are much leaner, and plastic beverage containers are far less heavy. Thoughtful product design can also extend the useful life of products by making them more durable, more easily recycled, and more easily repaired.

Dematerialization can also be accomplished by substituting services for products. Public libraries have long been viewed as a socially useful way to share books. Similarly, private video rental and car rental businesses pool customers to extract more use per unit of product.

All of these strategies share the objective of reducing the amount of extracted material and/or the amount of material discarded as waste per unit of social or economic value. In some cases, the flow of materials in the economic system is continued, but is no longer linear, resulting in many useful passes and much less waste per use. Market forces drive various current examples, and it is possible that in the future market forces will encourage even more dematerialization as materials prices rise. However, the recognition that today the relative price of many materials has been falling suggests that market pressures alone will not be enough to generate dematerialization at the scale needed to achieve a sustainable future.

8.4 Detoxification

Reducing the flow of materials and their dissipation is a critical component of a sustainable materials economy, but it is not sufficient. Often neglected in the literature on dematerialization—and too often ignored by those who promote industrial ecology—is the character of the materials themselves. In its carefully mediated materials cycles, nature is quite selective about the materials that are employed. Barry Commoner has noted that living organisms do not metabolize lead or mercury, never manufacture PCBs, DDT, or dioxins, and avoid polymers that cannot be enzymatically decomposed.[17] Such care is not manifested in human economies. Toxic materials are ubiquitous today and because many are quite persistent, they constitute a legacy of hazards for future generations as well. Thus, the transition to a sustainable materials economy requires a qualitative as well as a quantitative change. Both the volume of materials and the kind of materials must change. This means detoxification as well as dematerialization.[18]

Toxic materials are used in industrial production because they often have properties that are valuable in producing commercially desirable products. The toxicity may be a consequence of the reactiveness or energy transfer potential of a chemical compound, or it may arise from performance requirements, where, for instance, natural degradation processes are to be discouraged. In some cases toxic materials are used in production because they are by-products of other processes and therefore of low cost. Toxicity may also develop as an unintended consequence of a chemical reaction process or because of the mixing and transformation of chemicals in an industrial waste stream. Or toxicity may simply be the result of neglect or ignorance as potent chemicals are used, even inappropriately, without knowledge of their effects or a concern for their long-term consequences.

Historically, toxicity has been too easily accepted. Only the most acutely dangerous materials were avoided in product design, and the toxicity of production materials was little considered in process design. Those who cared about the chemical hazards of production were too willing to settle for exposure controls in the workplace, and the hazards of products were usually recognized and addressed only after harm was well documented and litigation or public reaction was substantial. Until quite recently, there has been surprisingly little evidence in industry, government, or universities of a preventive or proactive approach to reducing the toxicity of materials, which would have compelled a much more visible search for safer or more environmentally compatible materials.

Because so many materials exhibit toxic properties, it would be easy to assume that less-toxic alternatives cannot be found, at least with comparable prices and performance. But as the previous chapter has noted, the full costs of using toxic chemicals are significant, and from a total cost perspective, price may actually be an incentive for conversion. Likewise, the performance limitations of highly toxic chemicals may be higher than conventionally considered if all of the special handling and regulatory compliance obligations are factored into the assessment of a material's functionality. Thus detoxification should offer a means to protect the environment and to promote economic performance as well.

Why should we care about detoxification now? This past half-century has witnessed growing community awareness and activism about toxic chemicals in water supplies, airsheds, consumer products and foods, and increasing worker concern about toxic chemicals in workplaces. There is substantial scientific evidence that persistent, bioaccumulative, and toxic materials are building up in the quiet niches of ecological systems. These include the lower depths of groundwater, the sediment of lakes and oceans, the ice layers of the planet's poles, and all of those species that are at the top of ecological food chains, including ourselves. The consequences of these buildups are beyond our current knowledge. We have struggled to control the most acute toxins and to learn more about the subtle and chronic effects of thousands of other chemicals. But we know so little about so many chemicals and because so many chemicals are newly synthesized, nature offers us few clues. With so much scientific work ahead, and so much uncertainty now, it would appear best to reduce the use of toxic materials as a precautionary strategy.

Detoxification requires that more serious attention be given to the toxicity and hazards of industrial materials. Much of the professional literature and almost all government environmental and public health programs begin with an assumption that in general toxic materials are necessary and professional attention should be focused on managing exposures. Huge efforts are made to dilute toxic chemicals in ecosystems, capture and treat toxic materials in wastes, and protect workers and consumers from exposure to dangerous doses. Yet in the absence of sufficient data on the basic mechanisms of toxicity for the large majority of toxic materials, these exposure controls leave substantial levels of uncertainty. In the face of this uncertainty, most government programs rely on the complex methodologies of quantitative risk assessment and cost–

benefit analyses to identify what are euphemistically called "acceptable levels of exposure."

Those who do argue for reducing or eliminating the toxicity in industrial materials often invoke a different approach to managing exposure in the face of significant uncertainty. Using a concept first developed in Europe called the "precautionary principle," these advocates argue that if the scientific data are not adequate, it is better to prevent exposure by seeking alternative materials, processes, or objectives. This principle was derived from the concept of *Vorsorgeprinzip* written into German water pollution law during the 1980s and adopted into the 1987 Ministerial Declaration on the protection of the North Sea, where it was stated that "a precautionary approach is necessary which will require action to control inputs of ... substances even before a causal link has been established by absolutely clear scientific evidence" of their harm. Central to this principle is a commitment to take protective action where the weight of evidence suggests harm even though clear scientific proof is absent.[19]

In terms of the technologies of production, the precautionary principle, like the preventive approach to pollution, avoids struggling over levels of risk in favor of a search for safer alternatives. Instead of debating about hazards of industrial materials where health or environmental impact data are missing or weak, a precautionary approach seeks to reduce hazards by reengineering, restructuring, and reconsidering technologies to find substitute materials or safer means of production, consumption, and disposal. Rather than controlling human exposure to toxic chemicals or improving the management of toxic materials in workplaces or communities, efforts are directed at reducing the production and use of toxic chemicals and increasing the production and use of environmentally compatible substitutes.[20]

Detoxification involves reducing the toxicity of the materials used in industrial production while maintaining or improving upon material and product effectiveness. It means seeking substitutes for the most toxic materials, or developing production processes or products that do not require them, or shifting to new sources of materials that are less likely to leave chemically active and biologically threatening residues. Detoxification offers a new goal for the materials sciences by promoting the search for safety and ecological balance in the development of the next generation of industrial resources.

There are many well-developed strategies for doing this. Material substitution is relatively easy where there are simple "drop in" chemical replacements. While chemical distributors market such easy replacements for certain applications, most substitutions require some process or product modifications. The international phaseout of chlorofluorocarbons drove a worldwide search for chemical substitutes, and in metal parts cleaning, many manufacturers found that simple aqueous cleaners could perform as well as CFCs, although this often required new processing steps.

Products can be designed to reduce or eliminate toxic chemical constituents. This may involve material substitution or chemical transformations that reduce the reactivity, balance the pH, or bind the toxic compounds in less bioavailable molecules. Reformulating processes to reduce the need for toxic chemicals addresses those intermediates that do not end up in the product. New catalysts and reagents, reactions that occur at ambient pressures and temperatures, aqueous-based processing, and reactions that occur in fully contained environments reduce the need for toxic chemicals.

Detoxification requires a thorough analysis of how chemicals are used, what properties lead to toxicity, where toxicity is not required by the necessary functions, and what substances are possible substitutes. Because industrial materials are developed in integrated production processes, the search for less-toxic chemicals will require a consideration of the full life cycle of a material and all of the consequences of a change. A sustainable material must come from a sustainable system of production, use, and disposal.

8.5   Pursuing Materials Strategies for the Future

The materials flow model presented here appears distant today; nonetheless, it provides a useful vision for a sustainable materials system. It will require a substantial effort to convert to a three-loop cycling system that is steadily dematerializing and detoxifying. Given the potential resistances to change built into the current system of materials management, efforts to convert would certainly confront technical, economic, political, and even personal barriers. Earlier efforts to create a broad national materials policy suggest just how difficult this task will be. Turning such a large and functionally integrated economy even modestly will require time, good incentives, leadership, and public commitment. A direct assault on the current materials management system is hardly wise. The system has produced significant benefits and it is important to recognize those benefits and ensure their continuity.

A more promising strategy would involve a careful look at the current directions of materials development and a thoughtful consideration of the opportunities that are created by these processes. The mission would be to seek tactical openings where the principles of precaution and industrial ecology could be grafted onto current patterns of development and use of materials. It is worth exploring opportunities for joining a commitment to protection of environmental and human health with new initiatives in the materials and biological sciences.

New materials and new management systems for materials are emerging in the inventions and experiments that are continuously appearing. Governments across the country are now touting the waste avoidance benefits of recycling and reusing consumer wastes. Research laboratories in universities and corporations steadily provide a host of new alloys, composites, petrochemicals, and polymers. Agronomists and agricultural specialists have been proposing the virtues of renewable resources for years. New production processes and new uses of current materials are also the subject of bioprocessing innovations. At first appearances, some of these new materials and processes show promise of greater safety and reduced environmental impact, although closer examinations reveal many complexities.

These inventions and innovations are motivated by a collection of widely varying objectives, including improved performance, reduced cost, market considerations, military significance, and politics. Environmental or public health considerations could also be determining factors, but this needs to be further examined. In the chapters that follow, several areas of development are reviewed, both to better consider the dominant avenues of current materials development and to look for the role that environmental and public health considerations do or could play.

III

Alternative Materials Strategies

9

Recycling and Reuse of Materials

We should recycle, but it is not the first thing we should do, it is the last. Redesign first, then reduce, and finally recycle, if there is no other alternative.
—William McDonough

Nature recycles everything. Physical and chemical forces constantly churn the earth in continuous cycles. Nature's organisms, from plankton to eagles, participate in an intricate web of nutrient cycling that changes the energy of the sun into food and waste that continues on as food and waste. It is not difficult to imagine that human uses of materials could be designed to mimic the highly effective recycling found in all natural systems.

This is the idea behind the nested cycling loops of the materials flow model. Rather than dumping valuable materials back into the environment, economic activities could be designed as closed systems that maximize the containment and recirculation of materials. As proponents of industrial ecology argue, one firm's wastes become raw materials for another firm's production.

At first glance recycling of materials appears an easy solution to the environmental problems that arise from the use of industrial materials. Those materials too dangerous to cycle in an open system could be contained in a closed economic system. Industrial wastes recycled into raw materials could substitute for virgin materials and reduce or eliminate the need to mine, harvest, or synthesize new materials. This is not a small vision. Well over 41 million tons of industrial hazardous waste (excluding wastewaters) are generated each year in the United States. Add to this the annual 209 million tons of municipal solid wastes, and there is an enormous resource base already moving through the economy. Most of this is merely "spilled" back into the natural environment as waste, but within this flow there is a kaleidoscope of already processed materials. Indeed, much of the municipal waste stream is richer today in metals than the ores from which the metals were initially refined, and these metals are simply waiting to be extracted and processed.[1] The large amount of energy and environmental disruption required to extract metals from ores, harvest fiber from crops, and distill and refine fuels to process chemicals could be reduced if already processed materials or partially processed wastes were recycled and reused. As energy demand and environmental disruption decrease so will costs, and the result would be a cleaner, safer, and more efficient economy.

This attractive first glance proves to be more difficult to achieve upon closer examination. While recycling surely provides both social and economic benefits, increasing the current recovery rates will require overcoming significant structural barriers; developing new institutional infrastructures; and redesigning products, processes, and materials. Indeed, it is a daunting task to restructure large sectors of the materials economy that are primarily linear and open and convert them into contained economic systems based on closed-loop recycling.

9.1   Recycling and Waste

The practice of recycling and reusing materials is as old as human life. Archaeologists have found ample evidence of the reuse of broken pottery as a common domestic practice among ancient cultures that had mastered pottery making. The frugal Europeans who first settled on the North American continent ardently recycled and reused tools, clothing, building materials, and domestic objects out of sheer economic necessity. They saved and reused hundreds of items, from buttons to nails, to bottles, to horseshoes, to farm implements, to doors and windows.

Early mining and forestry practices were enormously wasteful because raw materials were abundant and technologies for processing them were crude. But once value was added through craft work or manufacturing, the resulting products were much less likely to be wasted. Secondary use and secondary product markets flourished. Most women mended clothing, many men repaired tools, and poorer children scavenged. Patchwork quilts, rag rugs, and hooked upholstery all began from recycled cloth. Peddlers, rag merchants, and scrap dealers abounded.[2]

Some used materials, such as scrap iron, tin and copper products, lead pipes, and cotton rags, were in great demand by nineteenth-century manufacturers. These materials could be traded or bartered with peddlers for products, or sold directly to manufacturers who would convert them into new products ready to be sold back again. Cotton rags were in particular demand by the voracious paper mills that used textile fibers to make quality writing papers. It is estimated that about 65 percent of the paper produced on the eve of the Civil War was derived from domestic cotton rags.[3]

Late in the nineteenth century, new charity enterprises developed that were funded by the sale of used and recycled products. Noteworthy among these were service societies such as the Salvation Army, Goodwill Industries, and Morgan Memorial Cooperative Industries, which united poverty alleviation and religious recruitment with the recycling of donated materials. In addition, various civic and social service organizations, particularly the Boy Scouts, took up used paper and metal scrap drives as a source of revenue. The economic depression of the 1930s created ample opportunities for private recycling enterprises that collected trash and salvaged materials for reuse. Among the more successful operations was Sunset Scavenger of San Francisco, which by midcentury was netting a half a million dollars a year.[4]

Each of the nation's major wars stimulated drives to collect and reuse materials. World War II launched the largest of such campaigns, with special appeals for paper, tires, iron, aluminum, and fats (used for glycerin to make explosives). Under the War Production Board, municipalities, businesses, junk dealers, and various civic service programs were organized into local salvage committees to gather iron, steel, tin, rubber, paper, and rags for wartime production needs. With slogans like "Slap the Japs with the Scrap," massive collection drives in 1942 and 1943 recruited thousands of women and children to go door to door seeking materials for war production. Although they were hugely successful at returning underused products and wasted materials to productive uses, most of these emergency efforts were abandoned after the armistice.

Driven by advertising and marketing, the efficiency of closed-loop cycling was abandoned in favor of the convenience of a disposable product economy.

Even a brief history of recycling reveals the relative nature of the concept of waste. Put simply, waste is a product of economic conditions. When the economy is doing well, wastes increase; when economic times are bad, there is much less waste. During the nineteenth century, household wastes were far less voluminous than they are today. Scarce materials are expensive and unlikely to end up as waste. The high price of gold and silver means that there is little gold or silver waste. Recycling follows these same conditions, decreasing as wealth increases and materials become abundant and cheap. Recycling can be stimulated by national emergencies and government initiatives, but a return to normalcy and prosperity has generally favored waste generation over recycling. The long-term upward trend in waste generation demonstrates how the economy has been moving progressively away from a relatively closed "two-way" system in which recycling and reuse were common, toward a more open "one-way" system in which materials throughput has accelerated and ever-growing volumes of wastes are dumped into the environment.

9.2   Recycling Today

Given today's economic prosperity, it is somewhat surprising to see how much recycling persists. In 1996, the EPA reported that waste recycling and composting included nearly 56 million tons of municipal solid waste, or 27 percent of the total waste generated that year. Data for 1997 show a recycling rate of 28 percent, which suggests that growth in the recycling rate for municipal solid waste may be leveling off. Measured by percentage of generation, the products with the highest recycling rates were lead–acid batteries (95.8 percent), corrugated boxes (64.2 percent), aluminum beverage cans (62.7 percent), major appliances (60.5 percent), steel cans (56.8 percent), and newspapers (53.0 percent).[5]

It is estimated that the current recycling market in the United States exceeds $15 billion in sales annually. There are nearly 100 facilities that process recycled glass. Iron and steel scrap can be processed in hundreds of small mills throughout the country, and more than 200 paper and paperboard companies rely solely on waste paper for their feedstock. In 1992, more than a million tons of steel cans were reprocessed; over one-third of glass containers were recycled; 27 percent of polyethylene terephthalate (PET) plastic bottles were recycled; and over 32 percent of the fiber required by U.S. paper and paperboard producers came from recycled paper. A recent study of materials use in the United States estimated that 55 percent of all metals are recycled.[6]

Recycling involves the collection of materials for reprocessing and remanufacturing. This differs from the concept of reuse, where a product is simply used again without remanufacturing. There are two forms of recycling. Primary recycling recycles the material back into the same or a similar product, with minimal dissipation in the quality of the material (low entropy generation). Secondary recycling involves reprocessing the material into a product of lower quality, which means that there is some material dissipation (some entropy generation). Some materials are dissipated during use and cannot be recycled. Fuels are an example. Other materials are degraded (dissipated) by use, but with treatment (added energy) they can be reclaimed and recycled. For instance, some spent solvents can be distilled and reused. Finally, some materials are only modestly affected by use (iron, aluminum, glass), and these can be readily recycled.

Metals

Generally speaking, metals recycle well and a substantial portion of metals are recycled back into production. The U.S. Geological Survey estimates that some 80 million metric tons of metals were recycled in 1998, accounting for some $17.7 billion worth of materials. Thus, recycling accounted for 53 percent of apparent metal supply by weight and 41 percent by value. This represents an increase of secondary metal consumption that rose from 37 percent of the market supply in 1970 to 47 percent by 1993.[7] Today, over half of all metal inputs to manufacturers in the United States come from secondary metals recovery. Table 9.1 presents data on the apparent supply and recycling statistics for common metals. The apparent supply is the amount of processed material plus the amount imported and the net amount drawn from existing supplies during the year. The amount presented as recycled involves both industrial

Table 9.1
Apparent supplies and recycling of common metals in the United States, 1997 (metric tons)

	Apparent supply	Recycled	Percent apparent supply recycled
Aluminum	8,880,000	3,690,000	42
Chromium	488,000	120,000	24.7
Copper	3,900,000	1,450,000	37.2
Iron and steel	127,000,000	73,000,000	57
Lead	1,660,000	1,090,000	65.7
Magnesium	235,000	80,200	34
Nickel	193,000	68,800	32.7
Tin	48,500	12,400	25
Titanium	[a]	26,400	45
Zinc	1,510,000	374,000	24.8

Source: U.S. Geological Survey, *Minerals Yearbook, 1997*, Washington, D.C., 1998.
[a] Data withheld.

scrap and reclaimed postconsumption wastes. As the data reveal, one-quarter or more of the most common metals are recycled, with recycling accounting for the greatest part of the apparent supply of lead, iron, and steel. A recent study of nonferrous metal scrap processors in New England found that 95 percent of the region's metals remain within the scrap reclaiming system.[8]

Recycling of metals usually requires separation, grinding, and heating to remove alloying elements and other contaminants. Although this may require significant amounts of energy, it is typically less energy than is required for extracting and excavating, processing, and refining virgin metals from ore. Aluminum is the best example. Preparing aluminum from recycled sources takes 5 percent as much energy as processing virgin aluminum from bauxite ores. Copper and steel are also good examples. Recycling copper takes seven times less energy than refining virgin copper, and recycled steel takes three and a half times less energy to prepare than virgin steel.[9]

Waste metals are referred to as scrap. Much of the scrap comes from inside metal processing plants, where croppings and trimmings are returned as "new" or "home" scrap. Additional scrap called "process" scrap is returned from the wastes of metal product manufacturing plants. Demolished metal structures and discarded metal products are called "old" or "obsolete" scrap. New scrap makes up about 40 percent of the scondary metals market while old and process scrap make up the other 60 percent. The recovery of old scrap has risen over 40 percent since 1970. The highest value is placed on home scrap because it is already at the point of production and it has never left the control of the manufacturer. Process scrap, like home scrap, is usually quite homogeneous and easy to recycle, but it must be worth the transportation costs and any risks associated with its past uses.[10] Old scrap is more likely to be of mixed materials and contaminated with coatings or oils or embedded in alloys, which makes it harder to recycle.

Among the metals, iron and steel make up the largest volume of recycled materials. They comprise nearly 90 percent by weight of the old scrap recycled in the United States, although this represented only about 46 percent of the total value of metals recycled. Today, about 57 percent of all iron and steel is recycled. Steel recycling rates were 92 percent for automobiles, 88 percent for construction plate and beams, 72 percent for appliances, and 64 percent overall. Steel mills consume about three-quarters of all iron and steel scrap, with iron casting foundries consuming the remainder. New steel hot off rolling mills may contain up to 40 percent recycled scrap steel. The infrastructure for collecting and reprocessing iron and steel is well developed. Recycled automobiles from some 12,000 automobile scrapping yards and 250 shredders provide nearly one-sixth of all ferrous scrap recycled in the United States. The remainder comes from demolished steel structures, steel consumer goods, and other transportation vehicles (boats, trucks, farm equipment, etc.). Much of the recycled steel is resmelted into the production of new mild and high-carbon steel. Basic oxygen furnace mills can use up to 30 percent scrap materials, while electric arc furnaces have been specifically designed to accept secondary metals, and some are charged today with 95 percent ferrous scrap.[11]

Copper recovered from scrap made up 36 percent of the total U.S. copper supply in 1998. About 87 percent of the copper recovered from new scrap was consumed at brass and wire-rod mills. Nearly 15 percent of the copper consumed that year was recycled obsolete scrap from conductive wire, electronic equipment, and various utility and defense operations, while process scrap accounted for nearly 67 percent of all scrap recovered. The difference between these figures is due to the falling price of copper during this decade and the absence of an effective infrastructure for recovering copper-bearing refuse.[12]

In contrast to copper, there is a very effective system for collecting and recycling aluminum wastes. Aluminum is recycled by the integrated primary aluminum companies and by independent secondary smelters who are often affiliated with the automobile and appliance manufacturing industries. The secondary aluminum industry has grown rapidly, increasing from a total metal recovery of 900,000 metric tons in 1970 to nearly 3.4 million tons of aluminum in 1998. Pure aluminum as found in aluminum foil can be nearly 100 percent recycled. The much larger volume of aluminum used in aerospace, automotive, and other industrial applications is alloyed with copper, zinc, magnesium, and other metals. Because of the value placed on high-grade aluminum, these impurities must be separated from the aluminum before recycling. This is typically accomplished by remelting the aluminum in a high-temperature reverberatory furnace. About 25 percent of the tin processed annually in the United States comes from recovered scrap. Because tin is typically used in plating, this requires special detinning facilities, of which there are seven in the United States. There are no commercial tin mines in the United States, so recycling is the only domestic source.

Metal beverage cans provide a sizable source of metal for recycling. Metal cans may be ferrous, bimetallic (ferrous cans with aluminum "pop-tops"), or aluminum. Bimetallic cans are the largest source of postconsumer ferrous metals. Aluminum cans sold back to the large aluminum production companies (Alcoa, Reynolds, etc.) for reprocessing into beverage cans make up nearly one-half of all aluminum recycled in the country. Indeed, domestically produced aluminum beverage cans are made of 51 percent postconsumer recycled aluminum, on average. Although ferrous, bimetallic, and aluminum cans are recycled separately, small amounts of cross-contamination by the other types is easily tolerated in preparing the metals for secondary markets.

Automotive and household batteries provide another source of metal available for recycling. Nearly 1.3 million metric tons of lead are consumed in the United States each year. Of this lead, 1 million metric tons, or 79 percent, are consumed in the production of wet cell lead–acid batteries, which are used predominantly for ignition in motor vehicles and boats. An average automotive battery weighs 36 pounds of which the lead anode and lead dioxide cathode make up about half the weight. The price of lead and various state laws and federal regulations have ensured a high rate of lead–acid battery recycling and today nearly 93 to 98 percent of the lead available from lead–acid batteries (or 76 percent of all lead consumed in the United States) is recovered for reclamation in an increasingly closed recycling infrastructure. These batteries are processed at some twenty-nine active secondary lead smelters in the United States, and these smelters rely on used batteries for more than 70 percent of their lead supply. In addition to the lead, the sulfuric acid solution is reclaimed for use in manufacturing fertilizers, and the polypropylene plastic housing is reprocessed into new housings or other recycled plastic products.[13]

Dry cell household batteries contain a host of metals that can be usefully recycled as well. These include nickel, cadmium, mercury, silver, lead, lithium, and zinc. Nearly 50 percent of the cadmium and 88 percent of the mercury consumed in the United States goes into dry cell batteries. While an individual battery may contain only a small amount of these metals, nearly four billion dry cell batteries are sold in the United States each year and, in aggregate, this adds up to a large amount of these metals. Recycling of dry cell household batteries is less well developed than the recycling of wet cell batteries and is hampered by the limited number of processing facilities. Currently there are only three U.S. facilities capable of recycling household batteries, so the large percentage of batteries recovered through collection programs is either shipped offshore for reprocessing or sent to hazardous waste landfills. Still, nearly 500 tons of cadmium are recovered from nickel–cadmium batteries each year.[14]

Some of the most effective metals recycling involves precious metals such as gold, silver, and the platinum group metals. The high price of these as virgin metals makes a strong secondary market possible. Precious metals are recovered from photographic film users such as hospitals and printers, electrical and electronic components, catalytic converters, metal plating operations, and the jewelry industry. Today, some 34 percent of the global supply of platinum and 87 percent of the rhodium are used in automobile catalytic converters. Prior to the use of platinum group metals in catalytic converters, they were recycled with efficiencies of over 85 percent. The few grams used to control automobile emissions resulted in widespread dissipation until the 1980s, when automobile scrapping yards began to collect and recycle catalytic converters. Gold and silver are commonly recovered from discarded printed circuit boards. In 1989, recycled gold made up 37 percent of national gold consumption and recycled silver made up 44 percent of silver consumption.[15]

Elastomers

Reclaimed rubber from used tires provides another material for recycling. Somewhere over 250 million car and truck tires are discarded annually, representing between 1 and 1.5 percent of all solid wastes generated in the United States. While recycling tires is relatively easy, less than 7 percent of the rubber tires discarded each year is recycled (about 11 percent is burned to generate energy). The majority of used tires—nearly 95 percent—are collected by automobile tire retailers who keep the used tires as a trade-in when they sell new tires. These retailers usually charge the customer a disposal fee and contract with waste haulers to transport and dispose of the tires.[16]

Those tires that do get recycled are often mechanically ground or chemically digested. Although there are some 120 firms involved in recycling tires, 90 percent of the market is supplied by the output of 20 large companies. Reclaimed rubber from chemical digestion produces a product that is quite competitive with virgin rubber—often 20 to 30 percent less costly than the primary material—but its lower quality limits its use to fillers and additives in primary rubber products. The crumb rubber produced by mechanical grinding is more expensive but displays greater materials properties and is used as new tire fillers, retreads, footwear, mats, roofing sealants, membrane liners for landfills and retention ponds, and various mechanical and stress-absorbing goods. Ground rubber is also used as a filler in roadway asphalt.[17]

Plastics

It is not easy to recycle plastics. The overall recycling rate is small: 1 million tons, or 5.3 percent of the nation's plastic production, were recycled in 1995. While plastics appear in many of today's consumer products, beverage containers provide the largest source of plastics available for recycling, followed by plastic films and rigid containers.

In 1985, 47 billion pounds of plastics were produced in the United States. Of the 39 billion pounds consumed domestically, 33 percent was used for packaging. While this represents only 8 percent by weight of the total municipal waste stream, it represents 20 percent of the waste stream by volume. By 1995, 38 billion pounds of plastics were disposed of and this made up just over 9 percent of the municipal waste stream.[18]

Recycling rates vary significantly by type of plastic. While the most common plastic used in beverage containers is high-density polyethylene (for milk and water bottles), polyethylene terephthalate (for soda bottles), polystyrene (for cups), polypropylene, and polyvinyl chloride (PVC) are also used extensively. Because of their high volume and ease of identification, high-density polyethylene and polyethylene terephthalate are the most commonly recycled plastics. In 1995 30.2 percent of polyethylene milk and water bottles, and 45.5 percent of polyethylene terephthalate soft drink bottles were recycled.[19]

Recycling of plastics requires that the discards be substantially clean and that the sources be as pure in type as possible. Plastic beverage containers are collected from local recycling centers, shredded and baled or granulated, and packaged for shipment to plastic product reformulators. Differences in the properties of various plastic resins requires that they be separated before shredding. In particular, polyvinyl chloride and polyethylene terephthalate are chemically incompatible in preparing recycled products. Indeed, one polyvinyl chloride bottle in 100,000 bottles provides enough contamination to prohibit polyethylene terephthalate recycling for most uses. As a result, municipal recycling programs that collect plastic bottles from households must separate the plastics by type. While this separation is facilitated by an industry coding system that appears on the bottom of the bottles and various new technologies that use laser identification for sorting, a good deal of sorting is still done by hand. Manual sorting is costly and prone to error. Costs for automated sorters can be substantial, ranging from $35,000 for simple PVC detection systems to $700,000 for multiresin and color identification systems.[20]

Because remelting of plastics for reforming tends to damage the polymer chains, most plastics cannot be recycled back to their original quality. Instead they are "downcycled" to less demanding products. For instance, a plastic lumber can be made from comingled plastic wastes that can compete with wood for park benches, roof shingles, fencing, deck flooring, and boat docks, although it has poor structural characteristics and cannot be used for tension or compression members.

Recycling of industrial materials offers important environmental benefits. It reduces the flow of materials into and out of the economy. The demand for primary materials is also reduced, and the generation of wastes is decreased. Materials that are recycled retain at least a portion of their value and continue to provide a social service. Recycling also reduces the dissipation of a material. Finally, it conserves materials for the future. By closing the loop of materials flow, recycling promotes dematerialization and provides a means to retain environmentally incompatible materials within closed economic systems.

9.3   Reuse Today

Some products remain in the market through an extended period of reuse. Product reuse is distinguished from recycling because reuse involves no reprocessing or remanufacture. The product or material is simply used again. Until the 1960s, many beverages, ranging from beer to milk, were sold in heavy, reusable glass bottles, which were either voluntarily returned or returned through the encouragement of a small deposit paid at the time of purchase. The advent of lightweight, inexpensive, disposable (often aluminum or plastic) beverage containers ended the use of most reusable beverage bottles.

However, there is a large and flourishing secondary market for "used" commercial goods, including automobiles, boats, airplanes, refrigerators, furniture, sporting equipment, guns, clothing, and books. In industrial settings there are brokers for used steel drums, wooden pallets, plastic containers, office furniture, construction materials, tools, and equipment. These markets may involve permanent used-product retail vendors, periodic flea markets, seasonal auctions, occasional garage sales, or weekly advertising magazines listing used products along with personal contact information. In most cases these products have not been designed to be reused; they simply have been discarded or offered for resale before their useful life is over, and they are purchased as less expensive substitutes for new products.

There is a substantial secondary materials market devoted to selling used discarded but yet functional products. It ranges from sophisticated retail chains to locally owned used-product shops to nonprofit charity and religion-based services. Given the supply of used products, this market could grow. In 1996, the EPA found that 15 percent (by weight) of the municipal waste stream consisted of appliances, furniture, carpets, and miscellaneous durables—all candidates for reuse. Yet in most areas of the country, there is an oversupply of products available for reuse, and a large proportion of these discarded but yet functional goods are sent to landfills or solid waste incinerators. A 1995 survey by the Washington-based Institute for Local Self-Reliance found that most reuse enterprises were small nonprofit operations with a few (an average of five) employees and several volunteers. These enterprises were frequently limited in their development by their inability to secure adequate capital for startup, operations, and expansion.[21]

There is also a reasonable secondary market for products that have been designed to be reused. Rechargeable batteries provide a good example. About 10 percent of all dry cell household batteries are rechargeable. They are used in flashlights, power tools, cordless telephones, appliances, personal computers, portable electronics, and children's toys. These batteries are designed to be frequently recharged and reused. The most common rechargeable batteries use nickel–cadmium dry cells, although there are also sealed lead–acid rechargeable batteries. While these batteries can be recycled, their true environmental value lies in the potential for repeated charging and reuse. It is estimated that one rechargeable battery can substitute for 100 to 300 single-use batteries.[22] Retreading used tires is another example of product reuse by design. While less than 7 percent of discarded tires are recycled in the United States each year, some 10 percent are retreaded for resale and reuse. Many tire dealers offer retreads along with their new tires.

Reuse is a fuzzy concept. From a materials perspective, there is not much distinction between continued use and reuse. For instance, kitchen knives can be discarded when dull, resharpened for continued use by an original owner, or donated to a used product charity for sharpening and resale. Resharpening, like recharging or repairing a product, allows it to be reused, sometimes for the lifetime of its owners or for use by several owners, and this reduces the aggregate flow of materials. The focus is on extending the useful life of the product rather than simply keeping the product out of the waste stream. Indeed, reusing products just as they are with little or no reprocessing by one owner or several is a fine example of dematerialization and a useful way to promote a closed economic system. Materials and energy are conserved and dissipation is forestalled.

9.4   Policies for Promoting Recycling

The rates of recycling of municipal wastes have varied considerably over time. During the 1960s, few U.S. municipalities recycled more than 10 percent of their domestic wastes, but following Earth Day 1970, a spontaneously organized, grassroots movement of community activists set up thousands of drop-off recycling centers that soon were formalized into local government recycling programs. Today, there are communities that report 35 to 40 percent recycling rates. (If organic waste composting is included, these rates jump to 55 to 65 percent.) Although these programs are touted as a way household residents can do something to protect the environment, most programs have been driven by the rising costs of municipal waste treatment and concerns over the future capacity of local landfills. In 1995 there were over 7000 municipal curbside recycling programs in the United States and nearly 9000 local drop-off centers for recyclables.[23]

Policies for promoting recycling can focus on either the supply of materials for recycling or on the demand for recycled materials. Focusing on the supply of waste materials from domestic households, more than 30 states have set some kind of quantitative recycling goal, which may range from 15 to 50 percent, depending on the target date. Some governments have introduced mandatory recycling laws for certain types of materials. Success with these laws has been mixed because they are difficult to enforce and because they encourage a shift to nonregulated materials. For instance, several states have passed laws prohibiting the landfilling of automobile tires with an expectation that this will promote a substantial supply of tires ready for recycling. Too often these laws have led to illegal tire disposal. Alternatively, some governments have imposed a disposal fee on certain products or materials to raise the price of disposal and thereby encourage recycling. It is difficult to find the appropriate fee; too high a fee encourages illegal disposal and too low a fee provides no incentive.[24]

One of the most visible supply-side programs in the United States involves returnable deposits on beverage containers. There was virtually no recycling of aluminum cans and plastic bottles before states began to enact mandatory "bottle bills" during the 1970s. Today, ten states have mandatory beverage container takeback laws, all of which depend on a small fee charged by retailers and refunded when the container is returned. These programs are generally very effective, with recovery rates of 90 percent or more. The ten states with mandatory bottle takeback laws represent just 30 percent of the U.S. population, but account for over 90 percent of all plastic soft drink containers recycled, over 70 percent of all glass bottles recycled, and nearly half of all aluminum cans recycled.[25]

Because the use phase of materials, especially those materials used in products for domestic consumption, involves a substantial amount of dispersal, recycling programs must reconcentrate the supply of materials through collection activities. Home scrap is readily concentrated, but obsolete scrap and postconsumer products require structured collection programs. Today, many municipalities fund drop-off centers and curbside pickup services to reconcentrate recyclable supplies. Most drop-off centers are multimaterial services that accept everything from beverage containers to newspapers, and some provide a small compensation if the recycled markets are doing well. The first curbside collection program for newspaper began in Madison, Wisconsin, in 1973. Today, curbside collection is often a municipal function integrated into the domestic refuse collection service. Such programs may require residents to separate the materials before they are set out, although most programs collect comingled recyclable products and transport them to recovery centers where the collected materials are separated by hand or machine.

From a social perspective, these local recycling programs have been a remarkable success, even though many of them have been less than effective in terms of their economic return. There are a multitude of economic studies on the effectiveness of the municipal recycling programs. Studies during the 1980s found that the costs of recycling recyclable materials in municipal trash were less than the costs of treating and disposing of those materials, sometimes by as much as one-half to one-third. This was largely owing to the high price of waste treatment and disposal, and not to the revenue derived from selling the secondary materials. Indeed, with the exception of aluminum, recycled materials were worth more in terms of avoided disposal costs than their actual material value.[26] But as secondary materials markets improved during the 1990s, revenues became more important. One major national study completed in 1993 found that recycling raised the overall costs of municipal waste management, but a recalculation made 2 years later when the markets for recycled materials had improved found the reverse. A second, often-quoted study of four municipalities in Washington state found that recycling materials was more cost-effective than conventional waste management options, but this study was based on waste management costs that were above the national averages. The current cost-effectiveness of recycling municipal trash appears to depend on waste management costs, the price recovered materials will garner on secondary materials markets, and the costs of collecting the materials.[27]

In focusing on the demand side, governments have sought to build the secondary markets that are critical to successful recycling. Setting out domestic wastes for recycling means little if there are no markets for recycled materials. An oversupply of waste newspapers collected by curbside recycling programs caused the used paper market to collapse in 1989. If recycling is to succeed at reducing the drain on virgin materials, then a large network of enterprises engaged in consuming recycled materials as feedstocks must be developed and encouraged.

A major barrier to these markets is the comingling of materials collected as wastes. There is a strong market for used polyethylene, and a good market for used polystyrene, but a much weaker market for a waste stream in which the materials are mixed. The key to high-value markets for secondary materials is quality, not quantity. However, it is difficult to implement a system for maintaining distinct postconsumer waste streams for an assortment of different materials, and centralized waste-sorting systems are likely to drive up the costs of secondary materials so that they compete poorly with primary materials.

Federal policy to promote recycled materials has been restrained. The Resource Conservation and Recovery Act gave the federal government direct authority to promote demand for recycled materials, specifically through requirements to purchase products made from recycled materials. The U.S. Department of Energy has also been charged with promoting recycling as a way to reduce waste and conserve energy. For several years the federal highway standards required that federally funded construction projects use recycled crumb rubber in the roadway asphalt. Yet, in general, the federal offices have been slow and underfunded, and the results have been limited. Indirectly, various federal regulations have encouraged the supply of recycled materials by foreclosing other waste management options. For instance, the supply of recycled lead has increased significantly owing to the EPA's tightening of environmental regulations on lead emissions and a special waste handling exemption that allows lead scrap dealers to store lead without acquiring the costly permits for hazardous waste storage.[28]

A fee on primary materials would make recycled materials more attractive. For instance, a pollution tax on sulfur emissions would raise the price of virgin copper and increase the incentive to use secondary copper. Market supports, technical assistance, subsidized collection and sorting facilities, targeted government procurement, and direct subsidies to recycling firms have all been tried. For example, a 1994 presidential executive order requires that all paper purchased by the federal government —some 300,000 tons per year—must be composed of 30 percent postconsumer waste paper. In addition, over a quarter of the states have set minimum recycled content standards for newsprint. Twenty-three states have set up grant or loan programs to assist the develepment of second-ary materials markets, and 27 states provide tax incentives for firms using recycled materials. Texas and California have established market-development zones in which firms involved in recycling or using recycled materials can get low-interest loans and other benefits.[29]

Consumers can support secondary markets by buying products made from recycled materials, if they can identify them at the point of pur-chase. Governments can encourage this by promoting product labels showing recycled content. Some photocopy, kitchen, and toilet papers currently sport such labels. But product labels can be counterproductive The federal Wool Products Labeling Act of 1939 required textiles to identify on their label the use of any recycled wool, and this declaration diminished the recycling of wool during a period when virgin material was popularly thought to be a sign of quality.

Proponents of industrial ecology have promoted a geographically based form of demand-side support for recycling variously called "envi-ronmentally balanced industrial complexes" or "eco-industrial parks." In the industrial parks where many firms are located it is possible that the by-products from one firm's production process could become the feedstocks for another firm's processes. For example, an often-cited in-dustrial park in Kalundborg, Denmark, demonstrates how the excess cal-cium sulfate generated by flue gas desulfurization equipment in power plants can be used as an input by gypsum plasterboard manufacturing firms, and waste heat from one process can be conducted to another facil-ity to preheat raw material inputs. The development of eco-industrial parks has caught the imagination of industrial developers and municipal planners, and several efforts have been initiated at existing industrial parks.[30]

Eco-industrial parks would increase the potential for recycling mate-rials beyond current market incentives, but largely because interfirm information flows are improved and resource transportation costs are minimized. The spatial factor is, in reality, a minor advantage, and the disadvantages may well outweigh it. By more closely integrating the waste-to-feedstock dependence of the firms, the flexibility of a more open materials market may be compromised. Where participating firms need to assure one another of a stable supply of waste or a continuing capacity to accept waste, they may curb improvements in each other's production processes. Thus, interfirm relations may discourage incentives to reduce wastes by improving production efficiencies and may inhibit the continuous search for process efficiencies. Such commitments may introduce a rigidity in both process and product novelty.[31]

Materials recycling requires the flexibility of open markets. During the 1980s some states and regions attempted to set up information networks called "waste exchanges" where firms with relatively pure wastes could connect with firms that might be able to use these wastes as feedstocks. One of the most ambitious of these waste exchanges was set up by the Chicago Board of Trade, but it was closed in 1999. These relatively open markets have proven difficult to maintain because quality assurance is absent, liability for wastes remains unresolved, and the price of virgin materials remains low enough to discourage risk taking in these secondary markets.

9.5   The Potentials and Limits of Recycling and Reuse

Recycling conserves natural resources, energy, and the assimilative capacity of the environment. It is estimated that the recovery of 1 ton of steel from scrap conserves nearly 1.5 tons of iron ore, 0.6 tons of coal, and 120 pounds of limestone. One pound of recycled steel saves enough energy to light a 60-watt bulb for more than 26 hours. Recycling aluminum saves about 95 percent of the energy required to process the same amount of metal from ore. Some analysts estimate that the energy saved in 1 year from recycling aluminum beverage cans (10 billion kilowatt hours) is enough to supply the residential needs of New York City for half a year. In addition, most recycling generates less pollution and less greenhouse gas emissions than primary production.[32]

The proportion of production supplied by secondary materials provides a rough indication of the degree to which a specific materials system is open or closed. Table 9.1 shows that both the lead and iron and steel industries are operating as partially closed systems. A historical analysis of the data shows that zinc recovery has been limited and unchanged over time; copper recovery ratios show a rapid increase in the early part of the century and a more modest increase after 1950; and aluminum recovery ratios rose, fell, and are rising again. The steel and lead recovery ratios demonstrate a continual upward trend. Since the 1970s there has been a gradual increase in the recycling rate for metals as a whole, with an estimated 45–55 percent recycled today. This led professionals in the U.S. Geological Survey to predict in 1995 that "if the current trend continues, the share of metal supply derived from scrap will surpass that derived from ores sometime in the next decade."[33]

In terms of the hard-rock metals (copper, silver, gold, etc.), recycling within the confines of the economy (Loop 3) has economic as well as environmental benefits over their continued mining and disposal (Loop 2). The importance of hard-rock mining in the U.S. economy is modest and shrinking, with employment declining by half between 1980 and 1990. Today, hard-rock mining employment in the twelve western states provides roughly 45,000 to 50,000 jobs, or less than 0.15 percent of those states' total employment.[34] Indeed, because these metals are a commodity traded on the international markets, their prices fluctuate widely and the economy of the domestic mines as well as their nearby communities is highly unstable. The collapse of copper prices in the early 1980s led to the shutdown of most U.S. copper mines, displacing thousands of miners. The rising price of copper reopened some mines by the end of the decade, but the price fell again during the early 1990s, only to rise again more recently. The effects of these fluctuations on local economies is often disastrous. With such a declining and unstable industry, there is plenty of potential benefit in reducing the reliance of the nation and these local economies on mining and instead investing in recycling infrastructures that reduce the need for mining metals and lower the volume of metal wastes.[35]

Still, there are limits to the effectiveness of recycling as a strategy for sustainability. First, success in collecting materials depends on the capacity of the recycling infrastructure. Recycling high-quality (low entropy) structural materials like the aluminum in automobile chassis is effective and efficient. Automobiles are easily recovered from dealers interested in selling new cars, and the composition of metals is large enough to warrant separation and reclamation. Higher entropy materials that are easily dissipated during use, such as solvents, detergents, flocculants, fuel additives, pigments, and lubricants, are all but irretrievable and well below economic thresholds for reaggregation and recycling.

Second, continuous recycling of some materials is never fully possible Technically, metals, ceramics, and glass can be recycled nearly indefinitely with some energy inputs. But these materials are in declining use. The recycling of the materials that are increasing in use, such as the polymers and petrochemicals, depletes the material and each successive product displays a somewhat lower quality than its predecessors. Recycling these materials does delay disposal, but after a certain number of iterations, their quality is dissipated. Such recycling tends to create a declining supply of regenerated products.

Third, because recycling remains a low-value process, dealing as it does with wastes, it is easy to find abuses. Some of the worst federally identified hazardous waste dump sites are old recycling facilities for industrial waste that went bankrupt. Recycling and scrap processing facilities can be fairly messy and polluting, and increasingly they are likely to be sited in countries with lax environmental enforcement. For instance, the tightening of regulations for secondary lead smelting in the United States has encouraged the export of used lead–acid batteries to recycling centers in Indonesia, Taiwan, Thailand, and the Philippines that are well below U.S. standards.

Fourth, the benefits of recycling can be overwhelmed if the rate of consumption is growing rapidly. In a fairly stable economy, recycling conserves natural resources and makes production more efficient. However, as the rate of consumption increases, the benefits derived from recycling are outweighed by the sheer volume of the materials used for production and consumption. A 20 percent recycling rate in an economy with a 1 billion-pound material throughput is more satisfying than the same rate in an economy with a hundred billion-pound throughput.

These limits do not negate the value of recycling or undermine its role as a contributor to dematerialization, but they do mark the outer bounds of a recycling strategy. Observing the limits, Donald Rogich of the U.S. Bureau of Mines, notes:

The term recycling as commonly used is not synonymous with continuous reuse, rather it is used for situations that only achieve life extension. Although that is beneficial from a resource conservation and waste disposal point of view, it is fundamentally at odds with a sustainable economy. In most instances recycled products are less effective than the primary feedstocks for the creation of new goods, and many current industrial processes can only accept a limited percentage of recycled feedstocks.[36]

Product and material reuse, unaffected by cyclical reprocessing, appears to be a more effective approach to dematerialization. Such reuse obviates the need for the materials and energy used in recycling programs to disaggregate and then remanufacture products. Long-life, more durable products may be more costly to manufacture the first time, but the payoff comes as the products achieve multiple cycles of use with little other inputs.

Community-based recycling programs are very popular among Americans; indeed, there is a common joke that more Americans recycle than vote. Yet the recycling of domestic wastes is difficult to justify purely in terms of waste management efficiency. Landfilling and incinerating are relatively inexpensive in the United States. The cost-effectiveness of recycling domestic wastes appears as hard to prove today as it did to those who struggled to create resource recovery centers during the early part of the century. Instead, recycling is better justified as an environmental protection strategy.[37]

Recycling reduces the volume of wastes, the dissipation of materials into the environment, and the exploitation of natural resources. It closes the loop on materials flows within economic systems, and for those materials that are not readily compatible with natural systems, it permits their continued use and delays their movement into the environment. However, an even better way to achieve these same benefits is to use fewer materials in the first place. Among waste management specialists this is often called "source reduction" and from their point of view, an ounce of material not used may do more for the environment than a pound of material recycled. This is not always well recognized by those who manage recycling programs and there are cases where recycling proponents have argued against the expansion of source reduction programs for fear that the supply of recyclables will fall below the rate necessary to support effective recycling investments.

Recycling promotes dematerialization and is a central component of a materials flow model, but it is not a comprehensive solution. There are current limits to the effectiveness of recycling programs and beyond those limits, further strategies will need to be employed to more completely implement dematerialization as a comprehensive strategy for a sustainable materials system.

10

Advanced and Engineered Materials

"I just want to say one word to you ... just one word: plastics."
—*The Graduate*, a 1967 film

Today, there are materials that can sense a human presence and respond by turning on the lighting in a room or sounding a security alarm. There are structural materials that sense the loading on a beam and transfer their greatest strength to that point. There are optical materials that are so pure that a light beam shot into one end of an ultrafine fiber filament can travel thousands of miles with almost no distortion.

From hundreds of laboratories throughout the world there flows a steady stream of imaginative, creative, and fascinating materials. Many of these novel substances are destined to be only curiosities, but some of them present an opportunity to literally transform the technological world. These new materials offer an enormous opportunity to create a safer and more environmentally compatible materials economy. By developing less toxic chemistries that are more consciously sympathetic to biological systems, a larger volume of industrial materials could flow between the economy and the environment without concern, and fewer materials would need to be managed within the closed confines of the economic system. These new materials offer such potential. And yet, for all that is impressive and intriguing about these materials, it is disappointing to consider how little attention has been paid to their effects on human health or the environment. In fairness, some of these materials are inspired by health problems (medical needs) or environmental issues (energy efficiency), but seldom are even the most obvious health or environmental effects of production or disposal considered.

There is nothing surprising here. Such effects were little considered in developing our existing industrial materials, so why should the development of new materials be any different? We do know much more about the health and environmental effects of industrial materials, but this knowledge is seldom transferred into the materials sciences or included in materials specifications. The reasons these properties are not considered in the design of new materials are embedded deep in the structure of the materials sciences, the incentives that promote new materials, and the government policies that cover these processes.

10.1 The Materials Sciences

The roots of the materials sciences can be traced back to the late nineteenth century when significant changes occurred in the organization of research, the development of science, and advances in instrumentation. Until the mid-nineteenth century, the development of materials relied on individual inventors such as Charles Goodyear, Charles Martin Hall, and John Wesley Hyatt, often working in isolation on highly empirical experiments. By the end of the century, a new form of research based on teams of specialists working in university or corporate laboratories began to appear in the new optical glass, electrical, automotive, and communication industries. Nowhere was this better demonstrated than the organic chemical industry; Hermann Stradinger and Wallace Carothers would never have achieved their successes in polymers without scientific theories, empirical research, and teams of scientists learning from each other's work. With the new century, the theoretical understanding of the atom and the development of quantum mechanics and fluid dynamics provided a basis for exploring materials in a more organized and directed fashion. This awakening knowledge was augmented by the development of X-ray diffraction techniques that allowed an ever-closer examination of the crystallographic structure of materials.[1]

Beginning in the 1930s, elements of several sciences began to converge into a loosely constructed network that became the basis for today's materials sciences. Much of the early work grew out of ceramics and metallurgy and focused on the composition and microstructure of metals and minerals and their mechanical, electrical, thermal, and magnetic properties. Following World War II, interest arose in finding materials capable of withstanding the intense heating and cooling cycles of jet engines, identifying the source of metal fatigue in aircraft frames, understanding the lattice structure of materials, and developing a replacement for the mechanical switches in the nation's telephone communication system. More recently, studies of nonlinear optics, crystallography, and molecular assembly processes have been added.

World War II provided a huge stimulus for advances in materials research and development. The requirement for critical materials that were in short supply galvanized research on substitutes, and the need for sophisticated weaponry and aircraft created specifications for materials not yet developed. These needs were matched by unending flows of government funds and closely integrated corporate collaborations that were prohibited during peacetime. The result was rapid advances in solid-state physics, semiconductors, high-temperature alloys, electrooptics, nuclear physics, and laser technology.

Following the war, the rapidly developing automobile, consumer product, aerospace, nuclear energy, and solid-state electronics industries created a voracious appetite for the new materials that had been seeded during the war. There were important constellations of scientists working at Los Alamos, New Mexico; Bell Telephone Laboratories at Murray Hill, New Jersey; Du Pont's polymer research center at Wilmington, Delaware; General Electric's research labs in Schenectady, New York; and various research centers at the Massachusetts Institute of Technology, the University of California, the University of Chicago, and elsewhere. These were multidisciplinary research centers where innovations arose from the interactions of different theories and bodies of knowledge. Dissatisfaction with the traditional disciplinary boundaries led to a desire for a more integrated nomenclature, and in 1958 Northwestern University renamed its metallurgy department as the nation's first academic department of materials science.[2]

The materials sciences involve the basic sciences, but link them directly to practical applications. In developing new materials, researchers have focused on three important design parameters—performance, processing efficiency, and the costs of production. Advances in performance have led to stronger, tougher, lighter, more durable, and more flexible materials. Production processes have become more sophisticated, materials efficient, and sensitive. The initial costs of production have typically been extremely high, but as markets have developed, mass production has led to rapid declines in costs. The semiconductor is probably the best example of a highly flexible material, involving incredibly sophisticated processing, that is increasingly available at a tiny fraction of its initial costs.

Today, materials science and engineering is a cross-disciplinary collection of intellectual enterprises that involve chemistry, physics, and several engineering disciplines. The field has arisen in a space that lies between basic sciences such as quantum and solid-state physics and practical disciplines such as mechanics, metallurgy, and chemical engineering. Unlike the earlier trial-and-error empiricism that led to many discoveries before the 1930s, materials scientists today search through a vast knowledge about the structure and properties of materials at the atomic and molecular level to "design" new materials that are tailored to meet rigorous performance specifications. In a bold statement on the future of the field, titled *Materials Science and Engineering for the 1990s*, leaders of the discipline wrote, "Without new materials and their efficient production, our world of modern devices, machines, computers, automobiles, aircraft, communication equipment, and structural products could not exist."[3]

10.2   Advanced Materials

During the 1980s, much professional and government attention focused on "new materials" (otherwise called "advanced materials," or "engineered materials"). These new materials exhibited high strength; great hardness; and superior thermal, electrical, optical, and chemical properties. Table 10.1 contains a selection of these new materials. They have dramatically altered communication technologies, reshaped data analysis, advanced space travel, restructured medical devices, and transformed industrial production processes. In attempting to define these materials, the U.S. Bureau of Mines noted the following characteristics:

- they are created for specific purposes,
- they are highly processed and have a high value-to-weight ratio,

Table 10.1
Some advanced and engineered materials and their desired properties

Material	Desired properties
Superalloys	Superior strength
Structural Ceramics	Superior hardness, corrosion passivity
High-performance polymers	Superior strength
Composites	
Laminated composites	Superior strength
Polymer matrix composites	Light weight and high stiffness
Metal matrix composites	Superior strength and stiffness
Electronic Materials	
Semiconductors	Capacity to store information
Superconductors	
Conductive polymers	
Photonic Materials	Ability to transfer light
Smart Materials	
Piezoelectric materials	Pressure-induced conductivity
Field-responsive polymers	Responsivity to environmental stimulus
Shape-memory alloys	Resistance to deformation
Nanoparticles	Capacity for self-assembly

Source: U.S. Bureau of Mines, *The New Materials Society, Vols. 1–3*, Washington, D.C.: U.S. Government Printing Office, 1990; and Anthony Kelly, ed., *Concise Encyclopedia of Composite Materials*, New York: Pergamon Press, 1989.

• they are developed and replaced with high frequency, and
• they are frequently combined into new composites.[4]

Like the petrochemical-based materials that preceded them, these materials are often synthesized from the by-products of conventional commodity materials. Gallium, indium, selenium, tellurium, and thallium—all by-products of aluminum, copper, iron, lead, and zinc processing—are important starting materials or catalysts in the development of advanced materials. Thus the supply of these new materials is closely linked to conventional mineral extraction and processing. The advanced polymers and many of the new composites are also largely developed from conventional organic chemicals. While the market for advanced materials has grown rapidly, the consumption of these materials in the production of finished goods is confined largely to the industrialized countries, with the United States often the largest consumer.

The production of advanced materials is carried out in labor-intensive, small batch processes, carefully tailored to meet the needs of prospective customers. Their production is not necessarily more or less energy intensive than the production of conventional materials, and the disposal of their scrap is little different than the disposal of commodity scrap. Advanced materials typically go through a protracted research and development period, and they are usually slow to penetrate existing markets. This slow adoption is due to several factors. First, being new, there is limited experience to draw upon and potential buyers may have little performance data for writing purchase specifications. Second, because the materials are designed for specific uses in high-value applications, customers are reluctant to risk a change from conventional materials. Third, with no industry standards, a given new material may vary in characteristics from supplier to supplier or from batch to batch. Finally, being a nonconventional material means that there is no guarantee that the material will succeed, which can raise concerns that the material will not be around in the years to come. These perceived risks limit the potential market for new materials and mean that it is often the military or other federal agencies that first pilot test new materials and provide a basis for evaluating performance and determining the market scale necessary to bring the costs down.

The new and advanced materials include metal alloys, structural ceramics, engineered plastics, and composites. The composites are typically classified by their matrix phase, as in ceramic matrix composites, polymer matrix composites, or metal matrix composites.[5]

Advances in powder metallurgy, mechanical alloying, and rapid solidification have produced a range of new alloys that demonstrate superior strength; lightness; stability; and resistance to corrosion, creep, and fatigue failure. The search for structural materials with higher strength-to-weight ratios has led to several new aluminum alloys. The aluminum–lithium alloys are 1–3 percent lighter than conventional aluminum alloys and substantially stiffer (higher elastic modulus). Aluminum–titanium alloys demonstrate excellent resistance to corrosion, creep, and fatigue, and better strength performance at high temperatures than conventional aluminum alloys. These new alloys are seen as improvements over the current alloys used for structural applications in the aircraft and aerospace industries. Between the 1940s and 1960s, significant research focused on the development of superalloys capable of withstanding the high temperatures (500° to 1000°C) and continuous cycles of heating and cooling that exist inside aircraft jet engines. Superalloys built from nickel, cobalt, and iron bases are produced in particles that range from tiny powders (less than 5 μm) prepared for super-plastic forming to microscopic columnar grains used in forming directionally solidified turbine blades.[6]

Ceramics, which are valued for their chemical stability, corrosion and wear resistance, and durability, have a long history in electronic components because of their insulation value, but their brittle nature and low strength-to-weight ratio has limited their further development. The Japanese have become world leaders in the development of "advanced" ceramics for electronics, bearings, and automotive combustion engine applications. Recently, new ceramics, such as aluminum oxide, silicon nitride, and silicon carbide, have been developed for advanced heat engines and coatings for metallic or polymer substrates. Currently, aircraft engines and cutting tools account for about 85 percent of the industrial market for advanced ceramic coatings.[7]

High-performance and advanced polymers are an extension of commodity polymers that demonstrate superior strength and resistance to thermal degradation or chemical corrosion. The advanced polymers include polyamide, polycarbonate, polyethylene terephthalate, polybutylene terephthalate, and liquid-crystal polymers. The lower density of these polymers makes them attractive in the structural elements of transportation vehicles, and in packaging and containers for products that are shipped. About one-quarter of the advanced polymers are used in the electronics industry, often replacing older polymers such as the phenolics. Another quarter are used as structural elements in transportation vehicles, and the remainder are used in a wide variety of consumer products, ranging from domestic appliances to videotape and photographic film. Electrically conductive polymers, such as polyanalene, appear to have substantial use in plastic batteries.[8]

Composite materials are either conventional materials (metals, ceramics, and polymers) that have a reinforcement (particulates, whiskers, fibers, wires) embedded in a matrix of the base material, or laminates made up of bonded layers of heterogeneous materials. The reinforcements enhance the mechanical or physical properties of the base material, making these materials attractive for their high strength-to-weight ratio, ease of fabrication, resistance to corrosion and weathering, and relatively low cost. The laminates may be thin films of polymer, metal, or ceramic compositions. These materials have found wide applications ranging from printed wiring boards to automobile hoods, bathtubs, and space suits.[9]

The layered sheets of reinforced elastomers used in automobile tires are among the oldest composites in common commercial use. While tire sidewalls are composed of natural rubber, polybutadiene, and carbon black, the treads are largely a butadiene–styrene copolymer. Over the years, the reinforcing cords embedded in the rubber have been made of a variety of materials, including cotton, rayon, nylon, polyester, glass, and steel.

Polymeric matrix composites have been produced for years. Of these, glass-reinforced plastics make up the largest volume on the market today. Many other advanced polymer resins are woven, with fibers of graphite, carbon, aramid, and boron comprising up to 60 percent of the weight. Nearly 90 percent of these composites are thermosets, with about three-quarters of these being epoxies. Those that are thermoplastics are usually polyolefins and terephthalate esters. When these thermoplastics are reinforced with whiskers or short fibers, they retain their moldability while in use. Because the reinforcing fibers themselves may have flaws, research has been directed at producing polymers reinforced with polymeric rods of the same chemical composition and some of these polymer–polymer composites offer even higher strength performance. Their high strength-to-weight ratio, excellent weathering properties, and good thermal and electrical insulation make polymer matrix composites attractive in boat-building, motor vehicle, sporting equipment, and aerospace applications. Nearly 60 percent of polymer matrix composites are used in the aerospace industry, particularly as replacements for aluminum structural elements, while the rest have specialty applications, including high-performance sporting goods.[10]

The primary metals used in metal matrix composites are aluminum, copper, magnesium, superalloys, and titanium. There are many reinforcements, including particulates of silicon carbide, boron carbide, titanium carbide, and alumina; whiskers of silicon carbide and alumina; fibers of boron, graphite, silicon carbide, and alumina; and wires of titanium, tungsten, molybdenum, stainless steel, and beryllium. Most of these have been developed to improve strength, stiffness, creep resistance, and resistance to fatigue failure in structural applications. In some cases metal matrix composites also yield improvements in thermal insulation, thermal conductivity, or acoustic absorption. Alumina–silica-reinforced aluminum is used in diesel engine pistons, and boron–fiber aluminum is currently used in high-performance bicycle frames, while silicon particle –carbon aluminum is used in precision machinery and highly sensitive instruments. At present the metal matrix composites are largely an interest of the U.S. Department of Defense, which provides most of the funding for their development.

The development of composites has done much to meet the engineering requirements for materials that are light, strong, durable, and easily molded. These features have made composites highly competitive with pure metals and metal alloys. For example, nearly 90 percent of all medium- and large-scale containers for bulk chemicals are now made from fiber-reinforced plastics rather than conventional steel or aluminum.[11]

10.3 Electronic and Conductive Materials

The solid-state semiconductor is the heart of the information age. In the 1960s, a semiconductor device was about the size of a dime and contained one transistor. Until 1980, the number of transistors integrated onto a single chip doubled each year, and today a single semiconductor about the size of a grain of pepper can hold thousands of transistors in a sea of tiny circuitry.

For practical purposes, the birth of solid-state electronics is set in 1948 at Bell Telephone laboratory, where William Shockley, John Bardeen, and Walter Brittain first demonstrated a transistor, although as far back as 1926 Julius Lilienfeld had patented a solid-state amplifying device based on copper sulfide. The Bell lab transistor was made of extremely pure germanium, and germanium continued to dominate semiconductor technology until the 1960s.

The production of a common semiconductor involves a huge assortment of materials and over 400 unique processing steps. Table 10.2 lists some of the prominent materials required to make a semiconductor. Most semiconductors today are based on silicon, which is the most common metal on earth and readily available everywhere as silicon dioxide. While silicon is a relatively weak conductor, substitution of silicon atoms with dopants at levels as low as 1 ppb in a silicon matrix permits the passage of electric current through negatively charged electrons or positively charged holes. Because precise conductivity is critical, the silicon is prepared as pure crystals.

In its common form, silicon is nontoxic, but the process of making high-purity silicon crystals generates significantly more waste than product, and the waste is highly hazardous. Metallurgical-grade silicon is a by-product of the steel industry and is prepared by reducing silicon dioxide (sand) with coke in an electric furnace. This material is further processed by reacting the silicon with hydrogen chloride gas to form trichlorosilane, which is decomposed in a chemical vapor deposition reactor to produce polycrystalline electronic-grade silicon. In order to eliminate any remaining impurities, the polycrystalline silicon material is cleaned with hydrofluoric and nitric acids and grown into crystals of ultrapure monocrystalline silicon, which are then sliced into thin wafers and after processing are cut into "chips." The waste products of the hydrochlorination step include silane, chlorosilane, and chlorides of phosphorus, boron, arsenic, and antimony. The cleaning and crystal growth steps generate waste acids and defective silicon.[12]

The microcircuitry of the wafer is generated by the selective diffusion of tiny amounts of impurities into regions of the silicon substrate that are then linked by metallic conductor pathways. The conductivity pattern is generated on the wafer surface by photolithographic and etching processes in which the desired pattern is first laid out on a mask laid over a photoresist layer (usually a polymeric film spin coated onto the wafer).

Table 10.2
Materials commonly used to manufacture semiconductors

Crystal Preparation

Silicon dioxide	Gallium

Dopants

Arsenic	Phosphorus
Boron	Antimony
Gold	Beryllium
Germanium	Magnesium
Tin	Tellurium

Etching

Sulfuric, phosphoric, nitric, hydrofluoric and hydrochloric acids	
Hydrogen bromide	Trifluoromethane
Carbon tetrafluoride	Fluorine
Boron trichloride	Helium
Hydrogen	Argon
Acetone	Xylene
Isopropyl alcohol	

Wafer Fabrication

Oxygen	Nitrogen
Hydrogen chloride	Trichloroethane
Hydrofluoric acid	Trichloroethylene

Photolithography

Photoresists

ortho-Diazoketone	Polyalkylaldehyde
Polymethyl methacrylate	Isoprene
Polycyanoethylacrylate	Polymethacrylate
Glycidylmethacrylate	Ethyl acrylate

Developers

Sodium hydroxide	Silicates
Potassium hydroxide	Ethanolamine
Isopropyl alcohol	Phosphates
Alkyl amine	Ethyl acetate
Methyl isobutyl ketone	Xylene
n-butyl acetate	Stoddard solvents

Cleaning Agents

Deionized water	Acetone
Isopropyl alcohol	Ethanol
Hydrofluoric acid	Sulfuric acid
Hydrogen peroxide	Nitric acid
Ammonium hydroxide	Chromic acid
Ethylbenzene	Xylene
Chlorotoluene	

Source: U.S. Environmental Protection Agency, Office of Compliance, *Profile of the Electronics and Computer Industry*, Washington, D.C., 1995.

The wafer is exposed to some form of radiation (ultraviolet light, X-rays, electrons, or ions) to transfer the pattern to the substrate. The substrate is then "developed" by solvent washing or plasma etching and the exposed areas are etched to produce the pattern in the wafer surface into which the impurities or dopants can be diffused. Once the wafer has been patterned, the surface is coated with thin layers of metal that connect the internal circuitry to external connections and then the entire wafer is passivated (made resistant to corrosion) with a thin layer of silicon dioxide.

There are a host of materials that exhibit semiconductor properties. They range from simple elements such as boron, tin, selenium, and tellurium to more complex compounds of silicon carbide, aluminum nitride, and gallium phosphide. An increasing percentage of semiconductors is based on wafers made from gallium arsenide. Gallium is produced as a by-product of aluminum and zinc refining, and the arsenic is generated from the flue dusts at copper and lead smelters.

Since the 1970s there has been enormous progress in producing inorganic crystalline conductors. Some of this has involved the search for superconductors, or materials that conduct electricity without resistance. A major breakthrough came with the discovery of a ceramic compound that achieved complete conductance at a temperature far above the superconductive temperature point near absolute zero ($-459°F$) that had been identified three-quarters of a century earlier. This has been followed by a wave of new research on superconductors that involves rare earth elements such as strontium, lanthanum, and yttrium, but so far these efforts have not found large commercial applications because of processing difficulties.

Instead, a wide range of research has been conducted on the optical transmission properties of glass. This has led to an utrapure form of silica that can be pulled into exceedingly thin fibers. By wrapping this glass fiber in a molecular layer of reflective metal, an optical fiber can be produced that can transmit light thousands of miles without distortion. Rather rapid commercial applications for fiber optics have emerged, including the large-scale data transfer from conventional copper cables to laser-activated fiber optic cables that can transfer billions of bits of information per second.

However, the real revolutions in conductivity may emerge from organic conductors. By the close of the 1970s, the first organic superconductors were identified and in 1977 the first conducting polymer was documented. As the conductive potential of organic films and polymers has emerged, researchers have become interested in the nonlinear optical properties of materials. Today research scientists are working with organic superconductors that can operate at very high temperatures, conducting polymers that are soluble and easily processed into fibers or thin films, and nonlinear optical polymers that can be used for high-speed electrooptical modulation.

10.4   Smart Materials

So-called "smart materials" or "functional materials" are materials that can change their properties in response to environmental conditions, such as changes in pressure, temperature, magnetic field, optical wavelength, electrical field, or pH. They are designed with the capacity to recognize, discriminate, and adjust in response to environmental changes. For instance, a smart valve can shut off the flow of a fluid through a tube if the flow exceeds a certain rate because the valve is made of a material that is capable of swelling and pinching off the flow when a critical fluid pressure is reached in the tube.

The history of smart materials is often traced to the work of S. Donald Stookey at Corning Glass. Stookey set out in the early 1960s to develop a glass that would darken when exposed to light and lighten when the light was withdrawn. Stookey's solution was partly derived from the light-triggered chemistry of photography. By dispersing silver and copper halides into the glass formula, Stookey created a photochromic glass. As light energy hits the glass, an electron from the copper ion is released and picked up by the silver ions, producing neutral silver atoms that block light just as they do in photographic exposures. Once the light is removed, the electrons are liberated from the silver ions and return to the copper ion, which does not produce light-blocking atoms. This simple glass formula became the basis for light-sensitive window glass and eyeglasses with self-adjusting variations in transparency.[13]

One branch of research involves "shape-memory" or "superelastic" materials that have the capacity to bend, stretch, compress, and deform, but always return to their original shape. Titanium–nickel and copper–aluminum–nickel alloys have been developed as well as various reinforced composites that have applications in protective coatings, monitoring fatigue, damping of sound and vibrations, and various medical prostheses. Work on so-called "intermetallics" has focused mostly on aluminides such as titanium, nickel, and iron aluminides that have applications as temperature-sensitive control and metering devices and as "shape-memory alloys." The ability of these materials to recover their form is a result of a balance between the external stress and the internal restoring force arising from the free energy difference that exists between the material in its deformed and normal positions. Many thermostats today use a bimetallic strip composed of two strips of two different metals connected together at their ends. Because the two metals have different thermal expansion properties the strip tends to bend as the temperature changes, thereby opening and closing electrical contacts. Such thermostats can also be made from single strips of shape-memory alloys that bend in response to temperature changes, and, today, some clutch fans in car engines are turned on and off by such alloys.[14]

There is a whole new specialty called "smart polymers" or "field-responsive polymers."[15] These polymers display photooptical, pressure, and temperature sensitivities. They are quite versatile, flexible, and reversible, and can be readily processed so they can be easily incorporated into structural matrices. Applications appear numerous in optical data storage, image processing, motion sensors, light-sensitive widow coatings, and radiation-responsive microwave shielding. "Smart structures" appear as beams or columns that are able to sense changes in point pressures and respond by realigning the polarity of the molecular structure so as to increase resistance at the point of pressure. Particularly interesting applications involve the development of polymeric "smart skins" that can integrate antennas, sensors, and transmitters into the surface of vehicles or buildings, and the creation of synthetic muscles that can convert chemical energy into mechanical responses.

Smart materials are increasingly of interest to the government. During the 1970s, the National Aeronautics and Space Administration started research on materials that could sense and display potential faults, cracks, fatigue, or excessive strains in structural materials. The idea was to create airplanes with smart skins that could sense aeronautical conditions, such as air speed, temperature, and pressure, and detect early indications of potential failure. Composite materials were developed that contained optical fibers through which changes in the passage of light could be detected. This early effort to develop skins with embedded sensors soon expanded to produce research into structural materials that could sense and respond to changes in light, temperature, or pressure.[16]

10.5   Nanostructured Materials

As innovative as composites, semiconductors, and smart materials may be, they are all produced through processes that reorder and recombine compounds or molecules. The increasing sophistication of analytical detection equipment such as the atomic force and scanning tunneling microscopes that allow us to "see" the molecular and atomic structure of materials has raised the prospect that we may soon be able to design and manufacture technologies by manipulating the molecular structure of materials. This is not microscale fabrication that involves processing by scaling down the parameters of production, but nanoscale assembly that relies on familiar chemical and mechanical principles to assemble atoms and molecules into super-tiny molecular machines. The nano (or nanometer) is one billionth of a meter and represents the scale at which atoms are the basic building blocks. At the nanoscale, the conventional laws of physics must be reinterpreted. At this scale, the valence electrons can be displaced within the atom, which can lead to different physical and chemical properties, depending on size. By combining several elements with particles of different sizes, a whole new range of magnetic, optical, and reactive properties may be developed.[17]

For some time nanoscale applications have seemed to be mere curiosities. Still, there is a growing commercial interest in nanoscale applications in the technology of ink-jet printers, in the microsensors that detect and react to small vibrations in machines or sudden impacts that trigger air bags in cars, in noninvasive blood analyzers, and in micropumps that permit measured doses of pharmacueticals to reach cellular targets. Recent developments have demonstrated that careful selection of monomers, catalysts, and nanoreactors can be employed to dramatically improve the properties of even such conventional materials as polyethylene.[18]

Polymeric nanostructures offer potentials in chemical processing and electronics. Superfine nanoscale membranes could be used to separate chemicals in fluids. The demand for ever-smaller electronic devices is driving research on the development of superthin polymer films with nanoscale channels that could be used as semiconductor matrices. Alternatively, copolymer nanofibers could be made with conductive polymer cores "wrapped" in insulating plastic layers.[19]

Beyond advantages of scale, nanotechnology offers opportunities to process materials at the atomic level and to assemble nanoscale machines that can manufacture products atom by atom according to "blueprints" programmed by electronic software. The idea of constructing machines at the molecular level has been around since the 1940s, but it was given renewed stimulus during the 1960s by physicists such as Richard Feynman. Indeed, Eric Drexler at the Massachusetts Institute of Technology proposed building small machines to build smaller machines that could build even smaller machines. As biologists have begun to learn more about how proteins are assembled and physicists have developed instruments that can move molecules with atomic precision, the idea of molecular assembly machines has grown more reasonable.[20]

Technologies that can be measured by nanometers blur the distinction between material and machine and present the possibility of materials designed to perform functions and carry out tasks—even to repair and replicate themselves. Minute replicating assemblers could radically alter mining and manufacturing, making it possible to make almost anything, any time, out of nothing more than dirt, air, and sunshine. In this view, the machinery of production will be increasingly less important than the software necessary to manufacture it.[21]

10.6   Environmental and Human Health Considerations of New Materials

Some new and advanced materials offer real environmental benefits. They provide substitutes for dwindling natural materials and more energy- and resource-saving products. For instance, the lightweight metal alloys and high-performance polymers used in aircraft and automobiles can reduce energy consumption. The same is true for many of the composites that maintain or increase strength while lowering the weight of structural or surface units. The use of ceramic parts in the heat engines of transportation vehicles can increase their thermal efficiency and reduce the need for oil lubricants. The processing of some advanced materials is less energy intensive than the processing of conventional materials. For instance, the energy required to produce a given volume of advanced polymers, both in terms of the energy embedded in the feedstock and the energy required for processing, is on average roughly one-third of the energy consumed by an equal volume of refined structural metals.[22]

Many of the new electronic materials increase energy savings in products. New fiber-optical and coaxial networks dramatically increase bandwidth in electronic communication. Where new electronic displays can store information without power and reduce the number of times an image must be refreshed per second, energy consumption is reduced. Elevators with new regenerative drives recover energy, reduce waiting time, and decrease heat losses that otherwise increase air conditioning loads in buildings. The "enhanced vehicles" planned by the Big Three automakers include many electronic features that increase fuel efficiency and permit communication with other vehicles or databases to modulate traffic congestion and improve trip planning.[23]

While many of these advanced materials are sophisticated developments, they are "new" largely because they are novel refinements or combinations of existing materials. In most cases the basic source materials, and even many of the intermediaries, are conventional materials. The metal alloys, composite materials, and semiconductors employ innovative processes with highly developed theories and sophisticated techniques, but the starting materials are traditional metals, polymers, and ceramics. The nanostructured and field-responsive materials provide opportunities for creating totally new materials, but at these early stages most of the source material is fairly conventional. Because they rely so substantially on conventional materials as raw materials, these advanced materials are often plagued by the same environmental and health problems that characterize conventional materials.

Although the hazards of many of the source materials and processing chemicals may be known, there is no long-term experience with these "new" materials to rely on in determining their hazards or effects on the environment. Most new materials, even those that appeared after the EPA established its New Chemicals Program, which requires the filing of a premarket notification, have little test data on toxicological or environmental impacts. It is commonly assumed that the small production volumes of advanced materials means that their environmental effects are modest. In the aggregate this may be true. Yet, even small amounts of new materials can have impacts on the immediate environment and exposed workers, and for those new materials that find rapid adoption, the small volumes may soon be replaced by much more substantial quantities. Polymers provide a telling example. While little environmental consideration was given to early plastics, the rapid rise in the use of commodity plastics, particularly in disposable packaging and containers, outpaced adequate recycling and disposal systems and led to widely dissipated wastes with substantial environmental impacts.

The opportunity for recycling and reuse of the advanced metal alloys has not been well considered, although the necessity for high purity in feedstocks suggests that much of the scrap must be used in downgraded applications. High-performance polymers, such as commodity plastics, cannot be recycled without being downgraded. Recent developments in plastic packaging rely on plastic films made of laminates of several different resins. Some of the new plastics are composites containing metallic or glass fibers. These new customized plastics are all but unrecyclable. Likewise, the inherent structure of composites and semiconductors that intermingle different materials makes them quite difficult to decompose into the starting constituents and therefore nearly impossible to recycle.

Many of the steps in semiconductor manufacture use large amounts of water and energy. Semiconductor production involves substances known for acute and chronic health hazards. Some, such as arsenic, benzene, and chromium, are well-known carcinogens, while others, such as hydrogen fluoride, lead, xylene, and n-butyl acetate are recognized as potential hazards to reproduction. Wastes are generated at each production step. Crystal preparation generates spent solvents, acids, alkaline cleaning solutions, and deionized water. Wafer fabrication produces waste solvents, acids, enchants, and developing fluids. A host of metals are released as wastes, particularly from crystal preparation and the final metalization steps. Acid fumes, volatile organic compounds, and dopant gases are released as air emissions from various operations.

While silicon is a nontoxic material, gallium arsenide is not. Over the past decade there has been growing concern about the occupational hazards associated with these substances and the hazards associated with the emissions from semiconductor production facilities. The industry has responded with increasing controls on emissions, improved exposure controls, and the elimination of some substances. The industry moved rapidly to comply with the international phaseout of chlorofluorocarbons, which were used extensively for cleaning and degreasing. Most semiconductor fabricators virtually eliminated ethylene glycol ethers because of concerns raised by workers over possible effects of these substances on their reproductive health, but there remain many hazardous chemicals used in the industry that are worthy of further consideration.[24]

Only in the past decade has there been enough waste from discarded electronic products to have created a recognizable problem of solid and hazardous waste. It is estimated that nearly 9.8 million electronic products were discarded in 1998 and that this included 2.3 million personal computers, 300,000 notebook computers, 54,000 mainframe computers, and 19,000 televisions. With millions more personal computers and cellular telephones in use today, waste problems from electronic components will only increase. The computer manufacturing industry has responded with a voluntary program for recycling returned personal computers, but a more serious effort may soon be required.[25]

The production of many advanced materials involves fine-scale processes with microscale feedstocks. The fibers and whiskers (as small as 5 μm in diameter) used in composite production can become airborne during manufacture or disposal (particularly through shredding or incineration), providing an opportunity for inhalation by workers or the public. Carbon and graphite fibers used as fillers in composites can irritate skin and aramid fibers (used to make Kevlar) can cause fibrosis in the lungs. Likewise, the fine powders of advanced ceramics provide opportunities for environmental or health problems resulting from air releases at manufacturing facilities, from products during use, or from discarded products during waste treatment and disposal.[26] Indeed, the minuscule scale and self-assembly potentials of nanotechnology carries these concerns to an ultimate level.

The production of composite materials typically involves compression molding, hot resin transfer molding, or spray-on forming, all of which require significant energy inputs. The resins and solvents used in composites present various occupational hazards. Epoxy resins are a common matrix in composite materials. Most are made from epichlorohydrin reactions and some epichlorohydrin, a probable human carcinogen, is left in all epoxies. Epoxies require a curing or hardening agent and bisphenol-A is the most common hardening agent. Polyurethane and phenolic resins are also used as composite matrices. The isocyanates (polyurethanes) are of reasonable concern, particularly toluene diisocyanate, which is listed as a possible human carcinogen. While the ketone solvents (e.g., acetone, methyl ethyl ketone) can cause central nervous system depression, it is some of the chlorinated solvents (methylene chloride, trichloroethane) that are considered more hazardous because of their carcinogenic potential.[27]

Some new materials involve highly reactive chemicals that raise concern about explosions. For instance, the semiconductor industry continues to process wafers with acutely hazardous gases such as arsine, phosphine, and diborane, which are quite flammable. The hydrofluoric acid used in the industry can cause severe burns. The titanium and lithium used in the aluminum alloys are quite reactive with water vapor. While the processing of these alloys can be accomplished in tightly controlled atmospheres, concern arises about later stages of the life cycle where secondary processors may remelt metal scrap containing small amounts of titanium or lithium.

In its otherwise enthusiastic review of new materials, the U.S. Bureau of Mines warns that "[w]idespread use of new materials may cause environmental problems unless health hazards associated with their production and disposal are addressed."[28] To date, these issues have been little addressed. The trade journals on alloys, ceramics, and electronic materials occasionally run articles on occupational safety issues, but these articles lack screening or test data and are largely confined to recommendations on industrial hygiene and exposure controls. There is increasing attention to advanced materials in the scientific journals on occupational health, but almost nothing in the environmental health journals.

Of more concern is the absence in the material sciences of formal attention to the health hazards or environmental effects of new materials. The materials science journals abound with papers on new substances and new production processes, but the number of articles on health or environmental effects is so small as to be almost impossible to find by consulting abstracts or performing keyword searches. There is little professional dialogue about the need for health or environmental impact analysis as a factor in developing or commercializing new materials. A brief look at curricula at leading schools of metallurgy, ceramics, plastics engineering, or polymer sciences reveals little indication of coursework on environmental or public health issues.[29]

As exciting as advanced materials may be technically, they reveal flaws in the way design and development processes are currently focused. While the three primary design factors for new materials—performance, processing efficiency, and cost—remain paramount, problems associated with the toxicity or dissipative effects have been partitioned off from the focus of the field and relegated to other disciplines. If new materials should be designed to meet all of these values, then it appears that the materials sciences rest too quickly; only half of the objectives for sustainability are being met.

The academic materials sciences have little interaction with those who study ecology, biology, or physiology, or with toxicologists or pharmacologists, for that matter. Knowledge about the environmental or health consequences of materials is deemed the responsibility of other disciplines. In industry, the arrangement locked in place by professional precedent, corporate procedures, and legal regulations that separates responsibility for environmental and health concerns from interest in material design and process engineering appears as rigid today as it did decades ago. In reviewing the institutional organization by which new materials are being developed today and will be developed in the future, there is no evidence that any new, more protective process or perspective is emerging. Left to follow current trends, the materials sciences will provide plenty of exciting new materials in the future and just as many causes for concern.

There is a need here for a change in the direction of the field and a new dialogue—one that engages materials scientists and their sponsors with those who study and respond to the health and environmental effects of development decisions and those who use and consume the products made from new materials. The materials sciences have proven to be full of creative and innovative talent, but the range of objectives has been too narrow. A sustainable materials base will require that this same creativity and innovation be directed toward the development of materials that are less toxic in their production, use, and disposal and that are less likely to dissipate the natural resources of the planet or pollute its ecological systems. The materials sciences are fundamental to the development of new materials and new production processes. Now, while so many new materials are under development, is an opportune time to ensure that these materials are also contributing to a truly sustainable materials economy.

11

Renewable Materials

Everything that exists is the seed of that which shall come out of it.
—Marcus Aurelius

Wood is renewable. So is cotton, wool, flax, silk, and the oils of corn, soybeans, and rapeseed. The most common examples of renewable materials are those that are easily derived from plant- or animal-based sources. Most are organic materials, although it is also useful to consider replenishable materials such as water, oxygen, hydrogen, and nitrogen as functionally renewable.

While it is estimated that the manufacture of industrial and construction materials in the United States consumes about 175 million tons of petroleum and 300 million tons of inorganic metals each year, only 10 million tons of plant matter other than wood go into such products. Plant-based material, or biomass, provides a plentiful and renewable alternative source of materials for many industrial uses, but it is little used for such purposes.[1]

Renewable resources could play much more of a role within the nested loops of the materials flow model. Because most plant-based materials are naturally degradable by ecological processes, they are likely to be more easily cycled between economic and natural systems. Where the rate of extraction complements the rate of renewal, economic activity could be maintained without drawing down stocks of natural capital. Although it might appear that the substitution of plant-based materials for metal- and petroleum-based materials could reduce overall toxicity, this cannot be assumed. Some plant material is toxic, and the current conditions of industrial agriculture require large infusions of synthetic chemical and energy inputs.

Throughout history most products were derived from renewable materials. Products made of wood, leather, wool, cotton, hemp, and straw have all been used for industrial production. Recent pressure from consumers and government regulators has spurred renewed interest and innovation in agriculturally derived materials. Indeed, the U.S. Department of Energy has set a goal of replacing one-quarter of organic industrial feedstocks with renewable agricultural sources by the year 2030.

11.1 Promoting Renewable Materials

The history of efforts to produce industrial materials from renewable resources is rich in experiment and debate. Much of the impetus for renewable industrial materials has come from those who have investments in agriculture. Farmers and government officials who represent farm-dependent districts have promoted agriculturally based fuels, structural materials, solvents, coatings, and plastics.

Early credit for promoting agricultural crops as a source of industrial materials goes to George Washington Carver, who developed more than three hundred products based on the peanut. But the idea really took off during the 1930s, when a vigorous movement emerged to promote agricultural feedstocks for industrial use. Developed under the term *chemurgy*, which was derived from the Greek words for chemicals and work, the concept was promoted by William J. Hale, an organic chemist who was chair of a prestigious National Research Council division on chemical technology. In 1934 he published a book called *The Farm Chemurgic* in which he argued for a closer link between the farm and the factory. Hale had two objectives in mind. First, he wanted to ensure a solidly domestic organic chemical industry that was not dependent on foreign imports. Second, he wanted to support American farmers, who were suffering from overproduction and inadequate markets. He believed that developing industrial materials from agricultural crops was the best way to meet both objectives.[2]

In promoting his ideas, Hale joined forces with Wheeler McMillen, a national farming activist, Thomas Edison, Irenee du Pont, and Henry Ford. Edison was interested in agriculture because he wanted to develop a domestic rubber supply, which he thought could be based on the wildflower, goldenrod. Ford was a particularly compelling advocate. He was convinced that he could develop industrial organic chemicals from the oil of soybeans, and in 1938 he set up the first of several soy processing plants to make soy oil-based enamels for automobile body coatings and the glycerin used in shock absorbers. In 1933 he planted more than 300 varieties of soybean on 8000 acres of his farms, and 2 years later he could boast that a bushel of soybeans went into the paint, horn button, gearshift knob, door handles, accelerator pedal, and timing gears of every Ford car. Over the next several years Ford supported research into crop-based polymers that he hoped could be used for constructing the auto body, and in 1941 a "farm-grown automobile" was displayed at the Ford factory in Dearborn, Michigan.[3] Ford's commitment was not simply a technological fix, for he saw "agrindustry" as an economic solution as well.

I foresee the time when industry shall no longer denude the forests which require generations to mature, nor use up the mines which were ages in the making, but shall draw its material from the annual produce of the fields.... I believe that industry and agriculture are natural partners. Agriculture suffers from lack of a market for its product. Industry suffers from lack of employment for its surplus men. Bringing them together heals the ailments of both. I see the time coming when the farmer not only will raise raw materials for industry, but will do the initial processing on the farm.[4]

Hale recruited the support of the Synthetic Organic Chemical Manufacturers Association, the Chemical Foundation of New York, and the Association of Official Agricultural Chemists. At a 1935 meeting in Michigan sponsored by Henry Ford, these advocates formed the National Farm Chemurgic Council and established pilot projects to demonstrate the value of various farm-based products, such as newsprint from southern slash pine and "agrol," a blend of ethanol distilled from grain substances that could be used as automobile fuel.[5]

During the early 1940s the chemurgic movement grew rapidly as the nation's Great Depression bankrupted large numbers of small farms. Of even greater importance was World War II, which created a huge demand for domestic substitutes for foreign materials. Chemurgy achieved a certain level of respectability during the war for its promotion of farm products such as cottonseed oil, which replaced imported palm oil for use in the steel industry. But the movement's continued efforts to pro-

Table 11.1
Industrial materials derived from plant matter

Product	1996 production (million tons/year)	Percent of total market, 1992	Percent of total market, 1996
Wall paints	7.8	3.5	9.0
Specialty paints	2.4	2.0	4.5
Pigments	15.0	6.0	9.0
Dyes	4.5	6.0	15.0
Inks	3.5	7.0	16.0
Detergents	12.6	11.0	18.0
Surfactants	3.5	35.0	50.0
Adhesives	5.0	40.0	48.0
Plastics	30.0	1.8	4.3
Plasticizers	0.8	15.0	32.0
Acetic acid	2.3	17.5	28.0
Furfural	0.3	17.0	21.0
Fatty acids	2.5	40.0	55.0
Carbon black	1.5	12.0	19.0

Source: Anton Moser, "Ecological Process Engineering: The Potential of Bioprocessing," in Robert U. Ayres and Paul M. Weaver, *Eco-restructuring: Implications for Sustainable Development*, New York: United Nations University Press, 1998, pp. 77–108.

mote the alternative fuel, agrol, may have hastened its collapse because this advocacy raised the ire of the petroleum industry. The oil companies were tough adversaries and they campaigned hard to criticize agrol as a fuel and discredit the chemurgy movement as a force. Eventually the surge of interest in petrochemicals and the astounding success of the early petroleum-based polymers eclipsed and stalled the farm-chemistry momentum. By the 1950s, few organic chemists were pursuing farm-based industrial materials and only the rare agricultural research station remained committed to the chemurgy vision.

11.2   Agriculture-Based Industrial Materials

There is a long history to the use of industrial materials derived from agricultural sources. Today there are products in many conventional markets.[6] Table 11.1 lists several of these products and their market share. Vegetable oils and plant resins can be used in the production of lubricants, paints, detergents, solvents, and plastics. Some oils, waxes, and fats are derived from crambe, rapeseed, castorbeans, the tungtree, corn, sunflowers, palms, coconuts, and soybeans. Terpenes and natural alcohols can be derived from pine tree resins and citrus residues. Wood, cotton, kenaf, and hemp can be used as a source of fibers. These agriculturally based materials can be divided into five categories: starches and sugars, fats and oils, resins and gums, alcohols and solvents, and natural fibers.

Starches and Sugars

Starch is a polymer made from a hydroglucose (long chains of glucose linked by glycosidic bonds). It is insoluble in water, so it acts as the principal carbohydrate storage product of higher plants. The glucose (sugar) bound together in starch can be hundreds of molecular units long and composed of simple linear chains or branched chains. Starch can be commercially derived from corn, wheat, sorghum, and potatoes. While there is a large market for processed starch in the food industry, over 60 percent of commercially produced starch is used as a sizing agent in the production of paper and paperboard products. Dextrin, a water-soluble derivative of starch, is used to produce adhesives and glues.[7]

Starch made from corn (*Zea mays*) is the least expensive source of plant-based starch and it dominates the natural starch market. Over 186 million bushels of corn were used in the harvest year 1996–97 in the production of industrial starch. The largest share of nonfood, nonfeed consumption of cornstarch is in the production of ethanol, with the remainder used in the paper and paperboard industry (in adhesives and sizing), in the textile industry (as warp sizers), and as thickeners and stabilizers. Cornstarch is the leading source of starch for natural adhesives. In 1990, U.S. adhesive consumption was about 5 million short tons, with an estimated value over $2 billion annually, and of this, natural adhesives accounted for about 40 percent of the market.[8]

Sugarcane (*Saccharum officinanum*) and sugar beet (*Beta vulgaris*) are the largest sources of sucrose (sugar), which is used extensively in the food processing industry, but also finds industrial applications as a source of ethanol. Sucrose esters make effective surfactants and emulsifiers;

sucrose derivatives are used in the production of polyurethane foams; and xanthan gum, a polymer of sucrose, is used as a lubricant in oil well drilling muds.[9]

Fats and Oils

Fats and oils (lipids) are water-insoluble compounds found in the cells of plants and animals that serve as structural elements and metabolic fuel. Lipids are composed of triglycerides that can be decomposed into fatty acids and glycerol. The glycerols and longer chain fatty acids can be used in making soaps and detergents. The shorter chain fatty acids can be used in making plastics. In 1992, 5.9 billion pounds of fats and oils were used in nonedible products (soaps, resins, plastics, paints, varnishes, lubricants, etc.) This accounts for roughly 30 percent of total consumption.[10] Soaps are a traditional fat-based product typically made of tallow (beef fat) and coconut oil, with palm oil recently replacing coconut oil for reasons of lower costs. The glycerin, or glycerol, that is a by-product of soap-making has other uses in industrial esters.

The soybean (*Glycine max*) is the largest source of plant-based industrial oil. The nation consumed 382 million pounds of industrial soybean oil in 1998. Of this, 31 percent went into resins and plastics, 17 percent went into inks, and 13 percent was consumed in paints and varnishes. Much of the soybean oil used in polymer resins is used to make plasticizers for various thermoplastics, including polyvinyl chloride.[11]

Soy-based printing inks are now fairly conventional. These inks were developed by the American Newspaper Publishers Association in response to the oil shocks of the 1970s. Once colored soy inks were commercialized in the 1980s, they were quickly adopted. Starting with only six newspapers in 1987, use of colored soy ink grew to cover over half of the nation's 9100 newspapers in 1994. This included 75 percent of the 1700 U.S. dailies. Colored soy inks rapidly replaced petroleum-based mineral oil inks because of superior performance—brighter colors and more efficiency in coverage—and today soy-based inks make up over 90 percent of the color ink market. Soy inks are also easier on presses, clean up easier, are less hazardous to workers, and reduce the potential for volatile air emissions. The price of colored inks is primarily determined by the cost of the pigments whereas black ink costs are more dependent on the price of the oil carrying the pigments; therefore the slightly higher priced soy oil has been less competitive with black inks.[12]

Oil from the castor plant (*Ricinus communis*) is quite a versatile material. The castor-oil plant is a non-native, broad-leafed plant capable of irrigation-supported cultivation in arid conditions. Castor oil is predominantly composed of ricinoleic acid (90 percent) and is one of the few naturally occurring, nearly pure glycerides. It is currently used in cosmetics and nonbacterial soaps. In industry it is used as a plasticizer in plastics and as a lubricant in two-cycle engine oils and metal drawing oils. Derivatives of castor oil are present in primers, binders, inks, adhesives, caulks, and sealants. Commercial production of castor-oil plants for lubricants and medicinal purposes began as early as 1850. By the 1930s there was substantial cultivation of these plants in the San Joaquin Valley in California, and during both World War II and the Korean War, domestic production was encouraged by the military because castor oil provided key ingredients in hydraulic fluids, greases, and lubricants. Higher value crops gradually replaced castor-oil plants and by the 1970s all domestic production ceased. Today the country imports 100 percent of its castor oil.[13]

Resins and Gums

Resins and gums are organic compounds derived from plant secretions. They are typically soluble in solvents and insoluble in water. Gums are polysaccharides composed of salts or sugars other than glucose, while resins are a heterogeneous group of compounds, all of which are polymerized terpenes. The most common resin is the natural rubber tapped from *Hevea brasiliensis*. The advent of synthetic rubber might have eclipsed the demand for natural rubber, but today natural rubber still makes up roughly 30 percent of the rubber consumed in the United States. The largest share of this goes into automobile tires. The high degree of strength, toughness, and durability required by radial tires requires the mixing of substantial amounts of natural rubber into the butadiene polymers.

The secretions of pine trees (principally, *Pinus sylvestris*, *P. palustris*, and *P. elliottii*) are used to make various resins, including rosin, a resin mixed with oils from chemical wood pulping. Rosin, which is used to increase the stickiness on baseballs and violin strings, has several industrial uses in the preparation of inks, rubber, varnishes, sealants, and paper coatings.

Many gums currently in use (e.g., xanthan, dextran, polytran, gullan, and pullulan) are derived from seaweeds and kelp. These substances are used primarily as thickeners, flocculants, and lubricants. The paper and textile industries use a large amount of natural gums as sizing agents, and natural gums are mixed with glycerin to produce the dried adhesive that is applied to the flaps of envelopes.

Alcohols and Solvents

Government air and waste management regulations on the use and disposal of halogenated solvents have promoted research into biologically derived substitutes. A very low volatile and easily degradable solvent made of dimethyl sulfoxide is currently being produced from wood lignin. The solvent competes well against polymer cleaners and solvents and specialty paint strippers such as methylene chloride, dimethyl formamide, and sulfone, although its cost is inhibiting its penetration of other markets. A nonlinear alcohol derived from rice hulls and corncobs has found a successful market in the electronics assembly industry and there is a new soy-based industrial cleaning solvent on the market as well. Purac and Cargill, two large agricultural product companies, have set up a joint venture to build a plant in Nebraska to produce lactic acid esters as a substitute for chlorinated solvents in cleaning and degreasing electronic parts. These high-quality, low-cost lactic acid esters are derived from fermentation of corn sugar and are nontoxic, biodegradable, and easily recoverable through distillation.[14]

The drive to replace chlorofluorocarbons during the early 1990s created a market for alternative industrial degreasing and cleaning solvents based on D-limonone, a terpene derivative of plant resins that sells well as a replacement for methylene chloride and acetone in paint stripping, degreasing, and resin removal. Several firms in Florida are making terpenes from the wastes of citrus juice manufacturers, and firms in New York and Washington are making terpenes from pine bark.[15]

Nearly 64 percent of the corn harvest that is used for industrial purposes goes into the production of ethanol, which is used as an additive in gasoline, where it competes with a synthetic additive, methyl *tert*-butyl ether (MTBE) as an octane enhancer and fuel extender. In the harvest year 1996–97, 435 million bushels of corn were used to make ethanol for fuel additives. Since the EPA has now moved to phase out the use of MTBE because of its negative environmental impacts, the market for corn-based ethanol is expected to rise.[16]

Natural Fibers

Natural fibers such as wool, flax, and silk have been used for thousands of years for textiles and clothing. During the past century synthetic fibers derived from petroleum replaced many natural fibers in industrial uses and found many new uses in composites and plastics. Natural fibers that may share in some of these industrial applications include cotton, flax, jute, hesperaloe, hemp, kenaf, and animal wool.

Cotton (*Gossypium*) is by far the most important natural source of fiber. Cotton is a 95 percent cellulose material that has a long history in textiles, clothing, bedding, and surgical bandages. It is valued for its flexibility, dyability, and ease of spinning and weaving. The harvesting and processing of cotton is usually less expensive than any other natural fiber. Cotton has found industrial applications in tire cords, tentage, sandbags, and press covers. It was cotton that was reduced to a soluble cellulose form and extruded to form the first synthetic fiber, rayon. A niche market for organically grown and harvested cotton has emerged in bedding, clothing, and children's products. Colored cottons have been around for some time, but were difficult to spin and weave because the fibers were short. A long-fiber cotton is now being grown in several earth tones on organic farms throughout the Southwest. The natural colors eliminate the need for dyes, formaldehyde, and harsh processing chemicals, and its fire-resistant qualities reduce the need for flame-retardant additives.[17]

Several natural fibers are composed of soft "bast" fibers that typically require retting (rotting) and scotching [beating] during processing. Flax (*Linum usitatissimum*) is the oldest bast fiber used by humans. The fibers are smooth and straight and two to three times as strong as cotton. Flax is used extensively for the manufacture of linen, but is also used in the production of hoses and mailbags. Jute (*Corchorus casularis*) is the second largest natural fiber grown worldwide. The fibers are short, inelastic, and tend to disintegrate in water. The brittleness of jute means that it finds use in rather coarse goods such as carpet backing, twine, sacks, wall coverings, and linoleum floor coverings. Hemp (*Cannabis sativa*) provides an oil rich in fatty acids and protein, and a long and rough fiber that can be used for cordage, canvas, and sailcloth. Although hemp production has been prohibited since 1958 because of its similarity to marijuana, some states now permit experimental farming, and Hawaii has encouraged a test program for developing hemp for industrial applications.[18]

Kenaf (*Hibiscus cannabis*) is a fibrous plant that can grow to 14 feet in less than 7 months and produces two to three times more fiber per acre than traditional southern pine. The plant stem consists of an outer bark of bast fiber (30–40 percent of the stem) and an inner core of shorter fibers. Currently four U.S. firms operate fiber separation facilities that generate bast fiber for use in packing materials, burlap, and non-woven seeding mats and core fibers sold for soil-less potting mixtures, oil-absorbent products, and animal litter and bedding. Recent large-scale pilot plants have demonstrated that kenaf produces high-quality newsprint and can be mixed with recycled wood pulp-derived newsprint to make recycled bond and coated papers. Bast fiber can produce high-strength, low-permeability papers, packaging, and wrappings, requiring fewer chemicals and about two-thirds the energy of comparable wood paper products. About 8000 acres of kenaf are now under cultivation in the southern and western parts of the country.[19]

11.3   New Agriculturally Based Industrial Materials

Research on domestic and foreign plants suggests the possibility of new agriculturally based industrial materials. Rapeseed, an oilseed that is used to make slip agents for plastic molds, is cultivated in many countries and is being explored more assertively in the United States. *Vernonia* has the potential to replace solvents used in paints. A new crop, quayule, could be a replacement for both natural and synthetic rubber. Table 11.2 lists several plants that could find industrial uses in the United States.

Table 11.2
Plants with potential applications in industrial materials

Crop	Compound of interest	Potential uses	Replacement
Oilseed			
Buffalo gourd	Oleic acid	Ethanol	Petro oils, soybean oil
Chinese tallow	Tallow	Paints, varnishes	Imported cocoa butter
Crambe	Erucic acid	Plastics, suppressants, lubricants	Imported rapeseed oil
Cuphea	Lactic acid, capric acid	Soaps, lubricants	Coconut oil, palm oil
Honesty	Erucic acid	Plastics, suppressants, lubricants	Imported rapeseed oil
Jojoba	Wax esters	Cosmetics, plastics, household wax	Banned sperm whale oil
Lesquerella	Hydroxy fatty acid	Plastics	Castor oil
Meadowfoam	Long-chain fatty acid	Cosmetics, lubricants	Petroleum derivatives
Rapeseed	Erucic acid	Plastics, suppressants, lubricants	Imported rapeseed oil
Stokes aster	Epoxy fatty acid	Plastics, coatings	Petro oils, soybean oil
Vernonia	Epoxy fatty acid	Plastics, coatings, paints	Petro oils, soybean oil
Gums, Resins, etc.			
Baccharis	Resins	Rubber	Wood rosins, tall oils
Grindella	Resins	Adhesives, varnishes, sizing	Pine resins, tall oils
Guar	Gum	Paper additives, cosmetics	Imported guar
Guayule	Rubber	Tires	Imported *Hevea* rubber
Milkweed	Latex	Adhesives	Petroleum derivatives
Fibers			
Kenaf	Bast	Newsprint, paperboard	Wood pulp

Source: Adapted from U.S. Congress, Office of Technology Assessment, *Agricultural Commodities as Industrial Raw Materials*, Washington, D.C.: U.S. Government Printing Office, 1991.

Jojoba (*Simmondsis chinensis*) is an evergreen native of the Sonoran Desert that grows in hot, arid conditions. The seeds contain 45 to 55 percent oil that can be processed into a fatty-acid monounsaturated oil. After processing, the meal that remains is about 30 percent protein. Jojoba is currently used in cosmetics such as face creams, lipstick, and shaving cream. Hydrogenated jojoba can be combined with polyethylene or polypropylene to yield mixed plastics that are harder and have lower melting points than the pure plastics. Jojoba waxes could be used in household waxes, soaps, candles, and crayons. The tannin in the seeds could be used in the leather industry.[20]

Crambe (*Crambe abyssinica*) and rapeseed (*Brassica napus*) are complementary oil seed crops that can be grown as cool season annuals. More than one-third of the seed weight is oil, of which 50 to 60 percent is erucic acid, a valuable triglyceride. The meal that occurs as a by-product of oil seed processing requires the dehulling of crambe and the decorticating of rapeseed to achieve commercially acceptable protein and fiber levels. The erucic oil is particularly valuable. Oils high in erucic acid offer industrial applications in textile, steel, and shipping industries as spinning lubricants, cutting oils, metal-forming oils, mold lubricants, and marine lubricants. Derivatives of the oils are used as lubricants in forming plastics; as plasticizers; as emulsifiers in soaps and detergents; as catalysts in polymer, rubber, and ceramic production; and as anti-static agents in magnetic recording tapes and films. Canada has aggressively developed rapeseed production and promoted a substantial market for an edible oil called "canola." Nearly 40 million pounds of high erucic acid oils are used annually in the United States and almost all of this is imported. Domestic production of rapeseed has fallen from 22 million pounds in 1988 to less than 3 million pounds today and most of this is exported.[21]

Plants of the genus *Lesquerella* are considered suitable sources of hydroxy fatty acids. The oil contains over 50 percent fatty acid, while the meal has over 30 percent protein and a good amino acid balance. Hydroxy fatty acids have a wide range of uses in resins, waxes, nylons, plastics, corrosion inhibitors, cosmetics, and coatings.[22]

Guayule (*Parthenium argentatum*) is a native shrub of the Chihuahuan desert that has historically served as a source for latex. In 1910 Mexican guayule provided 10 percent of the world's rubber supply. During World

War II it was studied as an alternative source of rubber. While it is rich in latex—20 percent of the plant's dry weight is rubber—it cannot be profitably harvested until its seventh year. Research on guayule production is currently being conducted at the University of Arizona and the Gila River Indian Community.[23]

Research is continuing to identify new industrial uses for traditional crops. A good example involves recent efforts to market a road deicer that could reduce the use of salt. Calcium magnesium acetate can be made by reacting limestone with acetic acid made from fermented corn. Estimates suggest that 60 bushels of corn are needed to make 1 ton of calcium magnesium acetate. While the production costs are not necessarily high, the low cost of salt currently limits the use of this alternative, although increasing concern about the effects of salt intrusion near roadways could open up the market.

Commodity organic chemicals that are currently made from petroleum can be made from agricultural materials as well. Indeed, synthetic gas has been made from crude oil, wood, pine tar, rubbish, and various agricultural by-products such as cottonseed and corncobs. Imperial Chemical Industries in Britain used molasses as a feedstock for organic chemicals up through 1944, making ethylene glycol, ethylene oxide, and polyethylene; and piloting the production of butanol, acetic acid, and acetone from other agricultural feedstocks.[24] Figure 11.1 identifies the alternative pathways for generating several common organic chemicals. The starch derived from corn or sugar beets and the cellulose derived from wood has the potential to make various first-order and second-order intermediates. Under the right conditions, agriculturally derived chemicals could become viable substitutes for petrochemicals. Plant-based chemicals already have a high oxygen content and this provides an advantage over many petroleum-based source materials that require oxygenization because petroleum contains little oxygen.

The prospect of converting industrial materials from hydrocarbon sources to renewable, plant-based sources has drawn David Morris of the Institute for Local Self-Reliance to propose a "carbohydrate economy" based on biomass. Morris writes:

The carbohydrate economy is still very much in its infancy, but our research indicates that it has clearly moved beyond the birthing stage. Plant-matter derived products now have toeholds in markets from which they previously were

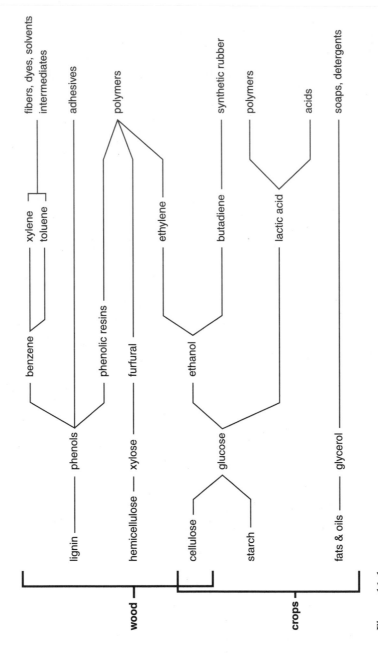

Figure 11.1
Routes of synthesis for organic chemicals from renewable resources.

excluded. And they have captured significant portions of other markets in which they previously had only a marginal presence.[25]

Morris's research demonstrates that a combination of green consumerism and environmental regulations is opening up traditional petrochemical product markets for plant-based products. A 1992 study found that plant-derived adhesives made up 40 percent of the national adhesives market; 35 percent of the surfactants on the market were plant-based; 11 percent of the detergents were based on plant matter; and vegetable oil-based inks made up 7 percent of the ink market. While the costs of the plant-derived products were falling over time, there remained a price premium of 20 percent for inks, 50 percent for detergents, and 100 percent for paints. Such cost differentials suggest that these products will need both technical and market-scale improvements if they are to grow out of niche markets.[26]

The scale of some of these markets may actually create their own limits. Twenty-five years ago it was estimated that substituting wood for petroleum to produce the organic chemicals and plastics then in production would have required 60 percent of the wood then used by the pulp and paper industry. Today, the production of a typical chemical intermediate at the rate of 500 thousand metric tons per year would require about 1 percent of the U.S. corn crop. At this rate it would take 50 percent of the nation's corn crop just to provide 100 percent of the annual ethylene market. In 1992 David Morris and Irshad Ahmed calculated that having plant matter capture one-third of the organic chemical market would require a 5-fold increase in the 1992 level of agricultural production for industrial purposes. Agricultural production at this scale would compete with the nation's food and forage requirements. Because of this, other plant materials would need to be used for industrial materials, or these chemicals might better be derived from crop wastes.[27]

Today there are many industrial materials made from renewable resources and there are many more opportunities available. But the relatively low price of oil, the dominant position of the petrochemicals, and the locked-in nature of today's materials on the contemporary market create formidable barriers to the penetration of many agriculturally based materials.

11.4   The Benefits and Limits of Agriculturally based Materials

The use of agricultural commodities as industrial raw materials offers a range of benefits, including diversification of agricultural markets, better utilization of land resources, reduction of farm commodity surpluses, rural development, increased international competiveness, and increased national security. Some of these are direct benefits to farmers. A more diverse market provides more opportunity for farmers and better insulates them against fluctuations in specific commodity markets. Some 60 million acres of potential farmland—nearly 15 percent of total national farmland—are removed from production under federal acreage conservation and reduction programs, in order to limit overproduction. Industrial demands that expand markets could put that land back into production and reduce current farm commodity surpluses as well. Some benefits directly affect the nation's materials flows. New agricultural products could increase the nation's trade balance and reduce its reliance on imported oil. Materials developed from domestic agriculture could provide substitutes for strategic materials that are currently imported.[28]

There are also direct environmental benefits. Some agriculturally based materials could lead to more environmentally friendly products. Of even more significance are the effects on agriculture. Because many of the new crops that are potentially valuable in industrial applications are adapted to arid conditions and are relatively salt tolerant, they could substitute for water-thirsty crops currently planted in irrigation-dependent lands of the West and Southwest. In addition, several of these crops are good ground covers that could reduce soil erosion; are nitrogen-fixing, which would improve crop rotations; and are reported to be relatively pest resistant, reducing the need for pesticides.[29]

Such environmental benefits need to be carefully balanced against the risks that may arise. Some of these new plants are not native to North America. Introducing non-native plants on a scale large enough for industrial cultivation can produce unwanted results, such as runaway invasiveness, unexpected effects on ecological balances, and unintended cross-fertilization with native crops and weeds. The relative pest-free quality of some of these new plants may depend on the genetic diversity found in their natural settings, and this resistance may rapidly evaporate when these plants are cultivated in single-species fields. This appears to be the case with jojoba, which is relatively disease- and pest-free in the wild, but appears less resistant under cultivation.[30]

Added to these concerns are the environmental costs associated with increasing agricultural intensification. Between World Wars I and II, agricultural production began a long-term transition from conventional family farms toward large-scale industrialized production increasingly reliant on petroleum-based technologies and chemicals. Like the chemical and mining industries, industrial agriculture demonstrates significant materials integration and corporate concentration. The small family farm has been dramatically affected by the growth of large corporate suppliers and customers; large tracts of farmland are now owned by mammoth conglomerates and are so mechanized as to be called "factory farms." These industrial processes have resulted in the dissipative use of large volumes of chemical fertilizers and toxic pesticides. By the 1950s, pesticides were used on 74 percent of vegetables grown in the country and on 81 percent of the fruits and nuts and 66 percent of the cotton. By the 1980s, herbicides accounted for 75 percent of the total volume of pesticides applied to agricultural crops. Corn and cotton are two of the most insecticide-intensive crops under cultivation. Cotton, which is defoliated with a herbicide before harvesting, accounts for about 10 percent of all pesticides used. Between 1964 and 1982, the volume of insecticides applied to corn and soybeans nearly doubled and even exceeded the amount used on cotton. Although pesticide use per acre has been declining since the mid-1980s, total U.S. pesticide use on agricultural lands runs at about 570 million pounds per year.[31]

The depletion of soil quality and quantity is a particular problem of industrial agricultural production. Intensive farming can lead to substantial deceases in soil biota, nutrient cycling, and organic matter. In the United States, nearly 8 tons of soil are lost per acre per year, which factors out at nearly 3.4 billion tons of soil being washed or blown away each year. The tillage of cornfields is particularly intensive and results in substantial amounts of soil loss (ranging as high as 2 bushels of topsoil for every bushel of harvested corn).[32]

Industrial agriculture is energy intensive. From tilling to harvesting to threshing and drying, machinery has replaced animal muscle, and the production of fertilizers and pesticides has further added to the energy intensity of every farmed acre. During the past century, energy intensity in U.S. corn production rose 8-fold. It is now estimated that U.S. food production requires two units of fuel energy for every unit of food energy produced, and this inefficient energy balance is made possible only by the relatively low price of fossil fuels.[33]

Intensive agriculture also requires substantial amounts of water. Although irrigation from surface water supplies a large amount of farmland, groundwater is often purer and more dependable. However, the excessive pumping of groundwater for agricultural uses is progressively depleting some of the nation's largest aquifers. Groundwater is dropping at least a foot a year under nearly 45 percent of groundwater-irrigated farmland. The vast Ogallala aquifer that underlies seven midwestern states supplies 20 percent of U.S. irrigated farmland, but unlike many aquifers of the eastern states, the Ogallala is a fossil aquifer that is not replenished by surface percolation. Thus this water is a nonrenewable resource and once it is mined and consumed, like any other nonrenewable resource, it is dissipated.[34]

Synthetic fertilizers and pesticides are a source of pollution. Nitrogen oxide is a common air emission from agricultural fertilizers spread on farm fields and may contribute as much as 20 percent of the total atmospheric load. Surface runoff from fertilizer or pesticides pollutes streams, and the leaching of these chemicals into soils contaminates groundwater. Something like 70 million pounds of fertilizers are used on U.S. farms each year and this has resulted in the growth of large blooms of nuisance algae on downstream surface waters and the eutrophication of coastal and freshwater aquatic systems. Nitrate concentrations in the major rivers of the Northeast have increased 3- to 10-fold since 1900. Various studies during the 1980s found 20 to 70 percent of farm wells contaminated with nitrates.[35]

Finally, the processing of agricultural materials also requires substantial amounts of energy, water, and toxic chemicals. For instance, the conventional processing of vegetable oils from oil seeds relies on extraction that uses a large amount of petroleum-based solvents. Substantial amounts of water and energy are often required to enhance organic oxidation and reduction processes. There are also occupational hazards. For instance, the ginning and carding of cotton creates fibers that can accumulate in workers' lungs, a hazard that has been long recognized.

It is too easy to assume that materials made from renewable resources would be environmentally friendly and that switching from nonrenewable to renewable source materials would greatly slow the dissipation of those materials most easily depleted. The scale of materials production required to meet even today's market requirements would put substantial amounts of land into agricultural production. Unless there were a significant transformation of the technologies of agricultural production, an increase in farming devoted to generating industrial materials would require a parallel increase in the use and dissipation of energy, water, and hazardous chemicals. Shifting from the petrochemical or mining industries to the agricultural industry would only substitute one set of environmental problems with another. The organization and technology of farming in the United States is hardly inspiring from an environmental perspective. Indeed, as regulations and policy have progressively reduced some of the worst environmental impacts of the petrochemical industry, the environmental consequences of agriculture have emerged as some of the nation's greatest concerns. It is just this dilemma that led William Hale, the father of chemurgy, to fear that his dream of a farm-based source of industrial materials might be little more than some plaintive "wail in the wilderness."[36]

11.6   Renewable Resources for Industrial Materials

Renewable resources may be environmentally preferable to nonrenewables, but the organization and technologies of modern agricultural production processes may be just as worrisome as those found in the mining and chemical industries. By considering the opportunities available for obtaining industrial materials from renewable resources, it is easy to understand that it is not enough to identify safer source materials; the processes of production must be environmentally sound throughout the material's life cycle. For a material to be sustainable, the raw materials, the production, the processing, and the disposal of the materials must be environmentally sound.

It is this kind of thinking that has led some critics of conventional agriculture to propose a different basis for farming. As the field of ecology has developed, it has had increasing impact on conventional agriculture. K. H. W. Klages first proposed an ecological approach to agriculture during the 1940s, and this has been further developed by scores of agricultural scientists writing since the 1960s. The effort to develop a new scientific specialty often called "agricultural ecology" or "agroecology" has brought together professionals from several disciplines, including agronomists, ecologists, entomologists, resource economists, anthropologists, and rural development experts.[37]

Proponents of agricultural ecology have demonstrated the importance of biodiversity in crop cultivation (often referred to as "polycultures"), which forms the basis for the recycling of nutrients, control of local microclimates, maintenance of local hydrogeological cycles, regulation of pests, and detoxification of noxious chemicals. The practical applications of agroecology include crop rotation, maintenance of ground covers, promotion of soil biotic activity, soil conservation, low energy-input farming, "no till" cultivation, integrated pest management, and organic farming. Miquel Altieri, a leader in this movement, notes that such practices emphasize long-term sustainability over short-term productivity because they:

• reduce energy and resource use;
• employ production methods that restore homeostatic mechanisms conducive to community stability, optimize the rate of turnover and recycling of matter and nutrients, maximize the multiple-use capacity of the landscape, and ensure an efficient energy flow;
• encourage local production of food items adapted to the natural and socioeconomic setting; and
• reduce costs and increase the efficiency and economic vitality of small and medium-sized farms, thereby promoting a diverse, potentially resilient agricultural system.[38]

Wes Jackson, a proponent of agricultural ecology with a research center in Kansas called The Land Institute, has proposed that the reliance on annual grains for the nation's primary source of protein has been misdirected. Perennial grains could be just as nutritious, and their cultivation would require far less energy consumption and environmental damage.

Instead of tilling vast tracts of the Midwest in order to plant monocultures of annual crops in chemically enhanced fields that require substantial amounts of pesticides to deter insects, Jackson proposes nurturing polyculture meadows of perennial grasses that are sustained by habitat interdependencies and that resist pests through natural processes. Eliminating tilling reduces soil loss and water evaporation. These prairie meadows could be harvested just like annual grain fields, but the plants would not be destroyed in the process. Instead, the preservation of the base of the plant and the root system would mean that fewer soil nutrients would be required and less waste produced in generating the desired seed heads of the crop.[39]

In much the same way, the oil crops required for an organic chemical carbohydrate source might be grown with far less energy input, soil loss, water consumption, and synthetic chemical additives. Indeed, Land Institute research has identified sunflowers and bundleflowers as two of the most promising perennials, and both are recognized as rich sources of oil seed. But the idea of a sustainable harvest of perennials goes further because many woody perennials, such as grasses, could be harvested without destroying the plant. The ubiquitous willow (*Salix*), for example, can be coppiced (cut back to the stump) to produce an annual source of lignocellulose without destroying the plant. Currently several pilot studies are investigating the harvesting of willow as a biomass energy source. Many common shrubs, such as sumac, privet, lilac, box, honeysuckle, bayberry, and forsythia, can likewise be cut to the stump to produce periodic harvests of lignocellulose.

Lignocellulose has attracted considerable attention as an alternative source of energy and materials because it is renewable and available in large quantities. With cellulose making up somewhere between 60 and 75 percent of dry lignocellulose, stem wood is the most abundant source of organic chemicals on earth. The total global supply of biomass is estimated to be 300 billion tons, and nearly 50 percent of this is lignocellulose. As a continuously growing material, it is estimated that the annual production rate of the planet's lignocellulose is between 20 and 50 billion tons and that nearly 4 billion tons can be harvested sustainably on an annual basis. In conventional forestry, much of this harvesting involves cutting the tree to the stump and often killing it. An alternative harvesting process designed to maximize a continuous yield of lignocellulose would cut branches from the tree, but leave enough of the plant standing to encourage rapid replenishment.[40]

Today, a huge volume of lignocellulose goes to waste. It is estimated that some 350 million metric tons of cellulosic waste are generated every year in the United States from pulp and paper mills, agricultural production, and the food-processing industries. The current waste lignocellulose generated by the world's pulp and paper industry facilities is estimated to be 30–50 million tons. Plant-based cellulose is more difficult to hydrolyze to sugars than plant-based oils, but the supply is large and the price would be low.[41]

Lignocellulose wastes are not the only wastes with organic chemical possibilities. Wastes from the agriculture and food industries and municipal organic wastes also provide feedstock opportunities. In 1996, 17 million tons of food wastes were discarded as municipal solid wastes (8.3 percent of the total waste), of which only about 4 percent was diverted for composting. Yard trimmings made up another 30 million tons of waste (14.3 percent of the total), with only about 30 percent recovered for composting. Composting of food and yard wastes for soil amendments is increasing in some communities, but the surplus wastes could as easily be collected for recycling into industrial chemical production.[42]

Manure wastes from farming offer another source for organic chemicals. During the past two decades, cattle, poultry, and hog farming have become highly centralized in the United States, generating huge volumes of manure that are well beyond the assimilative capacities of local regions. Like petroleum by-products, manure wastes are well concentrated organic wastes and in need of alternatives to environmental disposal. Indeed, the natural degradation of these wastes generates methane, which, if it is not captured for use, can become a problematic greenhouse gas. Finally, human wastes from municipal and domestic sewage also offer opportunities as chemical feedstocks. The nation generates 5.4 million dry metric tons of sewage sludge a year. Twenty-eight percent of the sludge from municipal wastewater treatment facilities in the United States is recycled back to farm lands as soil amendments, but because this sludge contains heavy metals and other contaminants from industrial and consumer product discharges to municipal sewer systems, farm application is often questionable. Recycling sewage wastes back into production of organic chemicals could provide incentives for improving the quality of sludge and provide a better means of closing materials loops. Where organic materials are currently going to waste, as in farm, food, or wood pulp processing, or in municipal sewage, there is an opportunity for an integrated materials production system that links food and fiber production with industrial chemicals.[43]

Renewable materials could facilitate a more open economic system by exchanging materials with the environment without harming natural systems. Indeed, in a sustainable economy, all organic materials should cycle, but this will require new approaches to agricultural and organic waste management. Through careful rethinking of the processes for cultivating and harvesting renewable resources, agricultural production could be better matched with the environmental compatibility of farm products. Without such a reorganization of agricultural production, the prospect of producing industrial materials from renewable resources will fall dramatically short of the vision of a sustainable economy.

There remains in the organic wastes of the forestry, agriculture, and food industries and municipal sewer systems an untapped reservoir of renewable resources available for industrial materials. By reorganizing the collection and management of municipal and domestic organic wastes, new opportunities could emerge for renewable chemical feedstocks. When these waste sources are added together with agricultural crops grown for producing industrial materials, there is a substantial volume of reasonably inexpensive renewable materials that could be used to support an organic chemical industry weaned from its reliance on nonrenewable sources. The nation started out well endowed with natural resources. Although the current economy has squandered its nonrenewable resources, a careful reorganization could once again tap the potential of the ample renewable resources of the land.

12

Biobased Materials

In a society accustomed to dominating or "improving" nature, [a] respectful imitation is a radically new approach, a revolution really ... [that] introduces an era based not on what we can extract from nature, but what we can learn from her.
—Jane M. Benyus

Nature is a marvelous manufacturer of materials. If we are patient, we can watch seedlings self-assemble into towering trees with lofty architectures of fractionate construction and fascinating beauty. This monumental task is carried out with little more than common polymers, sugars, and water. If we cut one of our fingers, we need wait no more than a few days before the tissue will disinfect and self-repair. The bacteria *Rhizobium* that burrows into the roots of legumes soon gratifies its host by converting atmospheric nitrogen, which is unavailable to the plant, into a highly usable form of ammonia that promotes growth.

Biological processes rely on a surprisingly few number of chemical substances. Plant structures depend on cellulose and lignin. Structural materials in animals rely on proteins (primarily collagen, elastin, and keratin) in combination with various polysaccharides, calcium minerals (for bone and teeth), or complex phenolic compounds (in insect shells). Although plants and animals are often taken for granted, the way in which living organisms manufacture materials is not trivial and there is an elegance and efficiency to many of the basic processes.

Learning how to make materials by observing the principles of natural processes is immediately appealing. After all, most of the materials created by nature, ranging from tree trunks and flower petals to bird beaks and human skin, are generated by living organisms at ambient temperatures and pressures, without chemicals hazardous to life, and most often in simple aqueous environments. These conditions minimize the use of toxic chemicals and eliminate the generation of toxic wastes. Products and wastes manufactured by nature's rules could be safely accommodated in open systems of materials cycling where materials readily flow between economic and natural systems.

The concept of learning from nature is not new; it was conventional practice until the rise of the analytical sciences. Leo Baekeland's work on developing a plastic from organic materials was inspired by natural models, as were the aerodynamic observations of the Wright brothers and the polymer studies of Wallace Carothers and Hermann Straudinger. Indeed, some of the newer fields of materials sciences have developed from studies biologists, chemists, and biochemists have made of the way in which organisms in nature make materials. There is a whole field of bioorganic chemistry that focuses on the structure and interaction of organic compounds of biological significance at the molecular level. The field draws upon thermodynamics, transition-state theory, acid–base theory, theories of stereocontrol, and the concepts of hydrophobicity and hydrophilicity, but the basic principles come directly from natural models.

Rather than the brute force approach to materials development that has for too long depended on increasing amounts of heat, pressure, waste, and life-threatening reactions, there is growing evidence of a milder, gentler path that is not only more compatible with nature, it actually mimics natural processes. This direction, which grows out of the life sciences, includes specialties with names like biosynthesis, bioprocessing, biomimicry, and biotechnology. Its not without risks, some well recognized and others only guessed at, but the opportunities for new materials that are safe and high performing appear abundant and inviting.

12.1   Biosynthesis and Bioprocessing

Biosynthesis involves making materials using natural processes. In nature, biosynthesis involves building complex cellular materials from simple molecules. Some organisms carry out the whole process themselves by building cells from carbon dioxide, nitrogen, water, and sunlight, while others consume nearby fats, sugars, and amino acids to make sophisticated macromolecules such as polysaccharides, proteins, and nucleic acids.

There is a long human history in using biosynthesis for food processing. For thousands of years people have been brewing beer, making wine, leavening bread, and culturing cheese with the aid of bacteria, yeast, and molds. The Egyptians were making beer and leavening bread with yeast some 6000 years ago. Cheese-making dates from 5000 B.C., and the preparation of soy sauce, which requires a complex biochemical reaction involving molds, yeasts, and bacteria, has a long history in Japan. Fruits had long been fermented to make wines and by the 1400s grains were being distilled to make alcoholic spirits in several parts of the world.[1]

The scientific understanding of these bioprocessing operations began to develop only after the 1870s when Louis Pasteur discovered that fermentation was caused by microorganisms. As the science of microbiology grew, its technical applications became increasingly important, and by the 1890s the first medical vaccines were commercialized. Beginning around 1900 there appeared a series of experiments designed to use large-scale fermentation processes to produce industrial materials. The development of acetone–butanol fermentation processes began prior to World War I, and the production of citric acid and various solvents by biochemical processes was spurred by the pressures of World War II. Both glycerol and acetone were produced by microbial processes as a raw material for explosives during the two world wars. The rapid growth of the petrochemical industry following World War II and the low prices of oil during the 1950s brought the development of industrial applications of microbiology to a standstill. However, medical interest in microbiology continued to grow. Stimulated in large part by the successful development of penicillin fermentation during the 1940s, the microbial production of antibiotics, vitamins, and enzymes grew steadily during the 1950s.[2]

Today, biosynthetic fermentation is used to produce a wide array of products. They range from organic feedstocks (ethanol, glycerol, and acetone–butanol) to organic acids, amino acids, enzymes, vitamins, antibiotics, and single-cell proteins. While the majority of these substances and processes find applications in pharmaceuticals, agricultural products, and food production, several have industrial applications.[3]

The production of industrial substances by fermentation requires a carefully managed environment. The heart of the process is a large vessel called a "fermenter," in which a population of organisms can be maintained at the required temperature, pH, dissolved oxygen concentration, and substrate density. However, there are a series of stages before and after fermentation that also require careful attention. First, the medium on which the organism is to grow—the substrate—must be formulated from raw materials and the fermenter must be cleaned, sterilized, and inoculated with a metabolically active culture of microorganisms. After fermentation, the culture fluid has to be harvested, the cells separated from the desired product, and the product extracted by filtering, flocculation, or centrifugation and then purified. The fermentation process is usually accomplished in large batches that may require a few hours to many days for complete production, although there are several forms of continuous fermentation processes in use today that rely on a steady feed of raw materials and a complementary withdrawal of finished product.[4]

Carbohydrates are the traditional energy sources in fermentation. Starch is the most important carbohydrate and may be used in the form of whole or crushed cereals or roots from plants such as corn, rice, wheat, potatoes, and cassava. The cost of pure carbohydrates, such as glucose and sucrose, limits their use, but beet sugar and sugarcane molasses are frequently used as raw materials. The malt extract from barley, sulfite waste liquors from the pulp and paper industry, and whey from the dairy industry are also used as sources of carbon. Plant oils such as soybean oil, palm oil, or cottonseed oil may be used as a carbohydrate supplement. Those fermentation processes that also require nitrogen are usually supplied with nitrogen in the form of ammonia, salts, or urea.

The products of fermentation of most interest as industrial materials are the organic feedstocks, acids, and polymers. While most of the country's ethanol is produced by the catalytic hydration of petroleum-derived ethylene, ethanol can be produced from the fermentation of yeasts (*Saccharomyces cerevisiae*) and bacteria (*Zymomonas mobilis*). Starch-containing roots and grains, and molasses are usually used to

make ethanol, typically in batch processes (although continuous processes are also available in the United States) in fermenters up to 600
cubic meters in volume. Biobased ethanol is used primarily as a fuel or
fuel additive or as a solvent in the cosmetic, pharmaceutical, and specialty chemical industry. Since most of the ethanol produced is used as
a fuel, the energy balance of the process determines its economic viability, and because there is usually a net energy loss in the production of
ethanol by biosynthesis, the market for biobased ethanol is weak.[5]

Glycerol, acetone, butanol, and butanediol can all be manufactured by
fermentation. Glycerol was produced by fermentation as a constituent
for explosives during World War II. Up through the 1970s acetone and
butanol were made through anaerobic fermentation for applications
in solvents, paints, and adhesives. The process relied on a matrix of
molasses and starch and various strains of the bacteria *Clostridium
acetobutylicum*.[6]

Many common organic acids can be produced through fermentation.
The production of citric acid by deep-vat fermenters began during the
1930s. Today, well over 350,000 metric tons of citric acid are produced
annually by fermentation for use primarily in the food and beverage
industries, although there is a growing use for citric acid as a substitute
for phosphates in detergents. The starting materials are conventionally
the starches from potatoes, sugarcane syrup, and molasses from sugar
beets. About half of the world's lactic acid is produced by fermentation.
The biosynthesis of lactic acid begins with glucose that is broken down
to glyceraldehyde and fermented for up to 72 hours. Total microbial
production of lactic acid amounts to some 200,000 metric tons per year.
Itaconic acid is a valuable intermediate in polymer chemistry. It is used
in the synthesis of pyrrolidones and as an emulsion in paints, and it poly-
merizes to produce a low-molecular weight plastic. Itaconic acid can be
produced using *Aspergillus terreus* in a media containing molasses as a
carbon source and ammonium salts as a nitrogen source. Except for citric acid, which is produced entirely by fermentation, there is significant
competition between biological and chemical processes for those organic
acids that can be produced by biosynthesis. For instance, lactic acid and
acetic acid are being produced today by both microbiological and chemical production facilities.[7]

Polysaccharides can be readily derived from fermentation. Most of the xanthan gum on the world market today is produced by microbial fermentation. The primary use of xanthan gum is as a stabilizer in gel suspensions and for viscosity in various lubricants. Research is under way on various other polysaccharides, such as pullulan, seleroglucan, and gellan, with some of the most interest focused on the production of polyhydroxybutyrate (PHB) based on the fermentation of glucose.[8]

Within the microorganisms used in conventional fermentation, the most significant activity is carried out by enzymes embodied in cells. The process of constructing complex molecules from simple ones inside a cell involves a series of reactions mediated by enzymes. Enzymes are large protein molecules composed primarily of amino acids. Their role in metabolism is quite complex. Essentially, they lower the energy required to speed up a reaction and thereby hasten biological activity. Under the right conditions, enzymes stimulate and guide the chemical reactions by which cells grow or gain energy. Because enzymes function as catalysts, they are not depleted by the process, but can often be reused several times.

The use of enzymes in cells to mediate chemical reactions requires catering to the needs of the living cell. Where an enzyme can be isolated from the cell, it has equal potential but requires less careful management, and a variety of enzymes have been isolated for commercial uses. Many of these enzymes are currently produced from the careful harvesting of bacterial excretions during fermentation. Rennet, a milk-curdling enzyme used in cheese-making, was one of the earliest commercially isolated enzymes. In 1915 a laundry powder using "tryptic enzyme additives" was patented and by the 1970s over 80 percent of all laundry detergents contained enzymes, particularly proteases. Amylases, which produce simple sugars from more complex ones, is used to hydrolyze starch for the production of dextrin and glucose. A bacteria-derived enzyme called "cellulase," which breaks down cellulose, the molecular base of cotton, is currently used to soften new blue jeans as an alternative to harsh stone washing.[9]

The processing of high-volume, bulk chemicals by biosynthetic fermentation offers various commercial opportunities, particularly in fuels, chemical intermediates, and acids. Several common chemical reactions could be controlled using enzymes as catalysts. Enzyme-catalyzed reactions offer advantages over conventional catalysts. They operate at ambient temperatures and pressures; they are relatively easy to control; they require no organic solvents; they are biodegradable; and they are very reaction specific. However, many of these bioprocessing methods are not currently economically competitive with established petrochemical processes, and today they remain more possibilities than realities.

12.2   Biodegradation

Biodegradation is the opposite of biosynthesis. In biodegradation, microorganisms break down rather than build up chemical and cellular structures. While many chemicals are naturally degraded in the environment by chemical reactions (e.g., hydrolysis and oxidation), microorganisms such as bacteria, algae, fungi, and protozoa are common biological decomposers of organic chemicals. These organisms, along with worms and some larvae, are the active digestors of the domestic garbage dumped into municipal landfills, and they can reduce the organic constituents to beneficial soil amendments and methane gas.

The biodegradability of a material is a measure of the degree to which it can be decomposed by biological processes. It is inversely related to persistence, which is a measure of the material's ability to resist microbial activity. While most biodegradation involves the transformation of a material to a condition that is environmentally acceptable, ultimate biodegradation leads to carbon dioxide, water, and inorganic compounds that are in or near elemental form.

Biological decomposition has long been used to detoxify human wastes. By 1914 microorganisms were being used to treat municipal sewage in order to reduce the biological hazards in the waste. This involved inoculating organic wastes with a tailored mixture of organisms that hastened the natural rate of decomposition. Soon thereafter biodegradation began to be used to treat agricultural wastes and wastes from food processing and other organic chemical industries. This treatment was used to metabolize the dissolved organic compounds which, if directly discharged to surface waters, consumed oxygen in the water and compromised aquatic habitats. Today, most municipal waste treatment facilities are based on aerobic (in air) biodegradation. The waste is fed into large tanks where it is trickled over filters or mixed with activated sludge to promote decomposition through a combination of microbial digestion and oxidation in air. Various microbial inoculants and enzymes may be added to the slurry to accelerate the breakdown of cellulose, hemicellulose, fat, and proteins.

Since the 1970s, composting (aerobic degradation) municipal wastes to produce agricultural soil amendments has been a growing practice. A recent U.S. survey found that in 1993 there were 128 such programs run by local municipal sewage treatment agencies, although concern over heavy metals in the sewage wastes limited the enthusiasm for the resulting sludges. Composting is also used by many municipalities to treat and reduce the large volume of yard and organic wastes collected through municipal waste collection services. Municipal composting of yard wastes has increased dramatically over the past decade, with more than 3800 operations now in service.[10]

Biodegradation is also finding uses in industrial processing. Advances in microbial processing offer opportunities to mine minerals that have remained uneconomical. A crude form of leaching metals from low-grade ores using living organisms can be traced back to ancient societies. Today, microbial processes that degrade inorganic ores to separate out desired minerals have been used to leach copper, nickel, lead, and zinc from their ores. At present, over 15 percent of the total U.S. copper production comes from bacterial leaching. Much of this leaching is dump leaching on mine tailings, but there is growing sophistication in *in situ* leaching, in which bacteria-laden waters are pumped into ore deposits and retrieved through separate pumping wells. Bacterial mineral leaching processes require relatively little energy inputs and offer accessibility benefits where ores are difficult to reach.[11]

Biodegradability can also be a virtue in product development. Paper cups and paper bags that compost easily have long been used where their service involves a high probability of single use and rapid disposability. Jute and low-grade wood pulp have found new uses in so-called "geotextiles" that are designed for temporary use in agricultural mulches or soil erosion control, after which biological degradation is a virtue.[12]

While there is no one-to-one comparability, products and processes based on biological principles have a potential to produce products and wastes that are likely to biodegrade in natural environments. Biosynthesis and biodegradation offer real possibilities for the benign cycling of materials from natural systems though economic systems and back to natural systems that is so often practiced today with products and processes that are far too toxic or persistent to behave appropriately.

In many ways, the cycling of biocompatible materials is simply an extension of materials management systems that have existed for millennia. However, the biological sciences have not been content just to extend conventional processes. Much of the new research is based on understanding natural processes in order to develop synthetic processes that function like or mimic natural processes. The most benign of these is called "biomimetics."

12.3   Biomimetics

Biomimetics is an emerging field of biology that focuses directly on studying and replicating the processes of living organisms. Research in the field ranges from studies of information processing in the brain to experiments involving biological healing and repair. Within this field there is a lively set of studies on the ways in which living organisms make basic materials such as teeth, hair, and skin, and process these materials into objects of use such as nests and cocoons.

An understanding of these processes offers a liberating set of possibilities because the processes are so compatible with biological systems. Most of them are low energy–low waste procedures that are elegant in their simplicity. The resulting products are impressive; not only are they environmentally compatible, they are also often finer, more intricately structured, and better performing than our conventional synthetic products. The inner shell (macre) of the abalone is twice as tough as synthetic ceramics, the rhinoceros can heal and rebuild its horn, and mussels adhere to underwater rocks with naturally derived adhesives.[13]

One area of productive research involves protein-derived fibers, such as silk, wool, and hair. Biological tissues are based on an array of chemical substances. Long-chain molecules of proteins and sugars are synthesized inside living cells. These macromolecules are then assembled under sensitive control mechanisms into complex networks called "microfibrils," which form the basis of cellulose, chitin, keratin, and collagen.

Silk is a fascinating, protein-based fiber. It is a solidified form of a viscous fluid secreted by special glands of several insects, namely, caterpillars and spiders. The manufacture of silk from the silkworm represents the only large-scale, human cultivation of an insect. Silkworm fiber consists primarily of the proteins sericin and fibroin. The high tensile-strength silk lines made by spiders largely involve glycine, alanine, and proline held together primarily by hydrogen bonds and ionic interactions among the amino acids. In terms of the strength-to-diameter ratio, these tiny threads are stronger than any steel wire we can make. The structures of hair and wool are much more complex than the structure of silk. Human hair is composed of sequences of fibroins and matrix proteins cross-linked by disulfide connections. In wool the fibroins are helical proteins composed of seven amino acids in repeating patterns.[14]

Another area of research focuses on shells, bones, and tusks, which are naturally assembled ceramic composites of proteins and various minerals (particularly calcium). Industrially manufactured synthetic ceramics are used for insulators, guides, bearings, coatings, optical devices, and electrical assemblies. They are conventionally made by high-pressure molding of ultrafine inorganic particles, but in industrial applications they are all plagued by brittleness. Natural ceramics (often referred to as "bioceramics"), such as the shells of mollusks, are formed of three-dimensional, polymeric lattices that are tough and resistant to breaking. Biologists at the University of Washington in Seattle are studying the crystalline architecture of the abalone shell to learn how the tough inner shell is structured. The shell is made of calcium carbonate (chalk) in an ordered hierarchial structure that uses polymers such as polysaccharides (sugars) and proteins as a mastic. The lattice limits cracks while the polymer allows for deformity, thus reducing the brittleness of the material.[15]

Researchers at the University of Arizona's Materials Laboratory in Tucson are studying the process by which mollusks grow perfect crystals and order them in functional hierarchies that are capable of absorbing ("biomineralizing") calcium from seawater. Biomineralizing involves two steps. First, the organism secretes polymeric fluids that assemble into three-dimensional lattice structures in which the open compartments encapsulate seawater, which is naturally rich in calcium and carbon. The organism then releases into the lattice proteins that provide templates for guiding the organization of the calcium and carbon building blocks. The ability of each protein to guide the assembly process is determined by a genetic code. These templating proteins initiate calcium carbonate crystal-building activity from specific locations in the lattice compartments, creating a complex structure of minerals that determines the resulting shape.[16]

Other studies involve the production of adhesives. Synthetic adhesives are a complex mixture of chemicals. They require cross-linked polymers to form the interlocking bond, initiators of the chemical process, and catalysts to speed the process up, and even then, the adhesion is often weak. However the common sea mussel produces a strong and durable adhesive for attaching to underwater surfaces. It can produce a cross-linked protein fluid that uses oxygen as the initiator and includes the catalyst in the interlocking proteins. Even more remarkable, this adhesive is produced and hardened underwater, a property that remains beyond our synthetic adhesives.[17]

Research on sea urchins has shown that they use electrical charges on their cell membranes to serve as the nucleation site for tiny calcite crystals to grow into a single, long crystal. Based on these findings, a group of researchers in Israel is experimenting with coatings of hydrocarbon molecules that inhibit the growth of crystals. These coatings may someday be used to control ice formation on airplane wings or to prevent damage to organ tissues when freezing them.[18]

Additional work involves the color characteristics of biologically derived optical thin films. The iridescent colors of insect wing membranes can be altered by physical means such as pressure, swelling, or shrinking. The colors of butterfly or moth wings are derived from the reflectance of sunlight. The wing surface is composed of a flattened stack of reflective ridges. Pigment granules are found at the base of each ridge; these ridges are pumped up or depressed as the wing tilts, leading to variations in the index of refraction on the ridges. A similar film could be used to sense alterations in the internal pressures of automobile tires or food packaging.[19]

The U.S. Department of Defense and the automobile industry have both been financial supporters of biomimetic research because qualities such as toughness, resistance to fracture, flexibility, and durability have significant value in the manufacture of vehicles and damage-resistant surfaces. Meanwhile, research sponsored by the U.S. Department of Agriculture and some large chemical, agricultural, and pharmaceutical companies has promoted somewhat different directions for these life sciences.

12.4   Biotechnology

Over the years, various efforts have been made to improve the performance of the microorganisms used in fermentation processes. Until the 1970s, most of these efforts relied on selective breeding or grafting of one organism onto another. Since these processes produce relatively random results, researchers had to painstakingly select for desired characteristics or rely on mutations to generate new organisms. While technologies have been developed to encourage and direct the mutations of desired cells, these were long and arduous processes. Natural hybridization by selective breeding follows quite narrow lines of development, and these processes had consistently run up against limits because natural microbes are difficult to modify in ways that can achieve highly efficient, large-scale production of materials. Beginning in the 1970s, the possibility of bypassing these limits emerged with the advent of the new technologies of genetic engineering, which provide a much wider latitude for modifying and tailoring organisms to achieve specific commercial objectives.

Many biobased processes and products are referred to under the general heading of biotechnology—a term used historically in a broad sense to mean any production process that uses living organisms. However, today most people use the term *biotechnology* in a narrower sense to refer to a new form of bioprocessing—"new biotechnology"—that involves the use of cell fusion and recombinant genetic engineering. They distinguish this form from "old biotechnology," which relied on plant breeding, fermentation, and conventional enzyme isolation.

The new genetic engineering techniques have been made possible only since the first cloning of a gene, during the 1960s. At that time researchers identified a pattern of information in the DNA (deoxyribonucleic acid) of genes which provides the code that directs a cell's synthesis of proteins. During the following decade the enzymes that cause DNA chains to join together or separate were isolated. Experiments at Stanford University and the University of California in 1973 led to the first introduction into the gene of one species' DNA that had been fragmented from the gene of another species—thereafter called "recombinant DNA."[20]

Most chemical processing based on genetic modification involves either recombinant DNA or cellular fusion. Those processes based on recombinant DNA involve the splicing together of pieces of DNA from different organisms and inserting this hybrid material into a vector (a carrier, usually a plasmid) that is then introduced into a host cell. Cellular fusion is another form of genetic engineering in which the walls of cells are stripped away by enzymatic treatment to create protoplasts that are then fused to form new cells with combined genetic material. By using specific reagents or electrical fields, these fusions can be induced between cells of species that are sexually incompatible. In either technique, the new cells, which now "express" the new DNA, are allowed to reproduce to produce organisms, plants, or cultures large enough to meet commercial requirements. By bringing together in one organism genes from two or more organisms, particularly where the genes from a higher organism are introduced into bacteria or yeasts, it is possible to increase the yield from conventional fermentation processes or to produce entirely new transgenic substances.

Genetically modified chemical production has opened up a plethora of new commercial opportunities. A large share of these products have found applications in the pharmaceutical and health care industries. During the past decade a host of new genetically modified agricultural products, including soybeans, corn, and tomatoes have come onto the market. In 1996, the first year that genetically modified agricultural seeds were widely available, farmers planted over 4 million acres with these seeds, and 3 years later the worldwide estimate was 124 million acres. In the United States, it is estimated that one-third of the soy crop is now grown from transgenic seeds. More is on the way. The Department of Agriculture claims that 4500 genetically altered plant varieties were tested by 1999, with over 1000 tested in that year alone. About 50 had been approved for unlimited use, including 13 varieties of corn, 11 tomatoes, 4 soybeans, and 2 squashes.[21]

The use of recombinant DNA technology to produce industrial materials has proceeded much more slowly. Some work has started on industrial applications for proteins, enzymes, and amino acids. For instance, genetic engineering has been used to improve the development and use of industrial enzymes. Enzyme-catalyzed processing requires inexpensive processes for enzyme production if it is to achieve competitive prices. Genetically modified enzymes can be effective in a variety of biochemical processes that offer industrial applications. The yield of natural fermentation processes can be improved by transferring the gene encoding the enzyme to a microorganism capable of producing the enzyme in large quantities. In addition, the enzyme itself can be modified by gene encoding to tailor it to meet specific functions. For instance, genetic variations of subtilisin, an enzyme used in detergents that is easily destroyed by common bleaches, can be generated using biotechnology to render the enzyme more resistant to bleach. In the future, genetically modified enzymes may be developed for use in harsher solvents and at higher temperatures than natural enzymes would tolerate.[22]

The development of new biotechnology based on genetic engineering offers a host of new opportunities, yet public reaction has been mixed. The range of new pharmaceuticals developed from genetic manipulations has been generally well accepted. However, public concerns have arisen about health and safety as genetically altered crops have been developed and commercialized and genetically modified ingredients have been used in the food industry. Particularly in Europe, these concerns have erupted into a lively and dramatic oppositional campaign. This opposition has led some large food producers and retailers in Europe to forgo the use and sale of all food products based on genetically modified organisms. How the issue of safety will affect the development of genetically altered industrial materials has been little tested because these initiatives are still modest in scale and little recognized by the general public.

12.5   Biopolymers

Conventional synthetic polymers such as nylon, polyethylene, polyester, polypropylene, and polyurethane are produced from nonrenewable petroleum products. Most are not biodegradable. The term *biopolymer* is used to describe a collection of materials that fall into three principal categories:

Table 12.1
Common biopolymers

Polyesters	Polysaccharides (plant/algal)
Polyhydroxyalkanates	Starch
Polylactic acid	Cellulose
Proteins	Agar
Silk	Alginate
Collagen/gelatin	Pectin
Elastin	Konjac
Resin	Polysaccharides (animal)
Polyamino acids	Chitin/chitosan
Soy, casein, whey gluten	Hyaluronic acid
Polysaccharides (bacterial)	Lipids/Surfactants
Xanthan	Acetoglycerides
Gellan	Emulsan
Curdian	Polyphenols
Polygalactosamine	Lignin
Dextran	Humic acid
Levan	Tannin
Cellulose (bacterial)	Specialty Polymers
Polysaccharides (fungal)	Shellac
Pullulan	Poly-γ-glutamic acid
Elsinan	Natural rubber
Yeast glucans	

Source: U.S. Congress, Office of Technology Assessment, *Biopolymers: Making Materials Nature's Way*, Washington, D.C.: U.S. Government Printing Office, 1993.

• polymers that are derived from biological starting materials, such as amino acids, sugars, or natural fats and oils;
• polymers that are derived through the use of biological processes, such as plants or microorganisms; and
• polymers that biodegrade in natural settings, such as compost.[23]

Table 12.1 lists several common biopolymers. In 1992 biodegradable polymer resins captured less than 5 million pounds (0.08 percent) of the plastic resin market and this market was expected to grow only modestly. Today there is increasing interest in biopolymers. The three largest potential markets for biodegradable polymers are food and nonfood packaging, personal and health care products, and other disposable products.

Much of the interest in biopolymers has been generated by concern over the environmental impacts arising from the disposal of petroleum-based plastics. For years, polymer research focused on increasing durability and chemical stability. The result has been an array of commodity plastics that resist microbial degradation. The design goals have now changed. Experience has shown that plastic wastes degrade slowly in landfills, add to litter in the landscape, and present a hazard to wildlife, so states and local communities have begun enacting bans on the use of nondegradable plastics in food and product packaging.

The rising concern about plastic wastes has been most acute in terms of wastes discarded at sea, and during the late 1980s the international Oslo convention on the discharge of wastes from vessels at sea was amended to require that all wastes dumped from ships at sea be biodegradable. This requirement compelled the U.S. Navy and several leading U.S. corporations to invest in research on biodegradable polymers. Most of this research involves corn, wheat, and potato starch, which produce a fully degradable product that is two to ten times more expensive than petroleum-based plastics.[24]

Research begun in the 1980s has branched into three distinct areas: photo-self-destruction, copolymers of plastic and degradable carbohydrates, and microbially derived polymers that degrade easily. Photodegradable polymers are used as plastic sheets on agricultural fields, where they degrade quite easily at a predetermined time when their photostabilizers become inactive. Polymers that are copolymerized with up to 50 percent cellulose or starch biodegrade relatively easily, but are difficult to process and demonstrate poor performance properties. Thin packaging films of polyhydroxybtyrate (PHB) derived from microbial fermentation and copolymerized with polyhydroxyvalerate, demonstrate significant deterioration in 10 days of anaerobic treatment, but the costs of production are relatively noncompetitive.[25]

Early research focused on the polymerization of lactic acid and developed specialty biodegradable polymers for medical uses such as reabsorbing sutures and biobased prosthetic devices, but the production costs were too high to compete well in commodity plastics markets. In 1996

Cargill began producing polylactic acid (PLA) as a thermoplastic, but the product was not cost competitive. More recently, polylactic acid has attracted new interest as a result of research carried out at Argonne National Laboratory hear Chicago. The lactic acid has been produced by the fermentation of a sugar derived from potato starch. Because the starch-based production costs are well under those for petroleum-based lactic acid, there is a substantial commercial opportunity for this biopolymer. Polylactic acid has been approved by the U.S. Food and Drug Administration for food packaging, so the range of applications is broad.

Du Pont and ConAgra have also been trying to market starch-based polymers. These polymers have been produced from wheat starch as an alternative to polystyrene packaging, but with limited success. Zeneca Bio Products (formally ICI Bio Products) initiated commercialization of a series of polyhydroxyakanoate (PHA) copolymers (specifically, polyhydroxybutyrate–polyhydroxyvalerate or PHVB) under the trade name BIOPOL. The prototype PHA, polyhydroxybutyrate, is a polyester homopolymer that is stored as an energy source in the cells of many bacteria, including a common organism called *Alcaligenes eutrophus*, which occurs naturally in water and soil. This bacterium can be easily grown in large fermentation vessels on a variety of substrates, and the polymer is easily harvested from the cells. By inducing the bacterium to produce a second polymer, polyhydroxyvalerate, the copolymer, PHVB, develops characteristics quite comparable to polypropylene.[26]

Polyhydroxybutyrate was first discovered in 1927 at the Pasteur Institute in France and first offered commercially by W. R. Grace in the 1950s. All synthetic PHA is produced through fermentation of a carbon source by bacteria, and most products biodegrade easily in microbially active environments. PHVB is broken down through bacterial secretions (depolymerases) into hydroxybutyrate and hydroxyvalerate, which are then consumed by cells as they grow. This results in water and carbon dioxide (and methane under anaerobic conditions). Zeneca has developed markets for BIOPOL as resins for films and paper coatings. Other PHVB polymers are being used in biodegradable cosmetic containers and, in Japan, in fishing nets.[27]

In recent years, the advent of genetic engineering and recombinant DNA technology has permitted significant improvements in biopolymer production and in the purity and performance of biologically derived polymers. For instance, the mechanical and chemical properties of structural polymers such as silk, elastin, and various adhesive polymers has been improved using genetically modified genes. PHA and PHB have been produced from a genetically engineered plant (*Arabidopsis thaliana*) under laboratory conditions.[28]

Biopolymers that are biologically produced and biodegrade may be more environmentally appropriate than synthetic polymers, but this is not a necessary conclusion. The manufacturing process, the disposal process, and the conditions of their use may generate significant threats to the environment that can only be avoided if they are considered as a part of the full life-cycle design process.

12.6   The Biobased Materials Industry

The largest sector of firms engaged in biobased materials consists of the agriculture and forest products industries. These industries are made up of thousands of small materials production firms and a much more concentrated number of large materials processing corporations. The largest food processing companies include Archer-Daniels-Midlands, ConAgra, and Cargill, while the largest forest products processing firms are International Paper, Georgia-Pacific, Kimberly-Clark, Boise-Cascade, Champion International, Mead, and Weyerhauser. Several of these corporations, particularly ConAgra and Weyerhauser, have industrial product lines, including alcohols, solvents, tars, and plastics, but in general the investments of the agriculture and forest products industry in industrial applications of biobased materials have been modest.

The chemical industry has been far more dynamic. The global restructuring of the chemical industry has sent many large firms into major mergers, acquisitions, and investments in the life sciences or biotechnology field. Monsanto, Du Pont, and Lubrizol have invested in agricultural and pharmaceutical companies. Du Pont recently divested itself of Conoco, which embodied most of its petroleum processing capacity, in order to focus on more service and information products in the life sciences. Monsanto spun off much of its chemical business into a new entity called Solutia so as to focus on biotechnology products such as genetically altered seeds. Dow and Du Pont have also invested heavily in biotechnology. Dow has developed a selection of more tailored and short-lived pesticides to replace its chlorinated pesticides. Du Pont's agricultural products division is also pursuing crop protection chemicals that are less toxic and have a lower use rate. Similar trends are occurring in Europe, where Hoechst and Rhone-Poulenc have sold off chemical businesses to concentrate on commercial opportunities in the life sciences.[29]

The new biotechnology industry in the United States grew rapidly after the first firm (Genetech) was founded in 1976. By 1988 there were 400 dedicated biotechnology companies, and over 70 major corporations had significant investments in biotechnology. Although the patenting of biotechnology inventions has been substantial and controversial, throughout the 1980s most of these firms had no sales and were losing money, which led to a significant run of mergers and acquisitions during the early 1990s. Today, there is a mix of corporate arrangements with some large, independent biotechnology companies such as Genzyme, Amgen, and Biogen; a host of small startup companies; and substantial biotechnology divisions in large chemical and life science corporations such as Monsanto, Du Pont, and Eli Lilly.[30]

The integration of the chemical and biotechnology companies offers a potent foundation for new materials. At present many of these products are pharmaceuticals, medical products, and foods. However, the basis has been laid for industrial applications and most of this market is simply waiting for a rise in the price of petrochemicals.

12.7   The Opportunities and Risks of Biobased Materials

In the long run, the biological sciences may have a major impact in shifting the production of fuel and bulk chemicals from a dependence on nonrenewable resources such as oil and natural gas to renewable resources such as biomass. Already, the bulk production of ethanol, acetone, and butanol from biomass has been pilot tested and technically refined. Developments in Brazil and the United States have demonstrated the potential to use biomass alcohol as a petroleum substitute, but these projects have been supported more as a means of finding uses for farm surpluses than because biofuels are likely to compete effectively with petroleum.[31]

Bioprocessed materials have many advantages over petrochemical materials. The feedstocks are typically renewable and ubiquitous. Most biobased materials are processed in simple aqueous environments at ambient temperature and pressure. Biocatalysts are highly active and selective; they produce no environmental problems, as do heavy metals; and they generate energy internally. Finally, bioprocesses generally produce biodegradable products and by-products. Among their disadvantages are the number of steps required to prepare feedstocks from the natural biomass and the low product concentration in the aqueous solution. Since organisms are not viable at high concentrations, fermentation solutions tend to contain a wide mixture of organisms and nonproduct compounds, making the recovery of the products and their purification difficult and expensive.

There is concern over the safety of biobased materials. In order to support microorganisms, most biological processes are nontoxic, which means that the solutions can be likely breeding media for bacteria and viruses of health concern. For example, the use of bacteria proteases in laundry detergents was rapidly reduced when production workers and consumers complained of allergies. Only after the introduction of microencapsulation processes were dustless proteases reintroduced into detergents.[32]

Much more concern has arisen over the safety and possible environmental effects of genetic engineering. These concerns have focused on the introduction of modified species, the spread of novel traits in existing populations, the creation or enhancement of pests, the unintended effects on population dynamics, and potential changes in ecosystems. For years, public interest in the risks of the new biotechnology has been balanced by the anticipated benefits of new drugs and life-enhancing medical applications. As genetic manipulation technologies have begun to be adopted into agriculture and result in genetically altered foods, public concern has risen rapidly. While this phenomenon has been more evident in Europe, the level of concern has also begun to rise in the United States.

European governments have been quite active in addressing the potential risks of genetically modified organisms. During the 1980s, Denmark and Germany passed legislation specific to genetically engineered products, and the European Union has enacted two directives that set standards for laboratory and industrial operations and for marketing and deliberate release of such material.

The United States has been more relaxed in its regulatory approach to genetically modified organisms. In 1976 the National Institutes of Health published guidelines covering safety precautions for research on genetically modified organisms. Over the following decade, these guidelines were revised and eased as more experience in laboratory work led to less scientific concern and today laboratory research is subject to minimal restrictions. However, genetically modified products have been subjected to regulation by both the EPA and the U.S. Food and Drug Administration. The FDA has been responsible for regulating and testing pharmaceuticals and foods produced through genetic modifications, while the EPA has sought to regulate the environmental release of genetically modified organisms, particularly those plants that produce insect-killing proteins, such as *Bacillus thuringiensis*. The FDA does not require premarket safety tests on genetically modified foods and has resisted calls for consumer labeling requirements. The EPA has been restrained in its approach because it does not have a clear statutory mandate. The agency has tried to use its authority under the Toxic Substances Control Act to regulate recombinant DNA products, but problems have emerged because altered microbes do not easily fit the statute's chemical definitions, and because the risks of genetic engineering remain more hypothetical than proven.[33]

The search for more biologically compatible materials and production processes is well on its way. Some of the technologies are old and well defined; others are highly experimental; and some raise significant concerns. It is one thing to make industrial use of materials that are already available in nature's bounty; that is the genius of using naturally occurring, renewable materials. It is quite another thing to learn the principles of natural processing and apply them in novel ways—ways that may be new and untried within the context of natural ecosystems. Science writer

Ivan Amato notes that biology works within small ranges of physical conditions that can continually support life, while human processing is unbounded by natural precedents and can occur in quite life-threatening environments. Furthermore, the logic of evolution, according to Amato, does not advance toward optimal forms of production, but rather toward "perpetuating evolutionary traits that are merely adequate for the survival and reproduction of a population under prevailing environmental conditions."[34]

The prospects of biosynthesis, biomass conversion, and the biocompatible cycles of bioprocessing and biodegradation offer avenues for developing materials that are as attractive as the prospects of sustainable harvesting of natural resources. If they are combined with the potential for recycling organic wastes from the forest, agriculture, and food industries or the effluent of municipal sewers, the possibilities appear endless. The vision of materials factories of the future that look like large greenhouses with long tanks of microbially infused carbohydrates in which tiny organisms manufacture well-tailored materials from recycled organic wastes is intriguing and attractive. There is an abundance of potential research investigations and production opportunities here. Indeed, the technological possibilities for new industrial materials based on natural processes are rich and varied enough that it appears unnecessary to leap to gene-altering technologies that raise unexplored risks. A rush to invest in genetically modified industrial materials or materials development processes appears cavalier when so many natural processes remain unexplored and untapped.

However, the prospects for biobased materials and processes and a materials economy based more on renewable and recyclable materials are dimmed by the current dominance of petrochemical materials. Without near-term commercial prospects, market interest and private investment are likely to be slow to encourage these alternative materials and pathways. The detoxification and dematerialization potentials of bioprocessing will require more than current market incentives and current government policies.

IV

Toward a Sustainable Materials Policy

13

Dematerialization

Less is More.

—Ludwig Mies van der Rohe

Following the publication of *Limits to Growth*, the Club of Rome continued to raise questions about the environmental effects of economic growth. Criticism of the first report was intense, but so were its praises. While the report met its objectives by raising a debate about materials scarcity and environmental capacity, there was always a note of gloom and foreboding within that public discourse. In 1980 a follow-up report was issued called *A Dialogue on Wealth and Welfare*, which proposed another kind of economic growth that would not be so dependent on resource consumption. However, it was not until 1997 that the Club of Rome would sponsor a report on resources and consumption that projected a more upbeat, solution-oriented approach. This report was drafted by two American leaders in energy conservation, Amory and Hunter Lovins, and Ernst von Weizsacker, the president of a prestigious German research center, the Wuppertal Institute. In their report, the Lovins and von Weizsacker proposed that sustainable levels of development could be achieved by increasing resource productivity by a factor of four. The report, which is titled *Factor Four*, provides some fifty examples of programs around the world that demonstrate that "we can live twice as well—yet use half as much."[1]

Resource use was back on the table, but with a different twist: attention had now turned to the efficiency with which resources could meet human needs. Avoiding the controversies about scarcity, the report promotes increased efficiencies through changes in energy technologies, materials use, and production and consumption practices. The tone is optimistic and pragmatic, and at times simplistic. But *Factor Four* marks a new optimism that was symbolic of the last years of the millennium; an optimism that businesses, markets, and community partnerships, working together can create a productive and sustainable resource management system.[2]

Increased resource productivity means getting more out of the use of materials. Intensifying the use of materials provides a powerful strategy for dematerializing the economy because it reduces the drain on the planet's resources. There are other strategies for dematerialization, such as recycling materials and decreasing our dependence on materials for achieving a high quality life. Before turning to those ideas, it is worth examining the progress in resource productivity already under way and the lessons that it offers.

13.1   Resource Productivity and Dematerialization

In a culture that values efficiency, resource productivity sounds like a worthy goal, and it is, because it offers an inexpensive way to get more value out of what we are already consuming. Resource productivity involves increasing the efficiency of materials and energy use, and this means that the extracted value per unit of material could be increasing while the amount of material consumed could be decreasing. If resources were progressively scarce and prices increased relative to wages, resource productivity would be an expected result of market adjustments. However, labor productivity has increased substantially and raw material prices in the United States tend to be falling relative to wages, so we might expect little resource productivity. Instead, there appears to be a long-term trend toward resource productivity. Resource costs are a factor in determining both interfirm competitive advantages and the global competitiveness of U.S. industries, and this is driving resource productivity and by definition, promoting dematerialization.

Early research on the intensity of materials use (dematerialization) was conducted by Wilfred Malenbaum for the National Commission on Materials Policy and used a ratio of material consumption to gross domestic product to study twelve major metal and mineral ores in ten world regions between 1950 and 1975. Following the work of Nicholas Kondratiev and Simon Kuznets, Malenbaum predicted that the intensity of materials use in a developing economy follows a bell-shaped curve, rising to a plateau and then falling. He was able to show that in the United States, the intensity of use for iron, steel, manganese, copper, zinc, and tin reached a peak during the 1950s; nickel, cobalt, and tungsten peaked during the 1960s; and chromium, aluminum, and platinum appeared to be peaking during the 1970s.[3]

This bell-shaped (or inverted U-shaped) curve has been much studied by resource economists. Some have argued that it demonstrates that as an economy matures and incomes rise, economic growth gradually "decouples" from materials use. For instance, Eric Larson and his colleagues studied seven materials (steel, cement, aluminum, chlorine, ammonia, ethylene, and paper) and noted a declining intensity of use as a function of increasing national income, leading them to conclude that the "era of materials" was over. Another resource economist, Michael Janicke, completed a study using four indicators of materials consumption in thirty-one countries for the years 1970 and 1985 and also found a substantial "delinking" between such consumption and economic growth.[4]

Still, not every study has demonstrated delinking. A follow-up study by Janicke that included more materials found substantial differences among them, with intensity of use inversely related to economic growth for some materials (steel, aluminum, cement), but directly related for other materials (paper). Other studies have found that dematerialization may be followed by "rematerialization" as economic growth continues. Thus, conclusions on the overall impacts of dematerialization are cloudy and this has led Boston University's Cutler Cleveland and Matthias Ruth to conclude that "although there is a wealth of studies that document improvements in the use of individual materials, we have a much less complete picture ... for broader classes of materials and for aggregate material use."[5]

Iddo Wernick and Jesse Ausubel of Columbia University have attempted to track trends in the intensity of materials use using data from the U.S. Bureau of Census and U.S. Bureau of Mines. They found that an analysis of the absolute consumption of all physical materials by weight reveals significant fluctuations, but no trend. However, when these data are normalized to gross domestic product, the consumption per unit of economic activity shows a dematerialization in the weight of physical materials of about one-third since 1970. Not surprisingly, trends in intensity of materials use vary by material. In terms of weight per unit of economic activity, the use of wood has declined at a steady pace since 1900. Since the 1950s, the intensity of steel, copper, and lead use in the overall economy has also fallen. However, the consumption per unit of economic activity of plastics and aluminum has substantially increased over this same period, as has the consumption of phosphorus and potash, both key ingredients in agricultural fertilizers.[6]

A different picture appears if the volume rather than the weight of materials is followed. While the weight of materials may be stable or declining, the volume of materials is gradually increasing. Wernick, Ausubel, and their colleagues show that the volume per capita of a combination of wood, paper, metals, and plastics has been increasing since 1970, with low-density plastics accounting for a substantial share of this expansion in volume. They conclude that "Individual items in the American economy may be getting lighter, but the economy as a whole is physically expanding."[7]

One recent study of five metals—copper, aluminum, zinc, lead, and iron and steel—found that the domestic markets are stagnant or declining as new materials such as plastics and ceramics and recycled materials from secondary markets consume increasingly larger market shares. With the exception of zinc, these metals show growing rates of recycling and consumption patterns that are tending toward less dissipative uses. This leads to a slower than predicted growth in extraction of virgin materials and a stable or declining rate of materials and energy use in the industries involved in processing and refining these metals. The result is that the reserves of these five metals are being drawn down at a diminishing rate. These trends contrast markedly with those of the 1970s, when the consumption rates for these same metals were growing exponentially.[8]

Another study of these same metals has tried to examine the effects that might result from decreasing primary production from mines and increasing the rate of recycling. This study shows that current rates of extraction and recycling will lead to a serious strain on reserves in the next 50 years, with the exception of aluminum, which is currently very abun-

Box 13.1
Operating Principles for Sustainable Development

> For renewable resources, such as air, water, and biomass, the sustainable rate of use can be no greater than the rate of regeneration.
> For nonrenewable resources, such as high-quality minerals and fossil fuels, the sustainable rate of use can be no greater than the rate at which a renewable resource, used sustainably, can be substituted for it.
> For a pollutant, the sustainable rate of emission can be no greater than the rate at which it can be recycled, assimilated, or degraded in the environment.
>
> Source: Herman Daly, "Toward Some Operational Principles of Sustainable Development," *Ecological Economics*, 2, 1990, pp. 1–6.

dant and well recycled, and zinc, which will be simply extracted to exhaustion by current trends. Using several scenarios, the study then notes that even if the extraction rate for these metals is cut to 2 percent of reserves per year, these strains will continue. Even with such a sharp reduction in extraction rates, the critical factor for achieving a sustainable rate of use of these materials involves substantially increasing the recycling rate.[9]

These studies reveal the importance of the rate as well as the volume of materials flow. The conventional materials balance model is usually presented in static form. The materials balance is computed at a still moment of time and does not account for rates of flow. The three-loop materials flow model introduced earlier provides a framework for how such flows might be managed, but no sense of the appropriate rates for the flows. To extend the model, we need some kind of guidance for the rates of flows as well. Herman Daly, a University of Maryland economist, has offered a relatively simple set of operational principles by which this might be accomplished.[10] These principles are presented in box 13.1. In order to translate these principles into operational rules within the materials flow model, they need to be adapted and restated as follows:

• For a biocompatible, renewable resource, such as air, water, or biomass, the sustainable rate of use can be no greater than the rate at which the resource can be recycled or regenerated.
• For a biocompatible, nonrenewable resource, such as ferric metals or fossil fuels, the sustainable rate of use can be no greater than the rate at

which the resource can be recycled or the rate at which another resource, used sustainably, can be substituted for it.

• For a biocompatible material, the sustainable rate of release to the environment can be no greater than the rate at which the material can be assimilated.

• For a nonbiocompatible resource (renewable or nonrenewable), such as organometallics and halogenated hydrocarbons, the sustainable rate of use cannot exceed the rate of recycling or the rate at which the resource can be treated to render it assimilative in the environment.

• There is no sustainable rate of release to the environment for a nonbiocompatible material.

This changes the idea of limits. It is not a limit of industrial materials, of unpolluted environment, of space, or of any other finite resource that is the governing factor of growth. The natural environment is a dynamic system that is continuing to cycle and recycle materials. Thus, it is the limit on the rate at which materials can flow, the speed of material extraction and disposal, and the velocity of throughput that are the true limits of capacity. Materials may be drawn from the environment, but not at rates that exceed their regeneration, or rates at which they can be replaced with regenerative materials; and biocompatible materials can be discarded into the environment, but not at rates that exceed the assimilative capacity of environmental media. Of course, volume is still important—the movement of huge volumes of materials has a significant impact on energy consumption and landscape destruction, but the rate of materials flow is what determines the longer term consequences of economic activity on the environment.

As we have seen earlier, these rates can be changed by technological innovations and improvements in resource productivity. The rate of regeneration of economically important renewable resources has been increased by advances in agricultural practices. The rate of use of nonrenewable resources such as iron and copper has been increased by the substitution of increasingly lower quality sources. Yet, even as these rates may be modified by the maturing of technology and practice, it remains important that they not exceed the capacities of environmental services.

This is not the case today. Peter Vitousek, a Stanford University biologist, has tried to compare the rate of materials use with the rate of natural regeneration. By computing the amount of solar energy captured by the photosynthesis of primary producers (plants) and deducting the amount of energy required to support those producers, Vitousek has developed a unit of "net primary production." Using this concept, he calculates that humans today are consuming nearly 40 percent of the planet's potential (terrestrial) net primary production and with one more doubling of the world's human population, nearly all of the net primary production would be consumed by this one species.[11] The current rate of consumption of some renewable (tropical and old growth woods) and several nonrenewable materials (lead, mercury, and zinc) is already unsustainable. And, while the rate of environmental release of some targeted toxic emissions is declining, the emissions rate for carbon dioxide, nitrous oxide, and methane is increasing, and the disposal rate of both biocompatible and nonbiocompatible wastes often exceeds the assimilative capacity of local ecological systems.

To achieve a sustainable economy, these flow rates must be reduced to sustainable levels. This will require increasing the rate of recycling and reuse and increasing the productivity of materials. These are two of the principles of dematerialization. Dematerialization can also be advanced by simply using fewer materials to meet the needs of daily life. This is often viewed as a form of substitution in which nonmaterial services substitute for materials. Conceptually, then, there are three general principles that characterize the dematerialization of the materials economy. These include

• closing the loop on material flows,
• increasing the intensity of material use, and
• substituting services for products.

The first requires changes in the way we design and use products, manage processes, and handle wastes. The second has to do with the efficiency with which we obtain value from our materials. The third transcends materials to seek a high-quality life less determined by the amount of materials we consume.

13.2   Closing the Loop on Materials Flows

Once materials have entered the economic system, there is great value in keeping them in use as long as possible through continued recycling and reuse. The ecological costs of extraction, particularly the impacts of moving large volumes of materials to claim and refine ores and fuels, and the costs of disposing of discarded materials back into ecological systems, are all reduced and potentially avoided by keeping materials actively cycling within the economy (loop 3).

It has already been noted that recycling and reuse cannot go on indefinitely. Some deterioration in materials quality is always a consequence of use and reuse, so there is always a baseline need for some material inputs. However, a concerted effort to adopt industrial materials that recycle well and then to maintain the infrastructure for continuing their cycling could significantly reduce the flow of materials between the environment and the economy and substantially reduce the dissipation of materials.

Recycling Materials

Recycling rates are on an upward trend in countries throughout the world. As nonrenewable materials grow more scarce, substitution is encouraged, but so is recycling. Copper provides an example. In 1910, recycled copper accounted for about 15 percent of production in the United States. As continued mining depleted the high-quality copper ores, copper prices rose relative to other metals and by 1996, nearly 35 percent of U.S. copper production came from recycled copper.

Recycling of process chemicals, structural materials, or finished products means that materials otherwise destined for disposal are continued in new products. Industrial scrap and postconsumer wastes contain large amounts of valuable materials already refined and processed. Up until the point of discard when the materials are mixed and comingled, they still retain a reasonable economic value and an attractive potential for recycling. We saw earlier the significant economic value that scrap recycling offers in iron, copper, lead, and aluminum.

Huge volumes of industrial wastes can also be recycled. Conventional industrial waste treatment processes often involve separation techniques that produce a filter cake, sludge, or slag that may be rich in materials with enough market value to justify refining and reclaiming them. Cadmium, chromium, copper, nickel, and zinc can all be "mined" from metal plating wastes and some spent processing solutions. Many secondary smelters accept wastes containing lead, copper, and zinc. Some processing solutions such as solvents, acids, and caustics are also reclaimable and can be recycled back into manufacturing processes.

As manufacturers come to rely more on recycled materials, they begin to alter their processes to facilitate recycling. Aircraft manufacturers stamp the alloy composition on parts during manufacture so that they can be easily identified during disassembly in order to facilitate recycling. Some leading automobile companies have been designing car components that can be easily removed and recycled when the car is discarded. These efforts have spawned a whole new area of product design called "design for disassembly" or "design for recyclability" in which products are designed with the intention of reclaiming them, disassembling them, and reusing or recycling their materials or components. Mitsubishi makes a washing machine that can be fully disassembled with only a screwdriver, and the German automobile maker Audi has developed a prototype for a fully recyclable sports car. The products of such design processes often exhibit a low number of constituent materials, a high degree of removable components, and an assembly process dependent on reversible snaps, slides, screws, and clamps.

Extended Producer Responsibility

In 1991, Germany enacted a bold new law called the *Ordinance on Avoidance of Packaging Waste*, which required that manufacturers and distributors of products be responsible for the reclamation and processing of postconsumer packaging wastes. Under the ordinance, the government sets mandatory targets for recycling and allows industry to set fees on packaging materials. Industry responded by establishing the *Duales System Deutschland*, a consortium of over 600 firms, that collects, processes, and recycles any of the members' packaging wastes, all of which are identified by a green dot on the packaging. Today, Germany requires recycling for 75 percent of glass containers, 70 percent of tin cans, 60 percent of aluminum packaging, 60 percent of paper and cardboard, and 60 percent of composites; over 75 percent of all packaging carries the green dot. The result has been a 13 percent decrease in packaging in Germany between 1992 and 1997 compared with a 15 percent increase in the United States for the same period.[12]

The German takeback system has encouraged a host of different programs in Europe loosely referred to as "extended producer responsibility" and all basically designed to require that producers assume responsibility for their products throughout their life cycle, or at least to the point of disposal. The Netherlands uses covenant agreements with product manufacturers to encourage integrated chain management that creates a set of product responsibilities all during the life cycle of a product. The Swedish Eco-Cycle Act of 1994 also sets out producer responsibility plans for a wide range of consumer products, including automobiles, electrical appliances, batteries, packaging, and tires. More recently, a European Union directive has been proposed that would require manufacturers of electronic products to carry responsibility for the collection and recycling of all used electronic products, including computers.[13]

In the United States, the most advanced producer responsibility, or "product stewardship," programs involve state-mandated beverage container collection programs and battery takeback and recycling. Minnesota and New Jersey led with laws requiring battery producers to carry the financial burden of recovery and recycling of rechargeable nickel–cadmium batteries. In 1995, manufacturers of nickel–cadmium batteries launched a voluntary national takeback and recycling program with the establishment of a nonprofit Rechargeable Battery Recycling Corporation (RBRC). Today, the RBRC involves 285 U.S. and Canadian companies (80 percent of the rechargeable battery market), who pay a license fee to participate, and some 26,000 retail stores and recycling centers willing to accept used battery-charged products and send them to a central facility in Pennsylvania for recycling. Minnesota, California, and Massachusetts have pioneered policies that promote producer responsibility on products ranging from car batteries, cathode ray tubes, paint, and pesticides to floor carpeting. In addition, some firms such as IBM, Du Pont, and Castrol have set up pilot programs for testing product takeback schemes. Behind these various initiatives and proposals is a desire to close the loop on consumer products in the hope of reducing their contribution to waste streams and of encouraging manufacturers to design products that are easier to recycle and less likely to contain toxic materials that are costly to handle in reprocessing.[14]

Reusing Products

Both recycling and reuse reduce material dissipation. However, whereas recycling permits materials to flow frequently and rapidly through many applications in the economy, the reuse of products slows the flow of material from production to disposal. Many domestic products are voluntarily reused as they pass from parent to child or as they are transferred among neighbors and friends. Children's clothing and toys are often passed from one child to another or one family to another, resulting in increased use, less material per hour of service, and less aggregate waste. Various nonprofit and religious charities are organized to pass products (particularly clothes and furniture) along from user to user at a cost only large enough to cover transfer expenses.

Secondary markets provide another form of dematerialization. Automobile junkyards, used-furniture stores, used-car dealers, and used-musical instrument retailers contribute to the commercial activity of most cities. A lively used-product market flourishes through weekly classified advertising in newspapers and special regional marketing magazines. Through these local advertising media, products ranging from used cars and boats to used tools, clothes, furniture, and architectural ornaments can be purchased by simple telephone contacts. Country auctions, flea markets, and estate sales also find new users for used products.

Although reuse of industrial products typically requires less energy and results in no downgrading of a product's material, recycling is often preferred because it more effectively absolves the producer and the user of liability concerns. A product that is reused by several different users may still carry some liability for the original producer if it somehow causes damage. On the other hand, the shredding and mingling of materials often involved in recycling basically relieves the original producer of most liability concerns. Simple changes in the liability laws could reduce this preference.

Zero-Waste Production

Although unheard of a decade ago, there is considerable recent interest in designing industrial production processes that produce zero waste. While the concept of a zero-waste process is thermodynamically impossible, the goal is worthy as a motivator. Those who promote zero-waste processes suggest three primary strategies: recovering and reusing wastes within the production facility, recovering and selling wastes to other facilities, or bringing waste producers together into common locations where one facility's wastes can easily become feedstocks for another facility's production. Much of the emphasis is placed on "environmentally balanced industrial complexes" or "industrial clusters" as exemplified by the integrated industrial complexes that have emerged incrementally at Kalunborg, Denmark, and Styria, Austria, where firms regularly recycle wastes to feedstocks. These integrated complexes are in keeping with the models proposed by proponents of industrial ecology.[15]

13.3   Increasing the Intensity of Materials Use

Surplus wealth allows products to be made and used with little regard for efficiency. Products are larger than necessary, heavier than required, more energy intensive than needed, and they break and fall apart more rapidly than they should. When a higher value is placed on resource efficiency, intense use of materials and energy becomes a design goal. The development of many of the new advanced and engineered materials has been prompted by a desire to obtain more functional service from less and less material. Efficiency of service per unit of material becomes increasingly important. In some sectors, competition within the market has provided substantial incentives for increasing resource efficiency, but other areas have lagged. Internationally, the connection between increased efficiency and reduced ecological impacts has been promoted by the World Business Council for Sustainable Development under its concept of "eco-efficiency."[16]

An increase in materials intensity is often best approached as a design problem. Product and process designers can play an important role in orienting their designs toward environmental values. A subspecialty of product design, called "design for the environment," emerged during the past decade and promotes products whose manufacture, use, and disposal is more environmentally friendly. Propelled by a series of professional conferences on environmentally conscious design and sustainable product development, and a federal program within the EPA, these initiatives have promoted a life-cycle approach to product design that promotes resource efficiency and dematerialization.[17]

Decreasing the Mass of Materials in Products

The most often-noted form of dematerialization involves designing products that are smaller in scale or lighter in weight. Product miniaturization and product "lightweighting" mean that less material is used to provide the same level of service. Bringing to market lighter, stronger, and more durable materials increases the amount of value per unit of material used. For instance, the bulky vacuum tube-based radios of the 1940s were replaced by much smaller transistor radios of the 1960s, which today have been replaced by radios with circuitry as tiny as a single semiconductor chip. The cellular phone today is pocket size, whereas the mobile phone of two decades ago required a shoulder strap.

Commercial beverage containers are increasingly lighter. At the beginning of the century, beverages were conventionally sold in glass bottles. The first steel soft-drink can was marketed in 1953. A decade later, aluminum cans appeared and grew from a 2 percent market share in 1964 to nearly 90 percent of the soft-drink market in 1986. During this period, the weight of the aluminum can was cut by 25 percent. Polyethylene terephthalate containers began to replace some conventional glass bottles during the 1970s. The weight of the plastic 2-liter soda bottle has dropped 27 percent since 1990.[18]

Transportation vehicles are also getting lighter. Although aircraft have increased in size and weight, the weight per unit of payload, or per passenger-mile has been falling. New composites based on aluminum and titanium have lightened air frames, and plastics have replaced most metals in interior compartments. Since the 1970s the amount of steel in the average U.S. car fell by about 600 pounds or 35 percent. Much of this reduction was due to the substitution of higher strength steel and the introduction of various plastic, aluminum, and composite elements. Tires are also less massive. The development of radial tires during the 1970s not only reduced tire weight by 25 percent but also doubled the use life of the tires. However, although cars and tires are lighter, the average weight of the U.S. automobile fleet has been climbing recently owing to the popularity of small trucks and sports utility vehicles.

The miniaturization of products and the lightweighting of materials does have limits. Products cannot become so small that they become incompatible with the size and capacity of the human body. No car will sell if it is too small to comfortably hold a person. Similarly, no material can be reduced in weight or density to the degree that it fails to perform effectively or becomes unsafe. A light-weight car raises safety concerns. Likewise, if lighter products are less durable or if smaller products are less expensive, this may increase the likelihood that these products will be discarded frequently and lead to the manufacture of more products to serve the same need.

Extending the Useful Life of Products

The intensity of use can also be increased by extending a product's duration of service and thus slowing the rate by which it cycles within or through the economy. Today's smaller and lighter products are less likely to be repaired and more likely to be discarded. For instance, nonrubber shoe production in the United States increased from 802 million pairs in 1970 (3.9 per capita) to 1.1 billion pairs in 1985 (4.6 per capita) just as shoe repair and refurbishing services have dwindled.[19]

Instead of designing products for rapid disposal, often after the first use, a product can be designed for continued service through multiple uses. Extending the useful life of finished products maximizes the amount of value provided during their useful life, reduces the need for new products, and slows the rate at which they are discarded. Some products, like antique furniture, fine art works, or family heirlooms are preserved, indeed, revered, for their age and generations of use. They are often durable pieces that have significant craft energy embedded in them, and their continued use diminishes the need for replacement products and slows the flow of materials. These more durable products may be relatively heavy and more complex than disposable products, because they are built "to last." The beautiful hand-crafted ceramic cup that can be used and enjoyed thousands of times is certainly heavier than a paper or polystyrene foam substitute, which usually has a single use and no particular aesthetic appeal.

Durable products are what Walter Stahel of the Geneva-based Product Life Institute calls "long-life goods," which he compares to goods that are maintained in service through "product life extension." Product life extension involves keeping products going through periodic servicing and rebuilding, through occasional refurbishing and refinishing, and through adapting and upgrading. These products are similar to those that are repaired for reuse in that their care may require skilled service people who know the products and are capable of attending to them.[20]

Designing Products for Upgrading and Adaptation

The utilization period of a product's life can be extended by designing it to be adaptable over time as the needs of the user grow and change. A house is a good example. Houses are large capital investments for most people. They are unlikely to be discarded frequently. Instead, as a resident's life changes, or additional wealth is acquired, the house may be upgraded and adapted to meet new needs. Rooms may be added, wall colors changed, kitchens modernized, windows and decks added, but the basic framework of the house is maintained.

Designing products so they are composed of removable components permits them to be changed and adapted as the needs of the user change. For instance, personal computers have been designed to permit modification with new components, peripherals, and software. The next generation of personal computers may be designed as "eternal hardware" with "mutable software" designed for regular upgrading. Razors are designed to accept different blades without replacing the handle. Cameras, likewise, have been designed to permit later lens upgrades and peripheral adaptation. Adapting and upgrading products may require increased use of materials, but not on the scale required if the product were to be discarded and replaced with a new one.

Designing Products for Reconditioning and Remanufacture

Remanufacturing converts worn products into like-new condition. There is a growing interest in the remanufacturing of products, but it depends on the return of used products that still retain enough viable parts and components to make it worth disassembling and rebuilding them. Remanufacturing is facilitated where an infrastructure exists to return worn products to manufacturers and where the products have been designed for easy disassembly, repair, and reconditioning. A steady supply of interchangeable components and parts is also critical. Xerox has exploited remanufacturing with a full-scale "asset management" system that repairs, rebuilds, and re-leases or resells photocopiers returned by customers seeking to change or upgrade their photocopying equipment. Among other products that are currently collected, returned, and reman- ufactured are industrial production equipment, automobile mufflers, jet engines, railroad cars, passenger boats, and office furniture.[21]

Designing Products for Repair and Reuse

Some products are more likely to be repaired and reused than others. Automobiles, farm tractors, and airplanes are all expected to require repair and rebuilding throughout their useful life. They are designed as a collection of interchangeable parts that can be readily disassembled as needed so that worn or broken parts can be replaced. Many products, ranging from home entertainment centers to personal computers are also designed as an assemblage of replaceable components.

Where products can be repaired, they can be reused, thereby decreas- ing the number of new products required and the number of old prod- ucts discarded. Both office photocopiers and automobiles are designed for easy repair and have components that are readily screwed or snapped in and out. A large share of the automated manufacturing and assembly machines bought and used by manufacturing plants are designed to be disassembled, serviced, and repaired as needed.

Such repair and restoration is generally labor intensive. A product can be made by machine, but it takes human skill and insight to repair it once it is broken or worn out. Susan Strasser quotes an 1896 manual on mending and repairing: "But all repairing must be done by hand. We can make every detail of a watch or of a gun by machinery, but the machine cannot mend it when it is broken, much less a clock or a pistol!"[22] Product repair and product servicing have long offered employment opportunities for immigrants and those beginning an ascent up the eco- nomic ladder.

13.4   Substituting Services for Products

Over the past half century, the U.S. economy has been in a gradual tran- sition to a service economy. In a service economy, an exchange of ser- vices is the primary economic unit. Social services, health care, education, recreation, hospitality, communications, transportation, maintenance, and finances increasingly make up the largest growth sectors of the economy, while farming, mining, and manufacturing are diminishing in relative size. Today, nearly 70 percent of the workforce is employed in service-producing industries. As a percentage of the total workforce, a deceasing number of people are employed in producing materials or products. This does not mean that materials and products are not important in a predominantly service economy. But a service economy, in comparison with a commodity economy, has far greater potential to use fewer materials to maintain its vitality.[23]

By reconsidering the need to produce commodities to generate income, corporations and institutions grow by selling services. Liberating the function of a product from the object, itself allows producers to focus on the quality of their service as a means of developing and maintaining a market. This requires rethinking the nature of market demand. Customers want clean clothes, not a washing machine; mowed lawns, not a lawn mower; snow-free walkways, not a snowblower. Throughout the economy today there are hundreds of examples where services are substituting for products, and these transactions are adding to the dematerialization of the economy.

Selling Services Rather than Products

Companies that measure their success by the number of products they sell are engines of material throughput, rapidly converting raw materials into products. Some firms have begun to reconsider their core mission by learning to sell services rather than products. Instead of selling office carpeting, Interface, the largest commercial carpet company in the United States, now offers floor covering services on a monthly rental basis. Carpeting is laid in an office, repaired if damaged, changed during redecorating, removed when worn, and recycled when discarded, all without being sold to the user. Another U.S. company, Carrier, now provides indoor climate management services as an alternative to selling air-conditioning units. Office furniture and indoor plants can also be rented as a part of an office furnishing service in most large U.S. cities.

Service contracts make up a part of some product's sale price. Purchased photocopiers are usually bought with a service contract. Computers, automobiles, household appliances, and various tools can be purchased along with a maintenance contract and various warranty agreements that ensure that if the product fails within a certain period of time it will be replaced by the seller. Where the manufacturer or supplier offers a service agreement or a warranty, there is an incentive for the supplier to produce and maintain the product in a condition that extends its life and reduces the need for rapid replacement.

Many companies rely on leasing products to acquire the services of the product without outright ownership. Most office photocopiers can now be leased rather than purchased from the equipment manufacturer. Xerox has pioneered a leasing system that leases and re-leases photocopiers to customers, who pay primarily for photocopying services rather than photocopiers. Delivery trucks are often leased from large trucking companies. Several computer companies such as IBM and Dell offer business and personal computers on a leased basis. Automobiles can be leased rather than purchased as well. The lease typically includes a contract for all repairs and periodic servicing. The car remains the property of the automobile retailer, who is actually providing the transportation service of the car rather than providing the car as a commodity. At the end of a customer's contract, the vehicle can be returned to the retailer, who may then re-lease it or sell it into a secondary automobile market.

Sharing Products

Many products are shared in industrialized societies. Public facilities such as roadways, bridges, buses, hospitals, schools, and recreational facilities are all shared products because their public and "nonrivalrous" character means that one person's use does not preclude the next person's use. Where the use pattern of a product involves long periods of disuse or the acquisition costs are high, products may be shared among multiple users. In residential settings, ladders, lawnmowers, snowblowers, and chain saws may be owned by one resident and loaned out when needed by others. Laundromats provide shared clothes washing and drying machines. Tool and equipment rental stores allow customers to share the services of hardware and avoid individual purchases. Video rental stores give customers a wide choice of films by sharing the services provided by the individual videocassettes.

Taxicabs and rental cars are other examples of shared products. Taxicabs and their drivers are "rented" by the minute and mile, but involve a sequence of multiple users throughout their day of use. At nearly any airport in the world, a traveler can de-plane and rent a car for a few hours to a few weeks. The rental agency provides the transportation service along with a guaranteed service contract assuring the driver that they can get where they want without owning a car. Upon return, the car is cleaned, serviced, and made ready for the next traveler.

By focusing on transportation rather than car ownership, some communities have developed vehicle-sharing cooperatives that make jointly owned cars available on a reservation basis. Many recreational communities have similar time-sharing residential units that provide scheduled vacation accommodation rather than second home ownership. These communities may also provide guest accommodations, swimming pools, and function rooms that can be used on a reservation basis.

In industrial settings, firms may borrow equipment, share facilities, or rent time in another firm's laboratory to avoid the costs of duplicative investments. Marketing specialists at Dow Chemical have pioneered a system they call "rent-a-chemical" to provide customers with the services of industrial solvents without a purchase. The solvents are supplied to a customer until they are spent, after which they are reclaimed by a company that treats and recycles the solvents for further use.[24]

Converting to Information Systems

The Internet has opened up the world's fastest growing service industry. In less than a decade this service has grown from a limited-access network to a powerful tool linking more than 170 million people worldwide. Already the Internet has changed educational, financial, and commercial services. In the next 20 years it is estimated that half of what people read will come in electronic form.

The potential to dematerialize information storage and transaction services appears enormous. While the system requires a personal computer and a service connection, beyond that, the service is all information. Increasingly it is the software, not the hardware, that is of most value to the user. People can shop, browse, play, learn, create, and communicate with no direct flow of materials. Institutions that exist primarily to process information, such as brokers, booking agents, and banks, may need far fewer branch offices. Manufacturers and retailers will be able to more carefully tailor supply to demand, reducing the need for large inventories and customer rejects. Federal Express is already offering a service that allows manufacturing companies to have product parts delivered by many suppliers to a customer's door ready to be assembled on site, eliminating the need for assembly plants, warehouses, and large inventories. Home Depot has eliminated many warehouses by having up to 85 percent of its products go directly from suppliers to retail stores. New opportunities abound. Consumers with unwanted products could use Internet auctions to find buyers or contact original manufacturers to facilitate product takeback, and firms with useful wastes should be more easily able to find potential buyers.[25]

Still, dematerialization is not a given in the new information economy. Information, education, work, and entertainment are all available with little material consumption once the infrastructure is constructed. Although a desk terminal could provide access to this vast network of resources, the excessive retailing, upgrading, and discarding of personal computers distorts the otherwise significant efficiencies of these services. Indeed, a large part of the information processing services could be provided on large mainframes with desktop access instruments, obviating the need for material-intensive personal computers. It is also useful to remember the early proponents of the computer, who predicted a "paperless office." Instead, in spite of widespread adoption of computers, office paper consumption has grown by one-third during the past 10 years. Indeed, electronic texts could increase paper use if people print texts out in order to read them.

Electronic commerce is certain to dramatically change retailing and reduce household shopping trips. Customers will be able to comparison shop with ease, and this may increase the potential to find the most energy-efficient and environmentally friendly products. One recent study predicted that the use of electronic commerce during the next decade could avoid the need for 1.5 billion square feet of retail space, 1 billion square feet of warehouses, and 2 billion square feet of commercial office space. The energy savings would be equivalent to the amount generated at twenty-one average power stations and would result in a 35 million metric tons reduction in greenhouse gases. However, home deliveries by shipping companies will still require large amounts of vehicles and fuel. Indeed, it is not clear that the Internet itself may not become a large consumer of energy. One estimate suggests that over 30 percent of the nation's electric supply may be required to power the Internet within two decades.[26]

Simply Using Less

There has always been a modest movement in the United States that eschews material products and advocates a lifestyle that is free of unnecessary dependence on material goods. Those who dedicate their lives to religious orders often renounce material property. The Amish and Mennonite communities try to live simply, avoiding the purchase or use of unnecessary products. There are rural communities in New England, Southern Appalachia, the West Coast, and Alaska populated by families that have voluntarily cut to a minimum their commodity consumption and refuse to participate in the consumer economy. Many elderly people retire to small accommodations that require few material possessions. Finally, there are many citizens who live simple, low-commodity lives because they do not have the economic resources to do otherwise.

Although living without a wealth of material goods is often regarded as a degraded condition by most of the population, there is a growing literature that supports the idea of living simply and consuming less as a value worth celebrating. Whether by religion, ideology, or economic condition, many Americans demonstrate the richness of lives that can be lived with little material throughput.[27]

13.5   Reducing Dissipation

Increasing efficiency through technological improvements permits an increasing amount of service from products even as stocks and flows of materials decrease or remain unchanged. Changes in social practices and commercial services offer an even more direct assault on the rate at which materials flow through the economy. In considering strategies that promote dematerialization, we cannot be overly simplistic or doggedly confident. Because the objective is the dematerialization of the entire materials economy and not simply one product or function, it is useful to consider a comprehensive analysis that includes an assessment of the life cycle of all of the affected materials or products. The energy and environmental consequences of the extraction and disposal of materials may not be intuitively obvious. For example, the energy and materials savings from the dematerialization of the primary metals industries can not be considered a "net savings" for the economy as a whole before the impacts of the recycling processes and the production of alternative materials (plastic, ceramic, etc.) are deducted from those savings. Indeed, a reduction in weight and size of individual products and their longer use lives could be counterbalanced by a growth in the number of products.[28]

Therefore it is important how dematerialization is measured. Frederich Schmidt-Bleek of the German Wuppertal Institute has argued that resource efficiency cannot advance rapidly in the absence of good metrics for measurement. Early assessments relied on weight as a proxy for a material's utility. As materials become lighter but more voluminous, and as products are used more, but also by more people, weight becomes an increasingly less effective indicator. In order to more effectively promote dematerialization, Schmidt-Bleek has developed a metric based on the intensity of materials use per unit of utility. He focuses on inputs as the primary unit and develops an index based on material input per unit of service (MIPS) as a measure of potential environmental impact. A simple index of the material efficiency of gasoline is the number of miles traveled per gallon. A more accurate index would require that the denominator include all of the fuel consumed and materials moved to produce that gallon of gasoline throughout its life cycle (the materials input), and the numerator would include the number of people moved across those miles (the service). On this basis, a multiple-person vehicle with a high torque-to-weight motor moving on a low friction surface would appear more efficient than a contemporary single-person automobile on a highway. Schmidt-Bleek argues that while they are still somewhat crude, inputs of materials are a rough measure of flow and that normalization against "units of services"—where common units of service can be defined—would permit the indicator to be used to compare the use efficiency of various materials.[29]

Using the concept of materials intensity per unit of service, the Wuppertal researchers have developed estimates of flow per unit of service for materials ranging from gold to gravel. They refer to the amount of material to be moved or processed and the amount of energy consumed per unit of finished material as the "ecological rucksack," or the ecological burden of each unit of material. Thus the catalytic converter used in cars, which weighs less than 20 pounds, has been found to carry a rucksack of some 2.5 tons, because the mining and processing of the platinum requires the movement of such large amounts of overburden and ore. Two faculty members at the University of British Columbia, Mathis Wackernagel and William Rees, have tried to translate this rucksack notion into the amount of land required to maintain various levels of human consumption. By calculating that roughly 10 acres of "appropriated carrying capacity" are needed to maintain the current consumption level of a typical Canadian citizen, Wackernagel and Rees conclude that Canadians (and their U.S. neighbors) are already consuming three times their fair share of the planet's environmental services.[30]

Once there are metrics, there is an interest in goals. Setting a goal for dematerialization could be quite straightforward. Von Weizsacker and the Lovins suggest cutting consumption by half and doubling efficiency (factor of 4) as a goal for the short term. Schmidt-Bleek goes further. Considering the total impact of human activity on the ecological systems of the planet, he argues that the total materials throughput of the world's economies should be cut in half. Since per capita consumption is nearly five times higher in industrialized countries than in industrializing countries, and the populations of the industrializing countries are growing at a rate of 1 percent a year, Schmidt-Bleek concludes that the materials intensity of the industrialized countries must be reduced by a factor of 10 within the next 50 years. The calculation is based on a principle of fairness that requires that the 20 percent of the world's population that currently consumes 80 percent of the world's materials cut the current flow of materials by three-quarters to achieve an equal distribution of consumption across the planet. This objective was endorsed by a meeting of government and business leaders at Carnoules, Spain, in 1994. The group issued a declaration calling for a reduction in materials flow in the industrialized countries by a factor of 10 by the year 2050.[31]

A 10-fold decrease in materials flow would produce a sharp drop in dissipation, but it would also require significant changes in the economy. Resource productivity would have to rise and a large amount of industrial materials would need to be recycled within the economy. This will require several bold strategies. The previous analysis suggests at least three:

- reducing the rate of mining of industrial metals,
- developing organic chemicals from biomass wastes, and
- promoting recycling and secondary materials industries.

Reducing the Mining of Industrial Metals

The first strategy would involve gradually reducing the rate of extraction of metals from the nonfuel mining industry (iron, copper, lead, gold, etc.). There is already a large enough volume of most metals currently flowing in the economy and/or sequestered in existing products and physical infrastructure (and in some cases in government and private reserves) that there would be little need for virgin inputs if recycling and reuse and material substitution were pursued more vigorously. A reduction in the nation's industrial mining (not supplemented by mineral imports) would have several important effects. First, it would cut the nation's total materials flow because it would take a large bite out of the amount of overburden moved by mining (even though coal mining produces a much larger amount of overburden than mining of industrial materials). Second, it would raise the price of primary (virgin) metals, encouraging the substitution of more lightweight organic compounds. Third, it would provide an economic incentive for encouraging the recycling of metals and the reuse, repair, remanufacturing, and life extension of metal products.[32]

As noted in previous chapters, the materials flows of the major metals have been slowly closing over the past decades, with more and more metal products made from recycled metals. The increasing use of metal scrap has been one factor in decreasing the mining of virgin ores. The continued consumption of lead is based almost fully on recycled material. Over half of the iron and steel flow occurs in a closed loop, and this is nearly true for aluminum and titanium as well. Given that the rate of extraction and disposal of several of these industrial metals is not sustainable according to the principles noted above, these metals are ripe candidates for materials that should be continuously recycled within the economy. Indeed, all of these metals have high technical potential for effective recycling, and they typically require far less energy and environmental impact to recycle than to mine. Finally, metals are often quite durable and likely candidates for creating long life and reusable products with highly productive uses.

Developing Organic Chemicals from Biomass Wastes

The second strategy involves developing an increasing share of organic chemicals from the wastes of the forest products and agriculture industries and municipalities. A serious examination of the lignocellulose wastes of landscape maintenance; logging, pulp and paper production; and harvesting of grain, legume, and forage crops could reveal a source of by-product material large enough to substantially supplement the production of petroleum-based organic materials. Estimates cited earlier indicate that these wastes make up nearly 350 million metric tons per year. Deriving chemicals from this waste biomass would increase the productive yield from farms, reduce the current forest product and agricultural waste streams, and result in more integrated farm, forest, and chemical production. In addition, the large volume of organic wastes from food processing and municipal sewage could also be used as a feedstock for organic chemicals. Food scraps and yard wastes make up roughly one-fifth of municipal wastes in the United States and if discarded paper is added, the biodegradable share of municipal waste is well over one-half, or roughly 120 million tons per year. In his 1999 national survey of garbage in America, Jim Glenn found "The organics portion of the waste stream is underutilized and ripe for picking. That is especially true for food and wood residues, and its even true for yard trimmings ... only a handful of states divert more than 50 percent of either yard trimmings, wood or food residues ... [t]he vast majority don't even come close to diverting 10 percent...."[33] Indeed, closing the loop on these organic materials already flowing in the economy by processing them into chemical feedstocks for organic chemicals and polymers could add significantly to overall dematerialization.

Although petroleum feedstocks remain abundant and cheap today, their use depletes a finite, nonrenewable resource. Promotion of an alternative source for organic feedstocks could slow the rate of depletion and encourage innovation and commercial opportunities in bioprocessing.

With some 4 billion tons of lignocellulose potentially harvestable at sustainable rates worldwide, this could be an enormous source of feedstock for biobased materials.

Promoting Recycling and Secondary Materials Industries

The third strategy involves an aggressive promotion of the existing secondary materials industries and the nurturing and development of new enterprises and corporate divisions dedicated to recycling and reuse. This would involve more industrial and postconsumer scrap recycling as well as developing manufacturers capable of taking back and reusing, remanufacturing, and recycling used products. Some products are already well served by sophisticated recycling infrastructures. For instance, about 8 million automobiles are recycled each year at some 12,000 scrap yards. There are more than 1500 facilities for shearing, shredding, and baling metal from construction and demolition sites. Another 18,000 recycling collection facilities exist for recycling over 80 percent of all durable household appliances and 60 percent of all metal beverage cans.[34]

The promotion of loop-closing programs such as extended producer responsibility and product leasing would offer new economic opportunities. In addition, this would require the development of economic enterprises dedicated to used products, repair and servicing, renting and leasing, and providing nonproduct-based services. The effect would be more materials cycling, more economic activity in secondary markets, and more employment in product repairing, servicing, and reselling.

These three strategies would have substantial economic consequences. Promoting and investing in the recycling industries and new biobased materials opens opportunities for new enterprises, expansions of existing industries, and new jobs. On the other hand, a reduction in metals mining could lead to increased unemployment in the mining sector and the economic depression of communities dependent on the continued operations of specific mines here and abroad.

None of these strategies are readily feasible today. Each would involve some dislocation and transitional stress. There are major interests at stake and they would certainly be resistant. While there would be new investment opportunities and new areas for employment growth, there would also be employment loss and economic insecurity, particularly in mining. Such localized costs should not be acceptable, even for the greater good. A commitment to reducing materials throughput will require some compensation from those who will benefit to those who will lose. The nation has a long history of such transfer payments. They are typically conducted through the tax system, although following World War II, payments were made directly to veterans for educational services. As reductions in mining are phased in, it should be possible to tax the wastes generated by manufacturers and consumers to provide the revenue for dislocation compensation while hastening the development of recycling industries and their potential jobs.[35]

It will take a substantial amount of effort and commitment to achieve dematerialization and an economy more sustainably balanced with the environmental services that supports it. The nation has grown complacent in a materials economy based on gluttony and waste. We seem unwilling to remember that inefficiency and wastefulness have historically led to economic stagnation. We have much to learn about managing a vital and sustainable economy. Reducing the economy's materials throughput to sustainable levels offers as many opportunities as it does problems. The challenge of dematerializing the economy and creating a truly sustainable materials system offers new opportunities for innovation and learning while protecting and conserving resources for the future.

14

Detoxification

The best way to reduce exposure to toxic chemicals is not to use them in the first place.
—Barry Commoner

In 1987, a wildlife biologist named Theo Colborn began a study of the impacts of environmental contaminants on the waterfowl of the Great Lakes. As she scanned the scientific papers and government reports for indications of adverse health effects, she began to recognize a similarity between poorly performing children in the region and the abnormal behavior of herring gull chicks. For nearly a decade Colborn compiled data that suggested that maternal exposure to low levels of certain toxic compounds might lead to physical and behavioral abnormalities in offspring, human or otherwise. In 1997, Colborn and two colleagues published a book that many saw as a sequel to Rachel Carson's *Silent Spring*. In her book titled *Our Stolen Future*, Colborn drew public attention to the largely unrecognized effects of endocrine disruption on both wildlife and humans, and like her predecessor, she pointed an accusatory finger at toxic chemicals.[1]

Whereas Rachel Carson sought to link synthetic toxic chemicals to loss of wildlife and cases of human cancer, Theo Colborn presented evidence that connected these same chemicals to reproductive and developmental disorders in wildlife and humans. In both cases the evidence was limited, but the results appeared worthy of more serious concern. Both books suffered because scientific literature can be painfully difficult to use as a secure foundation for firm convictions, especially where phenomena are new and studies are limited. Early advocates of causal factors for public health concerns are often criticized, and Colborn, like Carson, received a fair share of negative press.

Achieving scientific consensus about chemical hazards is a long and frustratingly critical process. It takes time to establish facts and for theories to become accepted. However, policy questions are often quite pressing. It is often difficult to scientifically determine a safe level of exposure for hazardous industrial chemicals, but it can be even more complicated because of the demands raised by social and economic conditions. The current debate over chemicals alleged to cause endocrine disruption is about the effects of minuscule doses that may pass through the placenta. As noted earlier, the conventional process by which an "acceptable" threshold of exposure to a hazardous substance is established often involves a political negotiation based on differing views of the scientific evidence. Such processes are long, complex, and contentious. Instead of investing in such protracted, costly, and too often unsatisfying processes, more of the resources of science might better be spent on finding or developing less hazardous substances. Such investments could point the way to a more sustainable materials system. This precautious approach to industrial materials would focus on reducing the production and use of toxic materials wherever possible and investing in the development of safer alternatives where none currently exist. This is the essence of a detoxification strategy.

14.1   Toxicity and Detoxification

Dematerialization involves reducing the amount of global flow of materials while maintaining a comfortable level of physical well-being. Detoxification involves converting from toxic to nontoxic the mix of materials that remains flowing. Whereas the former involves the amount and rate of materials moving through the economy, the latter involves characteristics of the specific substances. Before considering the different strategies for detoxification, it is useful to remember why materials are toxic in the first place.

Some materials are toxic by intent; others are toxic because their desired function makes them likely to be toxic; and others are toxic because no effort has been made to make them otherwise. Pesticides, fungicides, and disinfectants are toxic by intent; they are designed to kill living organisms. Various processing intermediates, such as catalysts, reagents, and solvents, are toxic because their chemical structure and functional properties are similar to those that make chemicals toxic. Reactivity, persistence, and fat solubility are functionally desirable properties of many of these chemicals and these are also the factors that contribute to their bioavailability and biological activity. The reactivity of an intermediate chemical is often an important indicator of its toxicity.

A reactive chemical inside a human body is likely to find hydrocarbons with which it can bond long before the normal cleansing functions of the lungs or kidney can transport and eject the exogenous substance. This is the case with some petrochemicals. The high solvency and significant reactivity of many petrochemicals makes them attractive in industrial processing, where they are valued for the manner in which they release the energy tied up in their chemical bonds. Yet once in the body, the solubility of these chemicals gives them relatively easy transfer across internal membranes. As they reach target organs, the reactivity that makes the chemicals so valuable in production leads to their relatively high degree of hazard. A hydrocarbon solvent reactive enough to remove grease from a steel surface is usually reactive enough to form bonds with the molecules in organic cells that may lead to alterations in their functioning.

Other chemicals used in industry or in industrial products are toxic, but their toxicity is not related to their desired function. Metals are good examples. Lead, cadmium, and chromium are used in many consumer products. Lead is used in solders, conduits, and plastics to provide flexibility and stability; in construction materials to resist corrosion; and in bullets to provide weight. Cadmium is used as a pigment in coatings, as a stabilizer in plastics, and as a conductor in electronic components, and chromium is used as a hardening agent in alloys and as a corrosion inhibitor in surface coatings. Each of these metals has well-recognized toxic properties, yet none of these properties contribute to the desired functions of the material. In each case, other materials that are less toxic could be used as a substitute, or some other less hazardous processes could be used to achieve the same performance. For instance, the lead used in soldering lids on metal cans could be replaced with a less-toxic metal like tin, or the metal parts could be joined by another process, such as welding or mechanical crimping. Often, in applications where a material is selected in spite of its toxic characteristics, this is due to unrelated performance virtues, or more commonly to cost advantages.

The toxicity of a material does not alone determine its degree of hazard. The hazard of a toxic material is affected by the conditions of exposure. An acutely toxic material requires little exposure time to be physically dangerous. Some quite toxic chemicals are also very unstable, and they naturally decompose to relatively benign substances in a short time. Indeed, chemical and biological processes common throughout ecological systems can be quite efficient at metabolizing and degrading many toxic materials. Where the rate of ecological assimilation is not overwhelmed, natural processes involving water, light, and microorganisms can manage many quite toxic materials, reducing them to chemically inert and biologically inactive compounds. (However, some degradation processes, such as the methylation of mercury, create even more toxic chemicals.) There are other toxic materials that resist natural degradation and are likely to remain toxic for years. It is those chemicals that are persistent and unlikely to degrade and those that, once inside living organisms accumulate in cells, that are of greatest concern.

These distinctions are often a matter of degree and condition, but through comparisons they can be used to form different groupings of the substances. By using these different properties, industrial materials can be divided into four groups according to their potential for harm:

Group one.   Degradable and nontoxic
Group two.   Persistent and nontoxic
Group three. Degradable and toxic
Group four.   Persistent, bioaccumulative, and toxic

Figure 14.1 is an illustration of the materials that fall within these groups. They range from the biobased materials that are often nontoxic and ecologically degradable to various heavy metals and halogenated aromatic hydrocarbons that are often persistent, bioaccumulative, and toxic.

This typology has direct application in the three-loop materials flow model developed in chapter 8. It provides the decision rules for determining which chemicals can be considered "reasonably compatible with environmental processes" and permitted to cycle between the economy

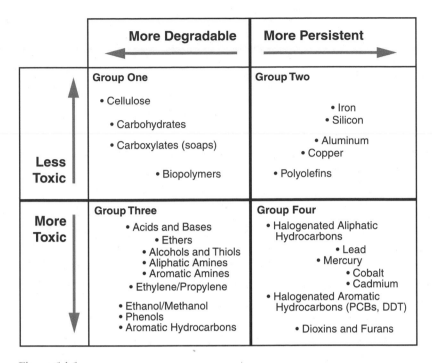

Figure 14.1
Industrial materials groups.

and the environment and which should not. It also clarifies the distinction among the different operational rules that regulate the sustainable rates of materials flow in the anthropogenic cycles. In other words, for materials to be "reasonably compatible with the environment," they should be degradable and nontoxic.

Using these groupings, specific principles can be developed for the management of different materials. Materials in the first group can be cycled between the natural and economic systems of the second loop with care and only modest concern. Many of these are renewable materials. Materials in the second group can be safely recycled inside economic systems, but should not be returned to natural systems, or at least not returned at rates that exceed the rates at which they are withdrawn. Materials in the third group must be carefully managed in production and use, recycled where possible, and effectively treated to eliminate toxicity before being transferred from economic activities back into ecological systems. Those materials that meet the criteria of the fourth group should be eliminated from production and use.

14.2   Detoxification by Substitution

Eliminating an industrial material from production and use is a final solution, but it cannot be carried out without considering what will fill the remaining vacuum. Most materials are used to serve a purpose. Left unattended, that purpose is likely to be met by other materials or processes, perhaps with new and equally dangerous hazards. If for no other reason, this is why it is best to drive a material out of use by promoting an attractive substitute than to pull it out of use with little thought to the consequences.

Material substitution is a fundamental driver in the history of industrial chemicals. Carbon tetrachloride (CCl_4) was first produced in Germany in 1839 and marketed as a grease remover under the name Katheryn. By the 1920s and 1930s, it was a commonly used industrial solvent, valued because of its nonflammability. During the 1940s, evidence began to mount concerning the toxic effects of carbon tetrachloride on the liver and kidneys, and it was slowly replaced in degreasing processes with trichloroethylene (TCE) and perchloroethylene (PERC). Increasing concern during the 1960s and 1970s over the possible carcinogenicity of TCE and PERC led to their replacement with Freon, a solvent trade name for chlorofluorocarbons, which was marketed as a safe and stable substitute. The conjecture about the destruction of upper atmospheric ozone by chemical reaction with CFCs and the empirical findings of ozone thinning encouraged the development of another round of safer substitutes. Indeed, it was the announcement by Du Pont of the availability of hydrochlorofluorocarbons (HCFCs) that eased the acceptance of the agreement for the CFC phaseout schedule established under the Montreal Protocol.

The replacement of a material through substitution can be accomplished through one or a combination of three procedures. *Material substitution* is quite straightforward and typically involves the direct replacement of one substance for another in a simple "drop in" process.

Common examples include the substitution of polymers for ceramics in coffee mugs, polyester for cotton in clothing, or glass fiber-reinforced composites for steel in structural components. *Process substitution* involves the replacement of one material by another by changing the process in which the material is used. Examples include the use of casting rather than forging, or the use of sand blasting rather than chemical treating to remove paint. Here the objective remains unchanged, but the process by which it is achieved is altered so as to introduce a new chemistry. The third type of substitution is referred to as *function substitution*, because here the objective or "function" to be achieved is changed. Examples include the substitution of a compact disk for magnetic recording tape, the substitution of transistors for vacuum tubes, or the substitution of electronic mail for postal mail. These three types of substitution—material, process, or function—can be used "at any of the steps in the resource, processing, and manufacturing cycle, from raw materials through primary products, parts manufacture and components to final design and assembly" as well as in the management of the end of a product's life.[2]

Thus, strategies for detoxification can be carried out at many points throughout a material's life cycle. Nontoxic materials can be used within economic systems and discarded into natural systems at sustainable rates. Dematerialization permits current rates to be reduced to sustainable rates. Disposing of biodegradable, toxic materials requires treating wastes to reduce their toxicity. Persistent, bioaccumulative, and toxic materials need to be reduced and eventually eliminated. Using less-toxic materials requires that products be designed with less-toxic constituents. Less-toxic production of materials means that raw materials should not be toxic and that materials processing should be more environmentally benign. In general, this means following three principles:

- reduce the dissipation of degradable, toxic materials;
- reduce the use of persistent, bioaccumulative toxic materials; and
- develop more environmentally appropriate materials.

Each of these points of intervention provides its own particular opportunities and these are considered in the following sections.

14.3   Reducing the Dissipation of Degradable Toxic Materials

Toxic materials should not be permitted to dissipate into natural systems. The federal Pollution Prevention Act, first enacted to reduce the generation of hazardous wastes, has become increasingly focused on reducing the environmental release of pollutants, particularly those monitored under the federal Toxics Release Inventory. When the dissipation of these materials is reduced, they become better managed, but a reduction in releases does not constitute detoxification, unless it is preceded by some kind of treatment that alters the toxicity of the chemicals. Those toxic materials that degrade readily to more benign forms can be treated and returned to ecological systems as long as rates permit safe assimilation. A guiding factor in the management of such processes is the time needed to convert the materials from toxic to nontoxic forms, and because this is a condition of degree, not a category, there will always be some uncertainty about the transition point. Therefore the precautious approach is to detoxify a degradable material before releasing it from economic systems into ecological systems.

Detoxification by degradation of an industrial material can be enhanced by designing the chemical for easy treatment or for rapid chemical or biological degradation. In general, persistence tends to be associated with lipid solubility, and water solubility is more likely to enhance degradation. Chemically mixed wastes are usually more difficult to treat than homogeneous materials. Effective treatment is closely related to concentration, so more dilute chemicals degrade more rapidly. There are many conventional and some new waste treatment processes that are useful for detoxification of a toxic material prior to release. Some technologies that simply separate the toxic material from the remainder of the waste stream can be useful in preparing materials for treatment. The most effective treatment appears to involve chemical or biological technologies.

Separation for Treatment
Many of the technologies used to separate toxic materials from effluents have been developed for air and water pollution control. Carbon adsorption uses activated carbon (carbon that has been processed to increase its

reactivity) in continuous-flow columns to absorb a wide range of soluble organic chemicals that can then be recycled or stripped and processed for treatment. Phase change-based separation technologies such as distillation, condensation, and evaporation often provide inexpensive and easily operated processes for separating miscible (volatile) liquids or liquids from suspended solids. Small stills are used for on-site distillation of fluids, particularly spent solvents, which can then be recycled or sent out for treatment. Ion exchange is a technology chiefly used to separate metals from waste streams. Because certain resins have a higher affinity for some ions than for others, the electrochemical differentiation can be used to separate ions between the resin and the wastewaters. Diffusion dialysis is a membrane separation process used extensively in the metals finishing industry to separate contaminants from spent pickling liquors and in the recovery of acids and caustic cleaners. Reverse osmosis and ultrafiltration techniques use pressure applied to concentrated wastewaters to force the water through semipermeable membranes in order to obtain a highly concentrated product, which is then available for reuse, reprocessing, or further treatment. Other methods used to separate toxic materials prior to waste disposal include freeze-crystallization, chemical precipitation, and electrowinning.[3]

Some separation technologies can yield materials that are as likely to be reused or recycled as detoxified. The metals content of an aqueous waste stream can be substantially decreased by complexation or adsorption methods to recover the metals, which can be recycled back into the economic system. Wastes rich in dissolved salts can be treated through ion exchange or reverse osmosis to clarify the wastewaters prior to discharge and provide a salt by-product that may still have economic value.

Chemical Treatment

Chemical waste treatment can detoxify some degradable wastes prior to release. A common example involves the neutralization of acidic wastes. Wastes rich in sulfuric acid can be neutralized by introducing a caustic chemical such as sodium hydroxide. With waste neutralization, one waste stream may be used to detoxify another. For instance, wastes containing cyanide and various heavy metals may be partially detoxified by mixing them with the more alkaline wastes from ethylene oxide or propylene oxide production processes. Sulfur dioxide can be used to dechlorinate chlorine-bearing wastes, which usually produces a much less toxic chloride.

Condensation polymers such as polyesters and polyamides are created by reversible reactions that can be used to convert them back to their immediate precursors. A depolymerization process called "chemolysis" uses water and alcohols at elevated temperatures to convert polymers back into monomers or short-chain polymers.

Chemical oxidation is a detoxification process that involves adding an oxidizing agent to chemically transform a compound. Organic chemical wastes are particularly prone to decomposition through oxidation. Various chlorinated hydrocarbons, phenols, mercaptans, and inorganic chemicals such as cyanide can be converted to carbon dioxide, water, and some residual inorganic chemicals. The most common oxidizing agents are ozone and hydrogen peroxide. Ozone is a powerful oxidant and readily reacts with many toxic organic substances, usually producing less-toxic compounds that are often biodegradable. Hydrogen peroxide and ultraviolet light are generally used together and demonstrate significant capacity to decompose trichloroethylene, trichloroethane, and many halogenated aliphatics. The oxidation rate of organic chemicals can be increased by a process called "wet air oxidation" that adds supercritically pressurized water, compressed air, and heat to the reaction process, although these processes require significant energy inputs. Using different processes, chlorinated compounds can be reduced to chlorides, nitrogen compounds to nitrates, sulfur compounds to sulfates, and phosphorus compounds to phosphates.[4]

Biological Treatment

Microbial detoxification of wastes involves the use of microorganisms to metabolize and decompose the wastes into more benign substances. The inherent biodegradability of a chemical depends on its molecular structure. For instance, pentachlorophenol is readily degraded by microbial processes, while a similar compound, such as hexachlorobenzene, is highly resistant to biodegradation. Microbial degradation of toxic wastes involves ambient temperature and pressure and enzyme-catalyzed reactions. As microorganisms metabolize the carbon to build cell structure, they release energy and consume and degrade the waste. This degradation may simply generate new compounds or, if complete, it is said to "mineralize" the wastes to carbon dioxide, water, and cell mass.

In general, and with major exceptions, many organic compounds are biodegradable. PCBs and dioxins are two organic compounds that are largely nonbiodegradable. Straight-chain aliphatic compounds are easily degraded through biodegradation. Unsaturated aliphatics are often less readily transformed than saturated aliphatics. The simple aromatic hydrocarbons are relatively easily degraded, but halogenated aromatic compounds are less readily biodegraded. Nitrogen and sulfur-containing compounds are also transformed by microbial action, although more slowly. Polymeric materials are among the most resistant to microbial attack, although microorganisms have been isolated that utilize nylon or polystyrene polymers as effective substrates.[5]

Inorganic materials may also be successfully detoxified through microbial action. For instance, ferrous iron in acid mine discharge can be converted to ferric iron by a bacterium, *Thiobacillus ferrooxidans*, and cyanide can be degraded by microorganisms to carbon dioxide and ammonia, which can then be subjected to bacterial oxidation to form nitrates. Microorganisms can also be applied to selenium compounds to alter their solubility, volatility, or bioavailability. Continuing research has identified naturally occurring organisms that degrade chemicals previously identified as fairly persistent. Much of this research comes from studying the microbes that grow on hazardous wastes that have been dumped into ecological systems without treatment.[6]

Physical Treatment
Conventional physical waste treatment technologies need to be differentiated by their potential for detoxification. Some, such as encapsulation and vitrification, may solidly secure a toxic material in a durable matrix which will reduce the risks of exposure as long as the matrix remains secure, which cannot be absolutely ensured. Others, such as incineration, which converts solid or liquid wastes into toxic ash and airborne emissions, may decompose some toxic constituents, but the processes are so dissipative, difficult to manage effectively, and often incomplete, that they cannot ensure detoxification. Likewise, waste treatment processes that

separate a toxic compound from stack emissions or wastewaters are of benefit only if there is an effective chemical or biological treatment that can follow, or some kind of reuse potential within the economic system. While the generation of a more concentrated toxic waste may improve waste management opportunities, it does not lead to detoxification. In general, waste treatment that still generates toxic materials is expensive and crude, and provides little incentive for detoxification in earlier stages of the life cycle of a toxic material; thus, its overall impact is limited.

14.4   Reducing the Use of Persistent, Bioaccumulative, and Toxic Materials

Barry Commoner is often credited with the observation that the best way to avoid exposures to toxic chemicals in the environment is not to use them in the first place. It is a simple and straightforward idea, but as the record of American efforts to build a productive materials economy demonstrates, the issue of toxicity was little considered in inventing and selecting materials for industrial production. The result has been a remarkable economy capable of producing a cornucopia of products and services at highly affordable prices, but with a shadowed legacy of health and environmental risks. The task ahead is not to create more and cheaper products, but rather to create safer products, produced with minimal workplace hazards and little damage to the environment. This will require new consumption and production strategies.

Consumption Strategies

Detoxification can be promoted by initiatives that reduce the use of toxic chemicals in products and in use. Although the federal government has a recent history of regulating toxic chemicals in pesticide, food, and pharmaceutical products, it has been less aggressive in regard to the chemical constituents of consumer commodities. The Consumer Product Safety Act provides the Consumer Product Safety Commission with the authority to set standards and ultimately to ban products found to present "unreasonable risks," but in practice, the Commission has focused primarily on the safety of products and used its power to ban toxics in products most sparingly. One strategy that the Commission has advanced more successfully has involved product labeling.[7]

Periodically, nongovernment environmental and consumer organizations have launched campaigns focused on the environmental and safety aspects of consumer products. Most of these efforts have required the mobilization of masses of consumers or at least enough public media coverage to permit the advocates to be taken seriously so that they might negotiate directly with government professionals and corporate managers. Recent examples have resulted in the phaseout of McDonald's polystyrene "clamshell" food packaging, Alar in apple production, and agreements to phase out polyvinyl chloride and certain phthalates in children's toys and hospital blood bags.[8]

Although there are many strategies for reducing the use of toxic chemicals in consumption, the following four are most commonly recognized:

• product labeling,
• providing information to employees and consumers
• environmentally-preferable procurement, and
• bans and phaseouts of particular substances.

Product Labeling   Product labeling to inform consumers about materials can range from simple identifiers such as the Society for the Plastics Industry's numeric polymer resin codes that appear on plastic bottles and containers, to the constituent lists that the Federal Trade Commission requires on many domestic product labels. Over the past two decades, consumer product labeling programs focused on environmental effects ("ecolabels") have appeared in more than twenty-five countries. Some of these, such as the German "Blue Angel" and the Scandinavian "White Swan," are government initiatives. In the United States, environmental product labels have been promoted by two private organizations, Scientific Certification Systems and Green Seal. Most of these programs provide a positive environmental label for products voluntarily submitted by manufacturers for testing and evaluation. Theoretically, these ecolabels could work as a detoxification strategy if the labels were clear and concise, the purchasing public well informed and sensitized to the hazards of the package contents, and alternative products were available at comparable cost and performance. For instance, a special hazard warning program initiated by ballot initiative in California during the 1980s requires that products containing known carcinogens or reproductive hazards be labeled. This program has led certain consumer product manufacturers to replace known carcinogens in products rather than label them. For instance, the Gillette Corporation removed trichloroethylene from its typewriter correction fluid, Dow Chemical reformulated its commercial spot remover to eliminate perchloroethylene, and Pet, Inc. accelerated the elimination of lead from its food cans.[9]

In practice, the effectiveness of ecolabels in altering product sales or marketing strategies has remained uncertain. For example, a label on the containers of many pesticides and industrial cleaners indicates that the contents are poisonous, but because purchasers assume that they can control the risks during use, these labels do not appear to affect purchase decisions, although they may improve disposal practices. Some studies suggest that ecolabels seem to have more effect as a deterrent, in getting producers to avoid certain constituents in their products.[10]

Even if there were a more intensive effort to provide product labeling for informing consumers, the absence of sufficient data on the health and environmental effects of many chemicals, even those with high production volumes, would make such labeling difficult. It has been proposed that indicating the absence of such data on a label might also be useful information for consumers, who might otherwise assume that an unlabeled product is a safe product. Not only might a label identifying non-assessed chemical ingredients be useful to consumers, it would most likely also encourage the development of more data.[11]

Employee and Consumer Information   Workers and consumers can become advocates for safer materials if they have the information needed to act. Production workers are often in an immediate position to notice the introduction of new chemicals into a process and to take note of the potential for exposure or environmental release. While workers, particularly those without trade union protection, often feel fairly vulnerable in raising questions about chemicals in the workplace, they are even less effective if they do not have access to information about production chemicals. Many states have passed so-called "right-to-know" laws, and the U.S. Occupational Safety and Health Administration has promulgated a "hazard communication standard" that provides workers with access to chemical information at their workplace. Although the information provided is not always complete or adequate, it has proved valuable for both workers and managers in better understanding what chemicals are being used and supporting a search for safer substitutes.[12]

Governments and nongovernment organizations have tried to assist consumers by providing lists of less-toxic products, or manuals on how to make simple nontoxic products that can be used as substitutes for toxic ones. Some authors have produced highly popular books that provide purchasing information for consumers on everything from home cleaning to house construction products. The shopping-oriented guides provide little more than alternative products for substitution, but some of the household manuals suggest process changes, such as using quilts instead of electric blankets, or even function substitutions, such as enjoying leisure time by bicycling rather than watching television. Many of these publications have been distributed widely and topped best-seller lists, attesting to the eagerness of many consumers for more information on environmentally friendly products; however, there have been few studies of the effectiveness of these lists and publications.[13]

Environmentally Preferable Procurement   State and local governments have pioneered environmentally preferred product procurement policies that tend to avoid products with toxic constituents. For instance, in developing its municipal program, the city of Santa Monica (California) has screened over 200 cleaning products with a detailed list of environmental specifications that includes avoiding hazardous chemicals. The state of Minnesota has established a scorecard system for evaluating multiple product attributes and screened over 400 products in some 33 categories.

Private companies have also instituted environmentally responsive purchasing programs. Daimler Chrysler, Volvo, Cannon, Sony, the Body Shop, and Ben and Jerry's (ice cream) use lists of chemicals to avoid in their purchasing. Companies such as Warner Brothers, Anheiser-Busch,

Patagonia, Volvo, Cannon, Daimler Chrysler, and Herman Miller also use lists of desirable environmental criteria (including the avoidance of toxic chemicals) to inform their purchasing decisions.

The federal EPA has produced lists of so-called "environmentally preferred" products as guides for government procurement departments. These programs, which are typically voluntary, have listed hundreds of products and been targeted at agencies that range in purchasing power from local school districts to the U.S. Department of Defense. For instance, the National Park Service now encourages contractors supplying park ranger uniforms to use recycled fibers, nontoxic dyes, and fabrics that do not require dry cleaning. A 1993 Executive Order of the President directed federal agencies to give preference to the purchase of products and services demonstrating the least burdens to the environment. Quotas on recycled pulp in purchased paper products are a particularly well-accepted requirement in some state and federal programs, although efforts to stipulate against paper products bleached with chlorine have met with substantial industry opposition.[14]

Bans and Phaseouts   Phasing out the production, sale or use of targeted toxic materials through legal prohibitions provides a fairly powerful strategy for detoxification, although it often generates implementation problems. Concern over the environmental or human health effects of various materials has led national government authorities to ban, withdraw, or severely restrict their production or use. The reductions of lead in human blood, phosphorus in surface waters, PCBs in human tissue, and CFCs in the atmosphere have been achieved through severe restriction or phaseout policies under national laws or international agreements. There is little disagreement that these policies are responsible for the observed declines in the use of these substances.

Sweden has made aggressive use of the phaseout approach, prohibiting the use of certain materials, including cadmium, arsenic, pentachlorophenol, and organotin compounds in most applications. In the 1980s the United Nations conducted an inventory of substances that have been banned or severely restricted by national governments throughout the world. The inventory identified some 600 substances on which prohibitive actions had been taken by at least 1 of 77 countries. Most of these substances are pharmaceutical or agricultural chemicals, but 82 are listed as common industrial chemicals.[15]

The EPA has used prohibitive regulations to phase out the use of various chemicals. Under its authority to register pesticides, the agency has banned or restricted the use of hundreds of pesticides, and under the Toxic Substances Control Act, it has prohibited the intentional production and use of PCBs (except in closed electrical equipment), CFCs in aerosols, and friable asbestos in schools. In 1989, Sweden put forward a proposal to establish an international phaseout—a "sunset"—of the use of thirteen commonly recognized problem chemicals. The Organization for Economic Cooperation and Development wrestled with the proposal for several years, but failed to reach a consensus and the initiative gradually died. More recently, a new international initiative has been launched by the United Nations Environment Program to negotiate a protocol for the global phaseout of twelve so-called "persistent organic pollutants."[16]

The use of government authority to ban a toxic chemical can appear appealing because it suggests clarity and finality. This can be quite illusory since many of these government restrictions are quite conditional and provide numerous exemptions. The very simplicity of toxic chemical bans can be misleading because they imply an overall reduction in risk. Indeed, there are cases where government efforts to phase out the use of a toxic chemical have opened up markets for substitute chemicals or production processes that are equally or potentially more hazardous. The simplicity of a single-substance government ban hides the complexity of our materials systems.[17]

Production Strategies

A more flexible and adaptive approach to detoxification is suggested by recent efforts to reduce the use or release of toxic chemicals in industrial production processes. Nationally, there are many pollution prevention programs that encourage less-toxic production processes or processes that generate less-toxic wastes. Internationally, these activities are referred to under the term *cleaner production*. Since the 1980s, many states have enacted pollution prevention laws. While several of these state laws, like the federal law, focus on reducing the release of pollutants to the environment, a few of the states enacted pollution prevention laws based on a concept called "toxics use reduction." Toxics use reduction programs employ various pollution prevention techniques (state technical assistance, facility planning, annual facility reports, etc.) to cut the generation of pollution by reducing the use of toxic chemicals within products or production processes.[18]

State programs to reduce toxics use, such as the Massachusetts Toxics Use Reduction Program, have demonstrated that these materials can be eliminated through two general processes: substitution of materials in product and process design, and reduction of volume through improvements in process efficiencies and internal recycling. Substitution involves using a less-toxic material in place of a more toxic one. This choice requires a comparative analysis of the hazard potential of alternative materials, including a characterization of their health and environmental effects over their full life cycle. Volume reduction strategies are usually based on yield improvements, modernization of equipment through upgrading or new purchases, and improvements in operations and maintenance. Many of these strategics have proven to be readily implemented, low cost, and surprisingly effective in reducing the amount of toxic materials purchased and used.[19]

Although toxics use reduction programs differ, most have promoted five general techniques. These include

- material substitution in the product,
- material substitution in the process,
- improvements in production efficiencies or yield,
- improvements in process operations or maintenance, and
- internal recycling, or closed-loop processing.

Identification of the appropriate technique often requires a reasonably thorough analysis of the available options, their technical and economic feasibility, and their environmental and health impacts. This may involve an audit of the entire production system and the preparation of a plan for implementation. Box 14.1 presents specific examples of toxics use reduction in industrial facilities.[20]

Material Substitution in Product Redesign   Material substitution in product redesign is a conventional method for reducing the toxic con-

Box 14.1
Examples of Toxics Use Reduction

Substitution in Product Redesign
Use water-based paints instead of oil-based paints
Use zinc-plated products rather than cadmium-plated products
Switch from hydrocarbon-based inks to water-based inks
Renegotiate customer specifications to eliminate toxic constituents in products

Substitution in Process Redesign
Replace chlorinated solvents in degreasing operations with aqueous cleaning systems
Replace organic paint strippers with mechanical blasting using plastic beads or Dry Ice
Replace caustic cleaning operations with alkaline-based cleaners
Substitute photoactivated catalysts for heavy-metal catalysts

Conservation through Process Improvement
Install automated temperature or pressure controls on reactor vessels to improve product yield
Install drip racks above electroplating tanks to reduce "drag-out" of plating solutions
Replace the nozzle on high-pressure paint applicators to better focus the paint spray stream
Upgrade tool and equipment quality to reduce off-specification products

Conservation through Improvements in Operations and Maintenance
Replace gaskets and valves to reduce fugitive emissions from chemical transport pipes and pumps
Install floating roofs on chemical storage tanks to reduce volatilization
Centralize material purchase responsibilities to reduce overpurchasing
Schedule production to reduce equipment cleaning

Conservation through In-Process Recycling
Reclaim, rejuvenate, and reuse rinse waters in metal plating
Recycle and reuse dedicated toxic chemical shipping containers
Distill and reuse solvent strippers
Rework batch process by-products back into the next batch

Source: Various case studies from the Massachusetts Toxics Use Reduction Program, Boston, Massachusetts.

stituents of the product. Many paints and coatings can be reformulated into aqueous(water)-based mixtures, eliminating ingredients such as toluene, methyl ethyl ketone, formaldehyde, and isocyanates. Water-based solutions can be used to replace toluene in the manufacture of adhesives. In Massachusetts, the gun maker Smith and Wesson replaced the lacquer used on its finished handgun stocks with a nontoxic carnauba wax, which yielded a higher luster and eliminated the xylene, 1-butanol, isobutyl isobutyrate, methyl ethyl ketone, toluene, and isopropanol contained in the lacquer and its associated thinner. The new process saved energy and generated a cost savings.[21]

Material Substitution in Process Redesign   Toxic materials can also be replaced by redesigning processes. For instance, coating products with solvent-based paints releases volatile constituents such as xylene, toluene, methyl ethyl ketone, methyl isobutyl ketone, and methylene chloride. Changing to a powder-based resin coating or a radiation-cured coating eliminates the need for the volatile organics. Hyde Tool, a Massachusetts hand tool company, reduced the sulfuric acid used to neutralize its waste streams by substituting citric acid and then eliminated the citric acid by controlling the pH in the production processes. The firm eliminated the chlorinated solvents used in one cleaning step by eliminating the cleaning step. Another Massachusetts company, the Robbins Company, was able to eliminate the use of sodium cyanide and hydrogen peroxide in preparing metal parts for electroplating by replacing the conventional bright-stripping of parts with an electrolytic process that uses the part as the anode and electrolytically removes the metal ions from the surface.[22]

Toxics use reduction programs have promoted water-based solutions where they have been a practical substitute for chlorinated solvents. In some cases this involves switching to hot water or water with a surfactant. In other cases there is a rapidly growing market of aqueous-based cleaning and degreasing products (e.g., alcohols) that are as effective and inexpensive as traditional hydrocarbon-based solvents.

Material Conservation through Process Improvements   The volume of toxic materials can also be reduced by improving the efficiency or yield of production processes. This may involve redesigning the physical plant layout, reprogramming operational schedules, tightening inventory controls, automating process temperature and pressure controls, and optimizing the operational parameters of equipment. Instituting integrated process management (such as statistical process control and "systems networking") can dramatically improve operations and reduce materials (including toxic materials) requirements. Process analytical chemistry is a practical monitoring and verification process that provides real-time information that can encourage rapid production corrections and higher levels of process optimization.[23]

Fugitive emissions can be reduced by installing equipment with minimum connections or so-called "sealess" or "leakless" valves. The efficiency of spray-applied coatings can be greatly increased by pressure and nozzle improvements, leading to much higher spray efficiencies and lower volumes of wasted paint.

Material Conservation through Improvements in Operations and Maintenance   Tightening up the operations and maintenance of production processes can reduce the volume of toxic chemicals used in industrial production. Surprisingly large amounts of materials are lost in production processes through leaks and spills, poorly maintained machinery, inadequately monitored equipment, redundant operations, and products sent to waste management because they did not meet quality standards. Unlike other techniques, improvements in operations and maintenance rely less on technology changes and more on improved management practices. Common improvements include improving employee training, preventing spills and leaks, instituting preventive equipment maintenance, increasing process documentation, and increasing product or process inspection.

For instance, routine maintenance and repair of equipment seals and gaskets on pumps, valves, pipe joints, and container covers can reduce fugitive emissions from fluid or gas conveyer and storage units. Leak detection and repair programs that use organic vapor analyzers have proven highly cost-effective in reducing fugitive emissions. In lithographic printing operations, the usable life of sulfuric acid baths can be increased by proper filtering, resulting in the purchase of less acid. Automated monitoring and timing controls can lead to less off-specification products in the batch production processes used to manufacture paints, dyes, adhesives, caulks, lubricants, and cutting oils.[24]

Material Conservation through In-process Recycling   Internal recycling procedures can reduce the demand for toxic feedstock materials by collecting and reprocessing by-products and reusing them in closed-loop systems. This may involve: collection of nonproduct residuals for direct reuse in a process; recovery of materials by chemical, physical, or electrochemical separation processes for reuse in a process; or treatment and removal of impurities for reuse. Separation (as described above) may be accomplished through distillation, evaporation, ion exchange, diffusion dialysis, reverse osmosis, ultrafiltration, or electrolysis. Vapor recovery units on chemical storage containers can capture, condense, and return potential fugitive emissions resulting from tank "breathing" losses. Spent catalysts can be regenerated and reused in some chemical processing operations, resulting in less demand for hazardous catalysts. Closed-loop processes are particularly attractive in metal plating, where the plating solution, rinse waters, and dissolved metals all can be recovered and reused.

Techniques for reducing the use of toxic materials must be carefully considered so that one hazard is not substituted for another. For instance, the volatile, petroleum-based inks used in printing on fabrics can be replaced with ultraviolet (UV) light-curable inks, although these inks still require hazardous photoinitiators and UV light is not without physical hazards. In some applications, powder coatings have proven effective substitutes for solvent coatings, but the reactive paint powders can produce respiratory hazards for workers. Liquid carbon dioxide can be used to replace acids in wastewater neutralization, but there remain hazards from the high pressure of liquefied carbon dioxide, and the carbon released adds to global warming.

Therefore, substitution processes for toxic chemicals require serious investigations of the impacts of the materials, energy requirements, and by-products of each available option throughout its full life cycle. Various procedures have been developed to guide such investigations. The Clean Air Act Amendments of 1990 required that the EPA develop a means of identifying safer alternatives for chemicals phased out because of their ozone-depleting potential. Under Section 612, the agency established a Safe New Alternatives Program (SNAP) for evaluating each substitute for its toxicity, hazard, exposure potential, energy impacts, degradation potential, and potential effects on air and water. The effort has been extended more recently with the publication of a *Cleaner Technologies Substitutes Assessment Guide*, which presents a step-by-step procedure for factoring in a wide range of data useful in comparative analyses of alternative materials and processes.[25] In more complex and integrated production processes, these auditing and planning activities themselves can prove quite costly and labor intensive. Because most programs to reduce toxics use have focused on individual production facilities, the more comprehensive impacts of the changes have seldom been well developed from a material life-cycle perspective.

These facility-level cleaner production programs have their limits as strategies for detoxification. Efforts to try to detoxify materials through changes in the product manufacturing industries can serve a useful purpose in creating a market demand for less-toxic materials, but they will always be constrained by the capacities of those materials that are currently on the market. A more fundamental strategy involves efforts to increase the supply of alternative materials that are inherently less toxic.

14.5 Developing More Environmentally Appropriate Materials

The state of chemistry, biology, and physics, and knowledge about physiology and toxicology has advanced enormously over the past half-century. We know far more about what makes materials toxic and how to make safer chemicals than we did. Herein lies the opportunity for the materials sciences. Many chemists and engineers today recognize that there is adequate knowledge to design chemicals and chemical processes that pose less risk to human health and the environment. Yet materials scientists are seldom asked to turn their design talents to minimizing or eliminating the hazards of materials. The domination of materials science by questions of functional performance, processing efficiency, and cost has provided little room for questions related to health or environmental impact. If materials scientists and synthetic chemists were challenged to come up with less-toxic materials, product designers and process engineers would be better able to produce more environmentally benign products and processes.

Already there is a growing number of chemists and chemical engineers interested in environmentally compatible chemical process design. Some are developing new chemical performance criteria that include environmental objectives for insertion into conventional chemical process design methods. Chemical engineers at the New Jersey Institute of Technology and the Massachusetts Institute of Technology are developing expert systems and new computer-assisted design tools that incorporate environmental factors into chemical product design. Some researchers have used group contribution theory and quantitative structure-activity relationships to set design boundaries for excluding undesirable properties in new chemical designs. Others have used a design technique called molecular-level reaction pathway synthesis to identify the chemical synthesis routes least likely to generate unwanted by-products.[26]

The EPA is one place where there is an increasing body of information about what makes chemicals more likely to be safe. Because the agency's New Chemicals Program must annually review data on hundreds of new chemicals for possible market entry, the agency has developed a set of procedures for encouraging safer chemicals. Beginning in 1987, it began to develop categories of chemicals based on those properties likely to be dangerous. Using the developing techniques of structure-activity analysis, the agency's first category was "acrylates and metacrylates." Today, there are forty-five chemical categories and the Chemical Categories List is generally regarded as identifying those substances least likely to be safe. Using this same experience more positively, the New Chemicals Program has developed a process called the Snythetic Method Assessment for Reduction Techniques (SMART) under which all chemicals submitted to the New Chemicals Program are reviewed to identify pollution prevention opportunities. EPA has also developed an interactive Green Chemistry Expert System (GCES) software program that allows manufacturers to conduct their own pollution prevention review, design more environmentally compatible chemicals, or review safer solvents and reagents.[27]

The idea of using existing chemical knowledge to design more environmentally friendly chemicals and chemical processes has opened a

Box 14.2
Twelve Principles of Green Chemistry

- Prevent waste (unconverted feedstock, spent reaction fluids)
- Maximize the incorporation of all process materials into the finished product
- Use and generate substances that possess little or no toxicity
- Preserve efficacy of function while reducing toxicity
- Minimize auxiliary substances (e.g., solvents, separating agents)
- Minimize energy inputs (process at ambient temperatures and pressures)
- Prefer renewable materials over nonrenewable materials
- Avoid unnecessary derivations (e.g., protection/deprotection steps)
- Prefer catalytic reagents over stoichiometric reagents
- Design for natural post-use decomposition
- Use in-process monitoring and control to prevent formation of hazardous substances
- Minimize the potential for accidents

Source: Paul T. Anastas and John C. Warner, *Green Chemistry: Theory and Practice*, New York: Oxford University Press, 1998.

---

rapidly developing new specialty in chemistry often referred to as "environmentally benign chemical synthesis," or "green chemistry." The "green chemistry" movement appeared early in the 1990s as the National Science Foundation and the nonprofit Council for Chemical Research sponsored a small program to support environmentally benign chemical synthesis and processing. The idea was to encourage and support research in alternative chemical synthesis and processing that might be more environmentally compatible. In 1995, the program was adopted by the EPA as a part of its national pollution prevention program and the next year the agency joined with the American Chemical Society in sponsoring an annual green chemistry and engineering conference that included a presidential awards program to recognize accomplishments in environmentally sound chemical methods.[28]

Green chemistry has been defined as "the utilization of a set of principles that reduces or eliminates the use or generation of hazardous substances in the design, manufacture and application of chemical products."[29] Paul Anastas and John Warner in their text on green chemistry lay out twelve principles (see box 14.2). There appear to be two complementary approaches in green chemistry. The first is based on chemistry and chemical engineering and focuses on developing alternatives to current production technologies and practices. This includes research on alternative feedstocks; environmentally benign solvents, reagents, and catalysts; aqueous processing; and safer and more readily recyclable chemical products. The second is based more on biology, toxicology, and ecology and focuses on developing approaches that reduce or eliminate materials or processes that are not compatible with biological integrity and ecological functioning. This involves research on nonpersistent, non-bioaccumulative, ecocompatible and renewable materials.

The chemical engineering approach begins with research on alternative feedstocks. This involves preferring renewable and biobased feedstocks over nonrenewable and petroleum-based sources and seeking starting materials that demonstrate the least hazardous properties (e.g., toxicity, flammability, accident potential, ecosystem incompatibility, ozone-depleting and global-warming potential). Both carbohyrates and lignan are viable feedstocks for organic chemical production. The supply is extensive and the separation processes are well understood, but the conversion technologies need further refinement. The use of polysaccharides as feedstocks for polymers has already been noted as an example of a renewable and nontoxic synthesis pathway. Likewise, glucose rather than benzene can be used as a raw material in the production of hydroquinone, catechol, and adipic acid, all of which are important intermediates in the production of commodity chemicals. Current research has shown that relatively nontoxic silicon is a useful replacement for carbon as a starting base for the synthesis of some industrial chemicals.[30]

Additional research focuses on alternative reagents and catalysts. This involves identifying catalysts that can be generated in the least environmentally destructive manner and that can function in the chemical transformation in a manner that is most environmentally beneficial (e.g., minimizes energy inputs, maximizes yield, minimizes waste outputs, generates the least occupational exposure and accident potential). For instance, addition reactions are preferred over subtraction reactions because they incorporate much of the starting materials and are less likely to produce large amounts of waste. Reactions that run close to 100 percent efficiency minimize the amount of unconverted feedstock and reagent.

Alternatives to the heavy-metal catalysts are sought because the common metal catalysts are often quite toxic. The use of liquid oxidation reactors replaces metal oxide catalysts with pure oxygen and permits lower temperature and pressure reactions with higher selectivity and no metal-contaminated wastes. New catalysis techniques that rely on enzymes, microwaves, ultrasound, or visible light may further obviate the need for harsh chemical catalysts.

Organic solvents with significant health and environmental impacts have been conventionally used as carriers in chemical reactions. Research on safer solvents includes investigations of aqueous solutions, ionic liquids, immobilized solvents, and supercritical fluids. Water has been shown to be an effective solvent in some chemical reactions, such as free radical bromination. Supercritical fluids, which are typically gases (e.g., CO_2) liquefied under pressure, are already commonly used in many applications, including decaffination of coffee and extraction of hops. Supercritical CO_2 can be used as a replacement for organic solvents in polymerization reactions and surfactant production. Future work may involve solventless or "neat" reactions such as molten-state reactions, dry-grind reactions, plasma-supported reactions, or solid materials-based reactions that use clay or zeolites as carriers. Box 14.3 provides examples of current green chemistry research projects.

By taking what toxicologists and ecologists already know, it is possible to generate a set of design criteria that could be used to guide the development of more environmentally friendly substances. Some of these include:

- use of inherently safer materials,
- use of nonbioavailable materials,
- use of physically benign materials,
- use of biodegradable materials,
- contained materials processing,
- on-demand generation of chemicals, and
- processing under ambient conditions.

Inherently Safer Materials The chemistry of a material should be designed to be less likely to damage a living organ or cell. Where possible, nonflammable, nonexplosive, nonvolatile, and noncorrosive chemicals

Box 14.3
Examples of Green Chemistry Research Projects

More Environmentally Benign Feedstocks
Polysaccharide polymers
Glucose-based chemicals
Biomass conversion

Green Reactions
Halide-free synthesis of aromatic amines
Nonphosgene isocyanate synthesis

Green Reagents and Catalysts
Carbon dioxide substitution for phosgene in isocyanate synthesis in poly-
    urethane production
Liquid oxygen reactions
Photocatalysis

Safer Chemical Products
Complexed photodevelopers
Degradable polymers
Degradable antifouling paints

Source: Paul T. Anastas and John C. Warner, *Green Chemistry: Theory and Practice*, New York: Oxford University Press, 1998.

---

should be preferred. Chemical reactivity not directly related to a material's function or performance should be avoided. If the reactivity of a chemical is important to its function (such as a catalyst), then a process should be designed in which a chemical is functionally reactive for the shortest time period. For instance, a highly reactive chemical should be shipped and stored in a bonded form, chemically "unzipped" into its reactive form only for those moments that it is required to be reactive, and then rebonded (blocked) to a less reactive state for postproduction functions.

Nonbioavailable Materials   The chemistry of a material should be designed to be less likely to reach a target organ or cell or to be stored there once it has been absorbed into an organism. If organisms create selectively permeable membranes to shield vital functions from exogenous chemicals, then chemicals should be designed that respect those barriers. Compounds that are composed of large, dissociated, nonlipophilic molecules are less likely to cross the cellular membranes of organisms. Water-soluble chemicals, for example, are less likely to pass through cell membranes than lipid-soluble chemicals. Large-molecule polymers are less likely to be absorbed than small-molecule polymers. Chemicals that cannot penetrate the membranes of cells or fat molecules are less likely to be stored and accumulate in organic tissue.

Physically Benign Materials   The physical properties of a material should be designed to be less likely to be absorbed into an organism. The physical properties of a material affect the possibility that it will be easily transported in the environment or readily absorbed into an organism. For instance, materials in fine powder form are likely to be transported in the air and easily inhaled. The same material in pellet, slurry, or solid mass form is less likely to be transported on air currents, dispersed onto food or water supplies, or inhaled into respiratory systems. Materials that easily volatilize are more readily transported and inhaled as vapors or gases than materials with lower molecular weights. Where a fine dust or volatile state is required, a material could be converted to that state for the minimum time necessary.

Biodegradable Materials   Chemicals could be designed so as to rapidly degrade in the environment or be converted to nutrients in an organism. Since synthetic materials are more likely to be hazardous if they persist in the environment or inside an organism, then chemicals should be designed that are likely to be easily biodegraded under natural conditions or readily metabolized into benign forms. It is possible to place functional groups into the molecular structure of a product so that prolonged sunlight or the action of microorganisms can degrade the material. The biodegradable polymers based on starch and other carbohydrates provide examples.

Contained Materials Processing   Production and use processes should be designed so that a material is less likely to leave a process and expose an organism. Production and consumption processes that are inherently or inadvertently dissipative are more likely to increase the probability of transport in the environment and plant and animal exposure. Materials that are in solid or liquid phases are more easily contained than materials in gaseous phase. Contained processing may require closed vats, sealed production lines, or process designs that reward recycling. Materials that have high economic value (gold, silver, platinum) are likely to be used in carefully controlled procedures to maximize their recovery and reuse. These same containment procedures could be used for materials with a lower value.

On-Demand Generation of Chemicals   Processes that require highly hazardous chemicals should be designed to generate them on demand. Some highly hazardous chemicals required by specific manufacturing processes can be synthesized on-site and on demand to eliminate storage and transportation hazards. The chemical constituents can be reacted as needed, and if there is surplus following use, reacted to a more benign form immediately after use. Some electronics firms have tried on-demand production of gallium arsenide and zinc selenide, both highly hazardous chemicals used in semiconductor production. This has involved a search for alternative precursors to arsine and hydrogen selenide to eliminate the on-site storage of these highly toxic gases.[31]

Processing Under Ambient Conditions   Production and use processes that occur at room temperatures and pressures and minimize water and energy use should be selected. Materials that are produced and used at normal temperature and pressure are less likely to dissipate, volatilize, or leak. Processing materials at ambient conditions also reduces the likelihood of explosions and fires. Biosynthesized materials provide a good example. Such materials are also more likely to require less water and energy over their full life cycle and thus to generate less environmental pollution or exposure.

In some cases, a careful consideration of the properties of chemicals makes it possible to "design out" the toxicity and thereby reduce or eliminate the hazard. EPA's Paul Anastas and Tracy Williamson make this goal clear when they conclude, "[g]reen chemistry seeks to reduce or eliminate the risk associated with chemical activity by reducing or eliminating the hazard side of the risk equation, thereby obviating the need for exposure controls and, more importantly, preventing environmental incidents from ever occurring through accident. If a substance poses no significant hazard, then it cannot pose a significant risk."[32]

14.6   Detoxification

A sustainable future should mean a less hazardous one. The materials that we offer the future should not be as dangerous as the materials that history has offered us. We have many high-performing and effective materials today that offer few dangers, but in the future we will need a much broader range of new materials to replace those that present too great a hazard. Our current research on advanced and engineered materials could offer many of these alternatives, but this will require some substantial redirection. We need to look more thoughtfully at the myriad of materials that nature already offers in its bounty of plant life. These are renewable materials that could provide effective substitutes for our diminishing stocks of nonrenewable resources and they are less likely to be dangers to life. The lignocellulose wastes of the forest and agricultural industries and the organic wastes of our municipalities provide an underdeveloped resource. There is an ample supply of forest and field biomass that could be grown with far less toxic chemicals and harvested sustainably to provide organic chemicals for a long time into the future. Finally, we are learning how an old process, like fermentation, could become the "soft path" for coaxing microbes to produce materials in simple aqueous solutions without harsh or dangerous chemicals.

This vision does require a sharp reduction in the production and use of ecologically unfriendly materials. If a reasonable goal of dematerialization is a 10-fold decrease in the flow of materials by 2050, then there should be a comparable goal for the use of toxic materials. Using the same target year of 2050, there should be a virtual elimination of the production and use of persistent, bioaccumulative, and toxic materials, and a 20-fold decease in the use of other toxic materials.[33]

The virtual elimination of persistent, bioaccumulative, and toxic chemicals in 50 years would mean that the materials used by future generations would be significantly safer and more ecologically sound than today's materials. Simply banning these materials could precipitate many unanticipated consequences. Instead, the commitment should be to phase these materials out of production and use by substituting more benign materials, changing the processes in which they are used, or changing the functions that they serve. This will require a large commitment from the materials sciences to develop safer substitutes and a sizable commitment from private industry to develop cleaner production processes. It will require a willingness on the part of consumers to identify and demand products and services that are not dependent on toxic materials, and it will require that governments promote and ease the transition.

A decrease in the use of the remaining toxic materials by a factor of 20 may appear ambitious, but there is ample reason to believe that this could be achieved. Data from the Toxics Release Inventory indicate a 43 percent reduction in the quantity of a core group of toxic chemicals released to the environment from 1988 to 1997. Data from the Massachusetts Toxics Use Reduction Program show a 48 percent decrease in generation of a similar core group of toxic chemical by-products from 1990 to 1997. Although these data are suggestive, they document reductions in the release of toxic chemicals and not in their actual use. A better indicator from the Massachusetts data reveals that the use of the core group of toxic chemicals in industrial production decreased by 33 percent (by weight) over the same 7-year period.[34]

The detoxification of the materials economy is an enormous challenge, but one we cannot avoid. To maintain a fair level of well-being, we cannot continue to pollute the planet and our own bodies with materials that are toxic and ecologically damaging. We will never have enough knowledge to be certain of all of the possible effects of such chemicals. We need to act now in the spirit of precaution, using what we know and can safely guess at today. We have learned from our experience with the chlorofluorocarbons that we can act in a precautionary manner, assertively, proactively, and internationally, even when all of the data are not on the table. Confronted with a complex array of toxic materials, many of which are largely unstudied, and the knowledge that we can find alternatives, we have an obligation to try to substitute where we can, and to focus our materials sciences on other solutions where we cannot.

15

A Sustainable Materials Economy

Can we move nations and people in the direction of sustainability? Such a move would be a modification of society comparable in scale to only two other changes: the Agricultural Revolution of the late Neolithic and the Industrial Revolution of the past two centuries. These revolutions were gradual, spontaneous, and largely unconscious. This one will have to be a fully conscious operation, guided by the best foresight that science can provide.... If we actually do it, the undertaking will be absolutely unique in humanity's stay on the Earth.
—William Ruckelshaus

In order to move toward a sustainable world economy, the industrialized world must progressively reduce its throughput of materials and energy. This involves a reduction in the volume and rate of flow of materials in the industrialized nations as well as a sharp reduction in the volume of toxic materials. We need to use fewer materials—we need to use even fewer industrial materials that are persistent, bioaccumulative, and toxic. As normative goals we should seek to reduce the flow of industrial materials in the United States by a factor of 10; fully eliminate the production and use of bioaccumulative, persistent, and toxic materials; and reduce the use of other toxic substances by a factor of 20 over the next 50 years. While there are many current initiatives that suggest the technical opportunities, the scale of this transformation and its social and economic implications rival the changes required to move from the Bronze to the Iron Age. That transition took hundreds of years; the transition needed here must be done in a far shorter time.

Converting the materials economy of the United States to a more sustainable materials system is a critical part of this global transition. The analysis presented here proposes that this could be accomplished by reconsidering the materials economy as a cyclical rather than a linear system. Based on a materials flow model involving three cycling loops, the economy should be viewed as an open, cyclical system nested within the larger ecosystems with which it exchanges materials at rates that do not compromise ecological viability. Some toxic and hazardous materials could be used within closed subsystems of the economy, but they must not be exchanged with the environment unless they are first detoxified. In order to achieve this model in the next 50 years, the six principles developed in the previous chapters have been proposed:

- close the loop on materials flows;
- increase the intensity of materials use;
- substitute services for products;
- reduce the dissipation of degradable toxic materials;
- reduce the use of persistent, bioaccumulative, and toxic materials; and
- develop more environmentally appropriate materials.

All of this must be accomplished while increasing the capacity of much of the world's population to enjoy a more secure, healthy, and comfortable life. This will involve reducing the quantity and composition of the materials flow through the more privileged parts of the global economy without substantially altering the quality of life for those accustomed to security and comfort. In order for consumption to reach a sustainable level, the rate of population growth must also stabilize, or all efforts to increase efficiencies will be overwhelmed by the growth in aggregate demand. Currently the population of the United States is just over 280 million people, but it is growing at a rate of 1 percent per year. The world population, which has just reached 6 billion people, is expected to reach 15 billion in the next 40 years. This growth must be slowed and stabilized. The subject is beyond the scope of this text, but it cannot go unrecognized as a determining factor in reaching materials sustainability.

15.1 Envisioning a Sustainable Materials Economy

The industrial materials systems of the future must be conceived of as cyclical systems of sustainably managed stocks and flows that conserve resources and minimize dissipation. Recycling and reuse must be a primary component of these systems, as well as new extraction, harvesting, and treatment practices. Those materials flows that are crucial to the maintenance of the planet's environmental services upon which life (at least human life) depends must be respected and protected. Of course this means that the rate of carbon dissipation from human sources must be sharply curtailed, but such reductions are also required for many persistent and bioaccumulative substances as well. The economic systems of production and consumption can continue to extract materials from the planet's reserves and dispose of them back into the planet's ecological systems, but the flow rate and the physical and chemical properties of those materials need to be monitored and managed responsibly.

Renewable, degradable, nontoxic materials can be readily exchanged between the economy and the environment as long as flow rates do not exceed the rates of regeneration. This includes the fundamental materials of natural plant life—cellulose, proteins, and carbohydrates—as well as water, oxygen, and nitrogen. This also includes a host of agriculturally derived materials, some of which are on the market today, many more of which could be developed and commercialized if their costs were competitive. For this, a new kind of plant management and harvesting system will be required that is less invasive and more in keeping with the diverse and interdependent processes of ecological systems. In addition, industrial materials could be developed from the cellulose and carbohydrate wastes of the forestry, agriculture, and food processing industries and the organic wastes and sewage of the municipalities. This includes a wide variety of bioprocessed materials that can be "farmed" through synthetic fermentation processes that rely on naturally derived organisms and substrates.

Nonrenewable, nontoxic materials can be used, but these should be recycled within economic systems and never consumed at rates that exceed the rates at which renewable materials can be substituted for them. This includes various minerals, many of which are already mined and refined. Extraction of minerals to produce primary metals should be reduced to favor an increase in the amount of metals being recycled within the economy. Some metals, such as iron and aluminum, already have well-developed recycling infrastructures that could keep the resources in continued use and reuse. Others, such as silicon, tin, and titanium, will require new recycling infrastructures. The materials sciences are well focused on many of these materials and they should continue to offer a wealth of new alloys and composites, but these innovations need to be better directed by environmental and health design parameters in order to ensure that new materials are safer and cleaner and more likely to be used at high levels of intensity.

Degradable, toxic materials can also be used, but these should be carefully recycled within closed economic systems and returned to the environment only after treatment has nullified their toxicity. Some of these, such as alcohols, acids, and caustics, can continue as viable intermediates in chemical processes, but their useful life should be extended through continuous reclamation and recycling.

The production and use of persistent, bioaccumulative, and toxic materials should be phased out through material, process, or functional substitutions. This should include well-recognized materials such as mercury, cadmium, cobalt, vanadium, and polychlorinated biphenyls, as well as less targeted materials, such as certain of the halogenated aromatic hydrocarbons. There is a broad range of challenging research opportunities in the substitution of these materials, and this should open new prospects for chemical process designers and materials scientists and engineers.

The development of new materials and the design of new production and consumption processes should play an important role in the conversion to a more sustainable materials economy. The new sciences of nanotechnology, crystallography, and biotechnology offer potent opportunities. The overriding objective should be high-performance materials that are less toxic than current materials and that can be produced with lower energy inputs and in less threatening environments. This suggests a new opportunity for renewable materials and bioprocessed materials. Whether genetic manipulation plays a role in this transition remains a debatable proposition. While it appears highly adventurous to create food sources from recombinant cells, there is some justification for using this technology to create pharmaceuticals. For now, genetically modified industrial materials should not be pursued. With so many natural organisms and processes unexplored, it appears unnecessary and unjustified at this point to invest in genetically engineered industrial materials.

15.2   Policies for Promoting Sustainable Materials

The transition to a sustainable materials economy requires new policies that promote human health and environmental quality as well as economic efficiency and product effectiveness. This requires broad policies that truly promote sustainability. Put conceptually, a sustainable materials policy would be one that optimizes value from the use of materials, adds no new risks to everyday life, increases natural capital, minimizes the transfer of risks from one generation to another, respects and enhances the natural functioning of the planet's ecosystems, and ensures no net loss of valuable resources. Such policies can be either public or private, adopted by governments, industries, institutions, or nongovernmental organizations, but they must be comprehensive and integrated and focused on the full life cycle of materials.

The market can be a significant factor in promoting this new materials system. Markets remain the most flexible, least intrusive, and most powerful mechanism for allocating materials and making changes. Prices and competition will need to continue to provide an opportunity for choice and a degree of efficiency in allocating the opportunities and costs of conversion. But the market cannot be relied upon alone. Earlier sections have documented the significant role that government policy and programs have had in developing the materials of the present and in protecting values that are not easily addressed by market signals, such as national security and protection of environmental and public health. Indeed, the rich array of materials currently available today is historically dependent on many now quite well-accepted government policies. Federal ownership of national forests, national materials stockpiles, government regulations for environmental protection, and federal funds for research in the materials sciences are all well-institutionalized programs. Thus there is no break with the past in suggesting that the transition to a new materials system will require government interventions and protections.

Nor will government policies be enough. There will need to be changes in what consumers want, what they understand, and how they behave. This will require more information and education, and also some shifts in cultural values. These changes will occur only to the degree that the public has more of a role to play in making decisions about the development and use of materials. This will require a new sense of accountability among materials producers and users. Government policies and economics can play a large role in promoting these social changes, but there is a need for personal leadership and campaigns by institutions and nongovernment organizations to influence public opinion and foster social change.

Each of these areas requires new initiatives. The market needs to be reconfigured through changes in tax policies and government expenditures to encourage resource conservation and discourage waste. Government policies need to be redirected to focus on reducing the use of the most dangerous chemicals and on the development of safer substitutes. Corporations need to adopt new missions that involve satisfying customer needs and managing production processes in ways that enhance public health and environmental quality. Citizens, workers, and consumers need to be better educated and informed about the materials in their lives and they need better avenues for expressing their interests and values in decisions about those materials.

There is no single policy remedy that can achieve all of this. It has taken hundreds of policy interventions to create and shape the current materials economy. Many of those policies need not be changed; they already promote environmentally sound directions. But other policies do need to be changed and entirely new policies will need to be designed. This will require a host of initiatives, only some of which are apparent today. The search for the best-directed and most effective policies will be an evolving and learning experience. Public sentiment, economic conditions, and political changes will create opportunities in the future that can only currently be guessed at. For now the best that can be achieved is to identify sound directions and begin to move where action is possible.

There are at least five broad themes that can guide these directions. These are:

• creating new information systems to improve tracking and management of materials,
• redirecting materials markets to encourage recycling and conservation,

• reconfiguring corporate culture and missions to embrace sustainability,
• redirecting government policies to promote safer and more efficient materials, and
• promoting public engagement to ensure longer term accountability.

Some of these initiatives are presented in this last section.

15.3 Creating New Information Systems

A first step toward a sustainable materials economy requires more effective materials and environmental data systems for collecting and managing information about materials stocks and flows and the state of the environment. As noted earlier, we lack adequate data. Data on materials stocks and flows are not integrated; data on environmental and health effects are often missing; and where procedures are in place to collect information, the data lack quality and the methodologies lack harmonization. To be useful, materials data are required at various levels. Production firms need tracking systems to identify wastes and increase resource productivity. State and federal governments need inventories in order to stabilize primary materials markets, ensure the effective functioning of secondary materials markets, and monitor the effectiveness of environmental, health, and economic policies. Internationally, data on materials stocks and flows are important to ensure that rates of cycling between economic systems and the environment do not exceed rates of material renewal or assimilative capacities. Finally, data are needed to ensure that material consumption supports a fair distribution of the quality of life.

For these various reasons, national and international materials tracking systems need to be created and maintained. This could be a government function, although there is evidence that private systems that provide publicaly available data can be equally effective.[1] A unified and reliable national database on materials that consistently monitors trends in supplies, demands, and flows would be of significant value in promoting resource conservation and new substitutions. Such databases could provide the information necessary to identify problems in supply and demand early enough to permit preventive actions. They could provide information currently lacking but needed to establish, implement, and evaluate federal policies. They could provide information useful for private sector planning and for monitoring market performance to improve corporate performance, stimulate international trade, and enhance the efficient allocation of investments. Such systems could also provide data for research and study to increase knowledge about materials flows and their consequences in the environment. Finally, they could provide the information necessary to better educate and inform the public about the role materials play in the economy and how this role could be improved.

To be effective, these data need to be interpreted through indices that permit monitoring over time. The data on toxic chemical releases reported by U.S. firms under the Toxics Release Inventory have proven to be one of the most effective environmental policy tools developed in the past 20 years. With relatively little public or private investment, data collection and annual publication of toxic chemical releases has been a potent force in promoting corporate environmental responsibility, because firms and the public have used the annual reporting as a benchmark for tracking improvements in pollution control.

During the past decade, there has been substantial activity in developing indicators of sustainable development. Among the most ambitious has been a program sponsored by the United Nations Commission on Sustainable Development and conducted by the international Scientific Committee on Problems of the Environment; it has produced a broad set of indicators now being tested by fifteen countries around the world. In 1998 a federal interagency working group on indicators of sustainable development developed a trial set of such indicators for the United States.[2]

In 1990 the Organization for Economic Cooperation and Development initiated a project to establish a core set of environmental indicators for its member nations that could be used to track environmental progress and promote the use of environmental data in sectoral and economic policies. The data include both resource use and the discharge of pollutants. The World Resources Institute and the Wuppertal Institute are jointly working on a project to develop national materials flow accounts for four nations that will permit cross-national comparisons and factor in resource depletion and environmental damage. The first step involved an inventory of each nation's materials input to the economy from the environment. In order to compute the global burden on the environment, each nation's draw on foreign resources was computed and added to the domestic input to compute the total materials requirement. A second step, which has only recently been completed, involves creating in inventory of each nation's material output in order to create national material flow accounts.[3]

This thinking is most advanced in Germany, where the German Bundestag established a special commission in 1992 to consider options for the "sustainable management of substance chains and material flows." The commission established a set of procedures for modeling national materials flows and gathering data; these were used to "map" the material throughput and total material input of the country. The widely reviewed German program demonstrated that the input of nonrenewable materials exceeds that of renewables by about 50 percent and that by far the largest amount of movement was in the overburden and wastes in the mining industries. During the past decade, several additional initiatives on national materials balances have been developed, with the more advanced programs established in Austria, Sweden, and Japan.[4]

Many new data reporting initiatives are being piloted at the corporate level. A small number of large corporations have begun publishing annual corporate environmental reports, and the International Standards Organization, the World Business Council for Sustainable Development, the American Institute of Chemical Engineers, and the Coalition for Environmentally Responsible Economics are promoting generic formats for corporate environmental performance indicators. These annual reporting systems are demonstrating the capacity corporations have for reporting on environmental performance. This experience should be used to promote data collection on corporate materials use and flows.

There are also initiatives to redesign the system that economists use to track national economic performance. The current design of the gross national product (GNP), the measure by which most nations monitor their economic performance, has been criticized as neglecting and therefore undervaluing natural resources. By neglecting to include natural assets and deduct environmental damage, the U.S. GNP misdirects those in the Federal Reserve Bank and the U.S. Department of Treasury who manage the nation's economy. In keeping with the World Commission on Environment and Development's challenge to nations "to take into full account [while measuring] growth, the deterioration in the ... stock of natural resources," ecological economists have been arguing for a new "green GNP." One model put forward by Herman Daly and John Cobb called an "index of sustainable economic welfare" deletes wasteful expenditures, adds the value of housework, weighs consumption for the degree of inequity, adjusts for the costs of pollution, and accounts for the depletion of natural resources.[5]

In 1993 the United Nations published an extensive system of integrated environmental and economic accounting for guiding national accounting schemes. The World Bank has also begun to explore national and global economic reporting systems that would account for three major capital components that determine a nation's wealth: produced assets (human-made capital), natural capital, and human resources. Although these indicators show dramatic differences in national wealth and "savings" from more conventional GNP figures, so far these accounts have only been used as adjuncts, not as substitutes for conventional GNP accounts.[6]

15.4   Redirecting Materials Markets

The market can be a central force in promoting a new materials economy. However, the current structure of industrial materials markets is not well designed to accomplish this task. By neglecting the long-term scarcity of natural resources, the limits to the absorptive capacity of the environment, and the damage caused by toxic materials, markets are permitted to optimize returns only on artificially short horizons. The full costs of the current materials mix are not factored into their prices. In economists' terms, these externalized costs (i.e., costs of product use borne by persons who do not consume the product) must be internalized to make markets truly effective. In order to conserve resources and avoid dangerous chemicals, the user must pay the full costs of products, including costs to society, the environment, and future generations.

Those who argue for the efficacy of market solutions are quick to point out the perversions created by government subsidies. Although many obsolete federal subsidies to materials markets have been eliminated over the past couple of decades, direct mining subsidies still authorized under the General Mining Act of 1872 remain a major source of inaccurate market signals that encourage virgin mineral extraction over recycling and recovery of secondary metals. Because the price of energy so directly affects the markets for materials, energy subsidies also distort materials markets. The underpricing of energy, particularly by state energy regulatory commissions, provides direct subsidies that discourage conservation of materials and energy, while the indirect subsidies that permit energy suppliers to avoid the costs of environmental damage and all but ignore global warming effects provide an estimated $32 billion in subsides a year, some portion of which subsidizes materials processing and use. If industrial markets are truly going to incorporate environmental and health effects into the price of materials, this will require the incorporation of increased costs to account for environmental damage and the elimination of those subsidies that support dissipation, waste, and environmental damage.[7]

Government tax structures provide another means for reshaping markets to account for environmental protection. This is not a new idea. In 1920, Arthur Pigou, a British economist, proposed that the economy would more directly account for the true costs of production if fair prices were paid for the consumption of open access resources (e.g., ocean fisheries, the atmosphere) and these price adjustments could be achieved by taxing natural resource use and environmental pollution. Following Pigou, various economists have argued that taxes on industrial emissions would provide an efficient means of allocating pollution control costs. Although the idea of using taxes to adjust for environmental damage has remained relatively dormant for years, during the past two decades there has been renewed interest by some economists and environmentalists in "green taxes." They argue that because natural resources and ecological sinks are common property and free for exploitation by anyone, private interests that exploit them receive a significant subsidy at a direct cost to the environment and an indirect cost to the public. By taxing the use of these environmental services, the true costs will be borne by those who benefit from their use. Furthermore, because these taxes will increase the price paid by those who use environmental services, these resources will be used more judiciously, leading to increased conservation and decreased environmental damage.[8]

Several European countries have advanced environmental taxes as a means to address air pollutants, solid waste, lead in gasoline, and excessive manure. Sweden, France, and the United Kingdom tax individual air pollutants such as carbon, nitrogen, and sulfur dioxide. The Netherlands has two revenue-raising taxes on energy use. Denmark taxes waste generation, and the United Kingdom has a landfill use tax. Denmark and Sweden levy taxes on batteries, with the revenues used to fund collection and recycling programs.[9]

Some of these countries have moved beyond these direct environmental taxes to experiment with a concept called "ecological tax reform," which includes a host of proposals for taxing mineral extraction, industrial wastes, fuel use, and environmental pollutants. At its broadest, ecological tax reform goes further than simply adding taxes onto environmental services. The proponents of ecological tax reform argue that the transition to a sustainable society requires economic incentives that reduce environmental damage but also increase employment. Therefore, ecological tax reform includes a revenue-neutral tax shift that increases taxes on environmental services and proportionately decreases taxes on earned income. Lower income taxes mean a reduction in the cost of labor for employers, who are expected to respond by opening up new jobs. Thus, by taxing things that we do not want—environmental destruction —we can remove taxes from things that we do want—jobs—while maintaining consistent government revenues. Such a rebalancing of the tax structure could also provide the means for compensating those workers and communities that would be affected by the reductions in mining suggested in earlier chapters.[10]

Although attractive as a concept, ecological tax reform has been slow to penetrate current government policies. Some European countries have experimented with minor tax shifts that follow the general idea. A 1991 Swedish tax on high-sulfur fuels has produced a marked reduction in sulfur dioxide emissions and a tax refund to the 25 percent of taxpayers who switched fuels. A Dutch water pollution charge introduced in 1970 has resulted in an 86–97 percent decease in cadmium, copper, lead, mercury, and zinc discharges, while it has served as the primary source of revenue for construction grants for waste treatment facilities. In 1994, Denmark introduced a new tax system that raised taxes on energy, water, waste, and shopping bags and at the same time, lowered taxes on personal income by 8 to 10 percent. A recent analysis conducted by the European Environment Agency examined sixteen tax programs in six countries and concluded that most of these "green taxes" had met their environmental objectives and generated tax relief for some taxpayers.[11]

In the United States, taxes that raise the price of gasoline by some 34 cents per gallon are the most common form of green tax. But because these taxes are used to build highways, which increases the use of automobiles and the consumption of fuels, gasoline taxes are not a form of ecological tax reform. They could be, if the taxes were higher and the revenue was used to offset the incomes of those who must depend on a car to get to work. A better U.S. example is the successful tax on ozone-depleting chemicals introduced in 1990 to hasten and ease reduction in the use of chlorofluorocarbons.[12] A federal carbon tax was proposed during the 1990s, with revenues to be directed to reducing the federal budget deficit, but it was withdrawn when significant resistance arose from potentially affected industrial interests. The lesson should be clear: green taxes will be difficult to enact without a broad constituency that sees more benefit than a cleaner environment. Taxes on polluting fuels, municipal wastes, and industrial pollutants are all needed and can serve as a source of revenue for compensating those negatively affected by new materials policies, but they will require a reduction in income tax rates as a fair bargain.

15.5   Reconfiguring Corporate Culture and Missions

Without broader market incentives, corporations tend to try to maximize profits only on cost and product performance. Traditional materials suppliers and product manufacturers alike have generally been satisfied to get the product out the door at the lowest possible cost. Competitive advantage has been based on product differentiation and price. However, there are new voices in the business community that are arguing for a broader array of factors for achieving profits. Many firms, both large multinationals and small regional enterprises, are embracing environmental factors as core business concerns and experimenting with new missions, new business strategies, new management systems, and new relations with suppliers and customers that go beyond regulatory compliance and conventional business dogma.[13]

To satisfy customers, appease regulators, and attract investors, these corporations are trying to become more open and transparent regarding the environmental and health impacts of their products. This requires a new sense of public accountability in a firm; one that holds that customers, employees, and citizens have legitimate interests as stakeholders in engaging corporate managers about business decisions. It is this sense of accountability that has motivated progressive companies to publish annual corporate environmental reports. While the early publications were often criticized as too heavily driven by public relations agendas, these reports have recently become more accurate on environmental aspects and they are now considered a legitimate means of assessing and rating corporate performance. Some European firms have gone further and begun preparing environmental product declarations for their customers that identify the environmental impacts of the product and its production (the Volvo declaration is eight pages long). As firms become more adept at corporate environmental reports, pressures are building for reporting on social issues as well.

New approaches to corporate governance have already embraced the idea that boards of directors are directly responsible for environmental performance and therefore require periodic environmental management reports from corporate managers. This has helped to spur the adoption of "environmental management systems" that parallel financial management systems and provide an opportunity for integrating environmental management into conventional business management practices. These management systems often, but not always, promote materials accounting within the firm.[14]

A life-cycle approach to materials management requires that firms forgo conventional identities as independent entities and come to view themselves as participants in a chain of value-adding and -subtracting suppliers and customers. Concepts like supply chain management, integrated product policy, life-cycle management, and life-cycle costing are being adopted from European practices as U.S. corporations become more sensitive to changes in global markets. By adopting a life-cycle approach, the Daimler-Chrysler Corporation was able to identify $2.5 billion in excess management costs (liability, training, occupational health costs, waste management, and lost opportunities for recycling) over 5 years that were due to the use of toxic substances; these costs are now being reduced through improved negotiations with suppliers. Large manufacturers in Europe and firms like Ford in the United States have made the adoption of environmental management systems a key requirement for all of their suppliers.[15]

The materials supply chains create the basis for a new dialogue among suppliers, customers, recyclers, and waste managers that is directed to increasing the efficiency of materials and energy use and decreasing the pollution, waste, and toxicity of materials throughout a product's life cycle. Extended producer responsibility and product takeback arc important principles for closing materials loops, extending product life, promoting materials efficiencies, and encouraging safer materials.

The business climate of the twenty-first century will depend on information and services far more than in the previous century. Reconfiguration of a company around information transfer and services has already been noted as a strategy for dematerialization. Selling information or selling "knowledge-rich products," such as software or products that can teach customers, respond to changes in the environment, correct for operator errors, or provide service themselves, can reduce the need for a constant flow of commodities. Companies should seek to build long-term relations with customers and supply "utilities" (product use) and "results" (satisfied needs) rather than physical products. There is a future here that relies on networks of customers and suppliers embedded in information-rich, customer-tailored services and well-managed loops of materials cycling.

All of this requires new management approaches that encourage and rely on continuous learning and organizational change. It is naive to assume that a sustainable economy can be created by the same corporations that provided the current unsustainable system. Corporations and other institutions will need to redesign their culture and reorganize their structure to meet the demands of a more socially accountable future. The solving of environmental and health problems needs to be seen as providing opportunities for learning how a company can institutionalize learning. The transition to a more open and socially responsive company will require that employees be able to contribute ideas without fear and share in the benefits that result. Changes in the way knowledge is managed and creativity is nurtured are important. In order to track materials flows and identify cleaner production opportunities, firms need to improve communication and information flows. Many management leaders find that rigid hierarchies inhibit information flow and often need to be replaced by more horizontal, team-based management structures to encourage innovation and learning. Such team-based workgroups facilitate the diffusion of environmental ideas more easily into manufacturing, marketing, and financing functions. Because change, even well-managed change, takes time, patience is key to organizational learning. It often takes 10–15 years for a new set of ideas to permeate a large corporation and it takes a generation to create such a shift in a society.[16]

15.6   Redirecting Government Policies

The transition to a sustainable materials economy will require the redesign and redirection of government agencies and policies as well. The focus must shift from managing the consequences of the production, use, and disposal of industrial materials, to promoting the development of safer and more environmentally compatible materials and systems of production and consumption.

There are several models conventionally used to describe government support for industrial innovation. The predominant model upon which much federal policy has been based since 1945 draws a linear connection between government-financed basic research, through industry-financed applied research, to the development of commercial innovations. The assumption that basic science is the starting point and that innovation is the result of such logical steps has been challenged by several other models. Some of these have emphasized the power of "market pull" over "scientific push" and the importance of customer relations, while others have shown that large firms may pursue several, even quite divergent, innovation strategies.[17]

Government investments should be used to foster research and support markets for new and advanced materials, biobased materials, agro-industrial chemicals, and green chemistry. The largest source of federal funds for agriculturally based materials comes from the U.S. Department of Agriculture. Much of this support is administered through the Cooperative State Research Service and the Agricultural Research Service, which provide funds to state experimental stations and universities. In 1990 Congress established the Alternative Agriculture Research and Commercialization Center to promote innovation in new uses for agricultural products, and this agency could become a central source for supporting innovation in biobased materials. Government support for agricultural prices decreased over the 1990s, but price stabilization could still be maintained if the government directed more public procurement to products made from agricultural materials.

The older model of government support for large-scale scientific programs has given way to smaller programs that promote private contributions and encourage collaborations. Recent changes in the legal structure of research have provided for more rapid transfer of innovations from laboratory to applications and among collaborating firms. The National Cooperative Research Act of 1984, which permitted cooperative research and development programs between competitors, and the Technology Transfer Act of 1986, which encouraged cooperative ventures between federal research laboratories and private firms, have provided more flexibility and communication as innovations are assessed and promoted.[18] Several of the federal laboratories could become focal points for applications of secondary materials. While the more applied and cooperative research approach needs to be maintained, there remains a need for large-scale projects designed to open up avenues for new development and production systems that close loops and conserve resources. The National Science Foundation, the U.S. Department of Energy, and the U.S. Department of Defense will continue to need to support the development of advanced materials, electronic and photonic materials, smart materials, and nanotechnology.

Government regulations can also be an effective stimulus to the development and use of industrial materials if the regulations are stringent enough and phased in with enough time to permit a reasonable period of transition. For instance, prohibitions on the disposal of toxic wastes have provided incentives to design products with less-toxic constituents. Strategies focused on eliminating toxic materials in products have provided motivation for seeking less-toxic materials in production, as illustrated by regulations on asbestos (in brakes), lead (in gasoline), urea formaldehyde (in insulation), and phosphates (in detergents). Efforts to detoxify process chemicals have encouraged innovations in feedstocks and raw materials. These regulations have indirectly promoted new materials by internalizing and thereby raising the costs of toxic and hazardous materials. In a similar way, increasingly stringent regulations on the release and disposal of persistent, bioaccumulative, and toxic chemicals would promote their substitution.[19]

Some government regulations need to be redesigned to promote sustainable material systems. For instance, federal regulations on the recycling of waste materials currently increase the compliance costs of transferring wastes for recycling. New regulations are needed that encourage recycling of secondary materials and reduce the burden of liability for those who can demonstrate that their by-products are being cycled into secondary materials markets.

Finally, government policy needs to be better coordinated and integrated so that there are consistent messages and common directions. This will require more than simply consolidating environmental regulations. To be effective in promoting more sustainable materials, government policies on technology research, intellectual property, technology transfer, taxes, and procurement will need to be coordinated and provided with a unifying set of objectives. Governments in other industrialized countries have used plans, agreements, and goals as a means to promote more sustainable economies. Sweden has prepared a national chemicals policy that requires that within 10 years all commercial products be free of lead, mercury, cadmium, polyvinyl chloride plastic, and other persistent and bioaccumulative substances. Canada has developed a broad plan for a more sustainable metals and mining industry that promotes metals recycling and a "safe use principle" in metals policy-making. The Dutch minister for the environment has proposed doubling gross domestic product while cutting wastes in half (factor 4), while the Austrian government has put forward a plan to dematerialize its economy by a factor of 10.[20]

This raises the old dream of a comprehensive and unified national materials policy in the United States. The failure to achieve a national policy on materials during the 1970s and 1980s need not be interpreted as a sign of its inherent irrelevance. Conversion to a safer and sounder materials base would be greatly facilitated through the development of a national policy on sustainable materials. It is important to recognize the reasons the earlier efforts were unsuccessful. The process needs to start with an open public dialogue and a wide-ranging search for alternative pathways. It needs to take account of the many significant interests that would be affected by such a transition, and it needs to put forward credible plans for easing the costs of change and dislocation. Leadership is required, and if that leadership is not available inside the government, perhaps the initiative for the policy must be begun and advanced outside of the government.

Some would doubt that the federal government in the United States is up to the task of promoting the transition to a more sustainable materials economy. The performance of the government over the past 20 years has been particularly uninspired in terms of major policy initiatives, most certainly in such areas as environment, health, or industrial policy. The transition to a more sustainable economy could certainly use leadership at the federal level, but that may be unlikely unless there is a growing constituency of informed and active citizens who press for these new directions. In the current absence of such leadership, it may be necessary to promote new policy work within state and local governments or within nongovernmental and community organizations. This would require a much more actively engaged public, and this is not so difficult to envision.

15.7 Promoting Public Engagement

An engaged public is a critical factor in promoting a sustainable materials economy. The history of the development of industrial materials reveals a crucial lesson. Concerns over material dissipation and toxicity were factored into decisions about production, use, and disposal of materials largely and often only when members of the public organized, raised awareness, and sought to be included in decision-making processes. In some cases these initiatives arose from the concerns voiced by public health and environmental professionals, but often they were propelled by civic activists, journalists, trade union advocates, environmentalists, and lay citizens. The emergence of government agencies charged with addressing these issues has been a mixed blessing. Government professionals certainly have more resources for studying the environmental and human health effects of industrial materials and more authority for requiring careful management. However, the record of federal and state government involvement demonstrates how professional jurisdictions, bureaucratic complexities, and political interferences can slow, if not stymie, well-intentioned efforts to protect the environment and human health.[21]

Such problems are an unavoidable result of a representative democracy that relies on administrative implementation of public policies. However, this same government design also opens opportunities for a certain amount of public participation and pressure. To be effective in shaping a more sustainable materials system, the public needs to be both informed and interested, and to be able to identify routes of access to private and public decision-making settings. The first requirement involves education and the second involves creating and publicizing new institutional openings.

Environmental education is the key to the first requirement. There is a growing movement for introducing environmental education into educational curricula ranging from primary schools up through undergraduate training. The scope of this training is broad, focusing on subjects ranging from ecosystems to domestic recycling, but it often stops short of teaching young people about the technologies of production and the costs and benefits of current practices. An environmentally literate public is a precondition to an active constituency willing to search for new modes of production and consumption, and an education directed by such a spirit must be empowering and solution oriented. Teaching children to recycle their trash is a valuable first step, but teaching people to understand the materials throughput of their households, their community, and their nation provides them with the tools to seriously challenge the assumptions upon which those materials flows are based. For this, adult education is also important. The country is dotted with thousands of local voluntary environmental action groups and lay boards of conservation, public health, and land use planning. Many of these people are good candidates for learning more about the production of materials, the generation of wastes, the functioning of the materials economy, and the environmental implications of these processes.[22]

A better educated public more attentive to the environmental effects of materials could be a powerful force for the environment. Environmental product labeling, targeted procurement programs, neighborhood recycling and reuse programs, and environmental literacy media campaigns can only be effective if consumers understand the environmental implications of the products they purchase and wastes they generate. Public participation can encourage better consumption practices. But, to become serious partners in promoting sustainable materials citizens need to be more than merely environmentally conscious shoppers.

An educated and empowered citizenry provides the energy for change, but the social and political mechanisms must be there as well. Citizen energy is often a factor in the enactment of new laws. Many of the current environmental and public health statutes were the result of extensive work by citizens. The Comprehensive Environmental Response, Compensation, and Liability Act of 1980 was propelled by a broad national grassroots campaign that demanded federal assistance in cleaning up the nation's hazardous waste dump sites. An active national constituency is still among the most powerful forces for achieving social or economic change through legislative initiatives.

Citizen-initiated litigation, or even the threat of litigation, continues to be a potent force for curtailing the worst abuses in producing, using, and disposing of industrial materials. While tort litigation is often criticized as too easy, too well rewarded, and too lucrative for attorneys, its role as a significant deterrent cannot be easily discounted. Litigation does promote a cautious attitude in corporate and government offices, even if it delivers justice to victims only on limited occasions.

There are several emerging new models for promoting more direct citizen participation in environmental policy. During the 1980s a broad array of initiatives generally referred to as "community advisory committees" encouraged small groups of citizens to organize and represent the interests of a wider constituency in providing advice to governments or businesses on environmental issues. Many of these were established to try to resolve difficult issues such as siting landfills or hazardous waste treatment facilities. In 1989, the Chemical Manufacturers Association recommended to its member companies that they establish community advisory panels as part of their Responsible Care Programs. By 1994 there were close to 200 of these panels set up and based on the concept of an "open and honest exchange of ideas and views." Over time, experience with these community advisory committees has led to substantial criticism. Participants often felt that the sponsoring organizations controlled the agendas and were not interested in advice that was counter to predetermined commitments, and sponsors often claimed that the citizen participants were ill prepared for technical choices and were too often "irrational and self-serving."[23]

Another model of citizen participation has been developed from so-called "planning cells" in Germany, which have been called "citizen juries" in the United States. These programs seek advice from a jury of randomly selected lay participants who are paid a nominal stipend to spend several days hearing the arguments of various proponents and who then meet behind closed doors to prepare a set of recommendations. The process, which relies on the judgment of presumably disinterested lay people, has been used in Germany and Switzerland for advice on the potential effects of new information technologies, the desirability of different national energy plans, and the siting of landfills. In the United States, the process has been used only on an experimental basis to advise on such issues as the impacts of agriculture on water supplies, the claims of competing electoral candidates, and the potential efficacy of national health care plans.[24]

A model more in keeping with domestic culture and politics has arisen around the advice given to federal agencies under the Federal Advisory Committee Act of 1972. Operating under well-defined procedures, public advisory committees made up of a divergent array of stakeholders have been set up to assist federal agencies with a host of technical and policy questions. The now-terminated Congressional Office of Technology Assessment used these advisory committees to assist with the production of its technology policy reports, and the EPA has used advisory committees under its National Advisory Council for Environmental Policy and Technology to address scores of issues ranging from classification of hazardous wastes to presentation formats for the annual Toxics Release Inventory. Federal agencys should create such multistakeholder advisory councils for overseeing government policies related to material resources, such as those involving national forest lands, strategic stockpiles, new chemical reviews, and new research and development programs on materials.

The Internet and the ubiquity of information-processing systems may provide an even more powerful way for citizens to express their interests. Already, environmental and public health advocacy organizations are using the power of electronic mail and Internet fora as a means to transfer information rapidly, expand participation in decision making, and coordinate activity among activists. As the public becomes more aware of the training and mobilizing capacities of interactive electronic media, new opportunities will open for participating in public and private decision making. Using the Internet communities are already experimenting with local opinion polls, and companies are seeking direct dialogue with stockholders about investment policies.

Increased public engagement provides a means to open information channels so that producers of industrial materials and their stakeholders —customers, suppliers, employees, stockholders, and facility neighbors —can develop more real-time interactions that offer the possibility of integrating environmental and health issues into business decisions. Citizens can negotiate directly with local firms, building partnerships where there are conducive opportunities, and creating pressures where such opportunities are less forthcoming. Public engagement is also critical to government innovation and reform. The development of new public policies on industrial materials will require the input of a wide range of professional and lay stakeholders.

15.8 Toward a Sustainable Materials Future

For too long the issues of public health and environmental protection have been viewed as costs to production and the economy. This is no longer a useful assumption. Producing a healthy human population living in an environmentally sound world could be as stimulating an economic development goal as increasing the production of disposable commodities. It takes an enormous marketing effort to convince consumers to buy several television sets for single-person households or four-wheel drive utility vans for picking up the groceries. People would be just as likely to pay good prices for ensuring the health of their families and the soundness of the planet if such values were the subject of commercial marketing. With equally clever efforts, a vibrant economy could be built to produce healthy people and healthy ecosystems.

In the final assessment, the current patterns of material production, use, and disposal cannot continue unaltered if we wish to ensure an ample and safe array of materials for the future. These patterns are simply not sustainable. We need to use fewer materials and we need to be more careful about what kinds of materials we use.

Materials are a fundamental determinant of sustainability. A focus on improving public health or protecting ecological systems without addressing the production, use, and disposal of industrial materials will prove inadequate and ineffective. We will not achieve a just or fair distribution of materials, an acceptable level of occupational and public health, or an adequately protected environment without policies that promote a sustainable materials economy.

For too long advocates for environmental and public health have settled for too little. In struggling to minimize environmental impacts and reduce human exposure, to set safety standards and establish regulatory infrastructures, these dedicated practitioners have attended to the victims and not to the solutions. Depletion of the resources of the environment and impairment of human health are the symptoms of a poorly designed and functionally flawed industrial production and consumption economy, not of an unprotected environment. It is time that we quit framing these consequences as environmental or public health problems, and face them directly as industrial problems. It is the materials of production and their management, not the effects of their mismanagement, that should draw our central attention. By focusing more on the composition and flow of the materials we use, we could focus less on the remediation of environmental damages and the treatment of human injuries and illnesses.

We have spent many decades (and billions of dollars) focused on the consequences of material use, without addressing materials directly. This could be viewed as a woefully regrettable misdirection. A more positive perspective would suggest that our slow start has left the field of materials design and use wide open and the opportunities for improvement abundant. Converting to a more sustainable materials economy will require significant commitment and effort. It will require a new system of materials management aimed not at environmental protection, but at the development and use of a new generation of safer materials and cleaner production processes. This will require a more socially directed approach to the development and use of industrial materials, one that seeks to generate a rich array of industrial materials able to satisfy the needs of the present without jeopardizing the opportunities for the future.

When called upon in the past, the government, industries, and people of the United States have responded with enormously ambitious projects such as the Manhattan Project, the Rubber Reserve Project and the Apollo Program. These projects substantially changed the economy and its materials. We need, once again, to find the will to think big and act purposefully. The work will need to be approached with respect for the prices that people and institutions will pay in order to carry out the transitions. However, action should not be delayed. Redirecting the materials economy sooner rather than later means that future costs and risks will be lower and fairer.

Nearly 50 years ago, the Paley Commission warned, "The actions that we as a Nation take or fail to take in meeting our materials problems in the period immediately ahead will affect profoundly the state of affairs many years hence. Upon our own generation lies the responsibility for passing on to the next generation the prospects of continued well-being. The first requisite for this is that we successfully meet the requirements that can now be foreseen."[25] Given that the Commission report paid no attention to environmental or health issues, it may be said that the nation responded well. Today, however, protection of the environment and human health are values that cannot be ignored. We have new challenges and we need new solutions.

This new century opens with a extraordinary set of opportunities for materials research, innovation, policy, and advocacy. A sustainable economy is an ambitious goal. The redirection of our materials economy must be a part of that vision. Materials do matter. We could develop safer and cleaner materials and we could use them more efficiently. We should be eager to get on with the task of creating a truly sustainable materials economy and a future more in harmony with the environment.

Notes

Chapter 1

1. See P. Jeanene Willcox, Samuel P. Guido, Wayne Mueller, and David L. Kaplan, "Evidence of Cholesteric Liquid Crystalline Phase in Natural Silk Spinning Processes," *Macromolecules*, 29 (15), 1996, pp. 5106–5110; and David Kaplan, W. Wade Adams, Barry Farmer, and Christopher Viney, *Silk Polymers: Materials Science and Biotechnology*, Washington, D.C.: American Chemical Society, 1994.

2. See Grecia Matos and Lori Wagner, "Consumption of Materials in the United States, 1900–1995," in *Annual Review of Energy and Environment*, 23, 1998, pp. 107–121; U.S. Interagency Working Group on Sustainable Development Indicators, *Sustainable Development in the United States*, Washington, D.C., December 1998; and Gary Gardner and Payal Sampat, *Mind Over Matter: Recasting the Role of Materials in Our Lives,* Worldwatch Paper No. 144, Washington, D.C.: Worldwatch Institute, December 1998.

3. Matos and Wagner, *Consumption of Materials*, p. 113.

4. U.S. Environmental Protection Agency, Office of Pollution Prevention and Toxics, *Chemical Hazard Data Availability Study: What Do We Really Know About the Safety of High Production Volume Chemicals?*, Washington, D.C., April 1998.

5. U.S. Environmental Protection Agency, Office of Air Quality Planning and Standards Research, *National Air Pollutant Emissions Trends*, 1900–1995, Research Triangle Park, N.C., 1996.

6. U.S. Environmental Protection Agency, Office of Pollution Prevention and Toxics, *1997 Toxics Release Inventory*, Washington, D.C., April 1999.

7. U.S. Environmental Protection Agency, Office of Solid Waste and Emergency Response, *Characterization of Municipal Solid Waste in the United States, 1997 Update*, Washington, D.C., 1998; and U.S. Environmental Protection Agency, Office of Solid Waste and Emergency Response, *National Biennial RCRA Hazardous Waste Report, Based on 1997 Data,* Washington, D.C., September 1999. The hazardous waste figures do not include industrial wastewaters.

8. John Young and Aaron Sachs, *The Next Efficiency Revolution: Creating a Sustainable Materials Economy*, Worldwatch Paper No. 121, Washington, D.C.: Worldwatch Institute, September 1994.

9. Less than one-third of these chemicals are used in industrial settings; the remainder are largely pharmaceuticals or pesticides. See United Nations Environment Program, International Registry of Potentially Toxic Chemicals, *Consolidated List of Products Whose Consumption and/or Sale Have Been Banned, Withdrawn, Severely Restricted or Not Approved by Governments*, 6th ed., Geneva, 1997.

10. See Robert Gottlieb, *Reducing Toxics: A New Approach to Policy and Industrial Decisionmaking*, Washington, D.C.: Island Press, 1995, pp. 170–176.

11. This was a highly controversial hearing. For a public health perspective, see David Rosner and Gerald Markowitz, "A 'Gift of God'?: The Public Health Controversy Over Leaded Gasoline During the 1920s," *American Journal of Public Health*, 75, April 1985, pp. 344–352. For an industry perspective, see Joseph C. Robert, *Ethyl: A History of the Corporation and the People Who Made It*, Charlottesville, Va.: University Press of Virginia, 1983.

12. Otto Hutzinger, Stephen Safe, and V. Zitko, *The Chemistry of PCBs*, Cleveland, Ohio: CRC Press, 1974.

13. Soren Johnsen, "Report of a New Chemical Hazard," *New Scientist*, 32, 1966, p. 612; Gunnar Widmark, "Possible Interference by Chlorinated Biphenyl," *Journal of the Association of Official Analytical Chemists*, 50, 1967, p. 1069.

14. Russell Train and Robert Risebrough, testimony before the U.S. Senate, Committee on Commerce, Subcommittee on the Environment, *The Toxic Substances Control Act of 1971 and Amendments*, Part 1, August 3 and 4, 1971, Washington, D.C., 1972, pp. 65, 104–106. Monsanto argued that there was no comparable alternative to PCBs for nonflammable lubricant applications. See Dan J. Forrestal, *Faith, Hope and $5,000: The Story of Monsanto*, New York: Simon & Schuster, 1977, pp. 196–197.

15. Midgley's unfortunate role in both the TEL and CFC stories is described in J. R. McNeill's, *Something New Under the Sun: An Environmental History of the Twentieth Century World*, New York: Norton, 2000; see pp. 62 and 111–112.

16. Mario Molina and F. Sherwood Rowland, "Stratospheric Sink of Chlorofluoromethanes: Chlorine Atom-Catalyzed Destruction of Ozone," *Nature*, 249, June 1974, pp. 810–812.

17. Philip Shabecoff, "Study Shows Significant Decline in Ozone Layer," in *New York Times*, March 16, 1988, p. A25.

18. See George Rosen, *A History of Public Health*, rev. ed., Baltimore: John Hopkins University Press, 1993; and Robert Gottlieb, *Forcing the Spring: The Transformation of the American Environmental Movement*, Washington, D.C.: Island Press, 1993.

19. See Barden R. Allenby, *Industrial Ecology: Policy Framework and Implementation,* Upper Saddle River, N.J.: Prentice Hall, 1999; and Robert U. Ayres and Leslie W. Ayres, *Industrial Ecology: Towards Closing the Materials Cycle,* Brookfield, Vt: Edward Elgar, 1996.

20. U.S. Environmental Protection Agency, Office of Pollution Prevention and Toxics, *Chemical Hazard Data.*

21. Some recent examples include Greg Eyring's work in U.S. Congress, Office of Technology Assessment, *Green Products by Design: Choices for a Cleaner Environment,* Washington, D.C.: U.S. Government Printing Office, 1992; and Tim Jackson, *Material Concerns, Pollution, Profit and Quality of Life,* London: Routledge, 1996.

22. Sustainable development has been defined by the United Nations and other international bodies as those forms of social and economic development that meet the needs of the present without jeopardizing the capacity of future generations to meet their own needs. See World Commission on Environment and Development, *Our Common Future,* New York: Oxford University Press, 1987.

23. The terms *detoxification* and *dematerialization* are used broadly in this text. Detoxification is conventionally used by toxicologists to describe enzymatic and biochemical processes that transform chemical toxicants to less toxic species. See Wayne G. Landis and Ming-Ho Yu, *Introduction to Environmental Toxicology: Impacts of Chemicals Upon Ecological Systems,* 2nd Edition, Boca Raton, Fla.: Lewis, 1999. The term dematerialization is provided both a broad and a narrow definition in Robert Herman, Siamak A. Ardekani and Jesse H. Ausubel, "Dematerialization," in Jesse H. Ausubel and Hedy E. Sladovich, eds., *Technology and Environment,* Washington, D.C.: National Academy Press, 1989, pp. 50–69.

Chapter 2

1. A detailed history of the early chemical industry can be found in Ludwig F. Haber, *The Chemical Industry in the Nineteenth Century,* New York: Oxford University Press, 1958.

2. See Williams Haynes and Edward Goody, *Chemical Industry's Contribution to the Nation; 1635–1935, A Record of Chemical Accomplishments, with an Index of the Chemicals Made in America.* Supplement to *Chemical Industries,* May 1935. A thorough history of the American chemical industry is presented in William Haynes, *American Chemical Industry,* Vols. 1–6, New York: Van Nostrand, 1945–1954.

3. The use of science as a basis for industrial production is one of the factors that differentiates the modern materials industries from traditional craft-based materials industries. See David Noble, *America by Design: Science, Technology, and the Rise of Corporate Capitalism,* New York: Knopf, 1977. From Braintree, Winthrop moved on to Connecticut where he became governor of the colony and

was granted wide-ranging mining rights. In 1662, Winthrop traveled to England to deliver the first American scientific papers on chemistry, titled "On the Manner of Making Tar and Potash in New England" and "Description, Culture and Use of Maize," a paper describing beermaking from corn. See Haynes and Goody, *Contribution to the Nation*.

4. The dates and specifics are taken from Thomas A. Rickard's rich, anecdotal history of U.S. mining up through 1900. See *The History of American Mining*, New York: McGraw-Hill, 1932, p. 10.

5. Rickard, *History*, pp. 15 and 177.

6. Rickard, *History*. Western iron mining is discussed on p. 11, lead on p. 155, gold on p. 19, and coal on p. 13.

7. For a lively history of coke, see Nils Anderson, Jr. and Mark W. DeLawyer, *Chemicals, Metals and Men: Gas, Chemicals and Coke: A Bird's Eye View of the Materials that Make the World Go Around*, New York: Vantage, 1995.

8. Haber, *Chemical Industry*, pp. 52–55.

9. This compares most ironically with the $15 million that the United States paid Mexico in that same year under the Treaty of Guadalupe Hidalgo for the purchase of the land that would become California, Nevada, Arizona, Utah, and parts of Wyoming and Colorado. See Rickard, *History*, p. 27.

10. Anderson and DeLawyer, *Chemicals, Metals and Men*, p. 32.

11. The Bessemer process was first introduced in 1865 at a steel plant in Troy, New York. See William T. Hogan, *Economic History of the Iron and Steel Industry in the United States*, Vol. 1, Lexington, Mass.: D. C. Heath, 1971, pp. 5–15.

12. Ashish Arora and Nathan Rosenberg, "Chemicals: A U.S. Success Story," in Ashish Arora, Ralph Landeau, and Nathan Rosenberg, *Chemicals and Long-Term Economic Growth: Insights from the Chemical Industry*, New York: Wiley, 1998, p. 73.

13. The early U.S. alkali industry is described by Haber, *Chemical Industry*, on pp. 142–151; and in Noble *America by Design*, 1977, on pp. 13–15. The history of Dow Chemical is covered by several authors, including Murray Campbell and Harrison Hatton, *Herbert H. Dow: Pioneer in Creative Chemistry*, New York: Appleton-Century-Crofts, 1951; and Don Whitehead, *The Dow Story: The History of the Dow Chemical Company*, New York: McGraw-Hill, 1968.

14. The history of electrochemistry is well documented in Martha Moore Trescott, *The Rise of the American Electrochemicals Industry, Studies in the American Technological Environment*, Westport, Conn.: Greenwood, 1981.

15. The first by-product ovens were opened in Syracuse, New York, by the Semet–Solvay Company in 1893. See Hogan, *Economic History*, p. 206. See also Joel Tarr, "Industrial Wastes, Water Pollution, and Public Health, 1876–1962," in Joel Tarr, *The Search for the Ultimate Sink: Urban Pollution in Historical Perspective*, Akron, Ohio: University of Akron Press, 1996, pp. 361–362.

16. Anderson and DeLawyer, *Chemicals, Metals and Men*, pp. 33–41.

17. Haber, *Chemical Industry*, pp. 92–93.

18. Haber, *Chemical Industry*, pp. 143–144.

19. The early history of Celluloid is documented in Susan T. Mossman, "Parkesine and Celluloid," in Susan T. Mossman and Peter J. Morris, eds., *The Development of Plastic*, London: Royal Society of Chemistry, 1994, pp. 10–25. The story of Hyatt's work is well told in Ivan Amato's highly readable book on the materials sciences. See *Stuff: The Materials the World Is Made Of*, New York: Avon (Basic Books), 1997, pp. 51–54.

20. For a history of Bakelite, see Wiebe E. Bijker, *Of Bicycles, Bakelites, and Bulbs: Toward a Theory of Sociotechnical Change*, Cambridge, Mass.: MIT Press, 1997.

21. The early history of Herbert Dow's work is well documented in Whitehead, *Dow Story*.

22. Amato, *Stuff*, p. 51.

23. Daniel Yergin, *The Prize: The Epic Quest for Oil, Money and Power*, New York: Simon & Schuster, 1991, p. 30.

24. Other large mills in the area were established by Jones and Laughlin (Pittsburgh, Pennsylvania), Bethlehem Steel Company (Bethlehem, Pennsylvania), and Wheeling Steel Company (Wheeling, West Virginia). See Hogan, *Economic History*, pp. 100–108.

25. The story of the discovery of California gold can be found in Rickard, *History*, pp. 18–39, and of Idaho copper on pp. 318–340.

26. Lead mining history is described by Rickard, *History*, on pp. 147–178 and that of copper mining on pp. 222–248 and 277–300.

27. Iron statistics are from Anderson and DeLawyer, *Chemicals, Metals and Men*, p. 121; oil and copper statistics are from Alfred E. Eckes, Jr., *The United States and the Global Struggle for Materials*, Austin, Texas: University of Texas Press, 1979, p. 15.

28. This plant was so hazardous that an explosion in 1918 destroyed the TNT works, killed 52 employees, and resulted in the eventual closing of the plant. See Anderson and DeLawyer, *Chemicals, Metals and Men*, p. 63.

29. Arora and Rosenberg, "Chemicals," p. 78.

30. Noble, *America by Design*, p. 16.

31. Arora and Rosenberg, "Chemicals," p. 81.

32. For the story of nylon, see Peter Spitz, *Petrochemicals: The Rise of an Industry*, New York: Wiley, 1988, pp. 274–279; and Amato, *Stuff*, pp. 62–62.

33. Eckes, *Global Struggle*, p. 52.

34. The rise and fall of the nation's most significant trust is well documented in Yergin, *The Prize*.

35. Noble, *America by Design*, p. 17. The aluminum industry offers an exception, where Alcoa remained a virtual monopoly up until World War II. See Steven Kendall Halloway, *The Aluminum Multinationals and the Bauxite Cartel*, London: Macmillan, 1988.

36. For a brief overview of these corporate histories, see Alfred Chandler, Jr., Takashi Hikino, and David Mowery, "The Evolution of Corporate Capabilities and Corporate Strategy and Structure within the World's Largest Chemical Firms: The Twentieth Century in Perspective," in Arora et al., *Chemicals and Economic Growth*, pp. 415–457. For specific corporate histories, see Alfred D. Chandler, Jr. and Stephen Salsbury, *Pierre S. du Pont and the Making of the Modern Corporation*, New York: Harper and Row, 1971; Dan Forrestal, *Faith, Hope and $5,000: The Story of Monsanto: The Trials and Triumphs of the First 75 Years*, New York: Simon & Schuster, 1977, Campbell and Hatton, *Herbert Dow*; Whitehead, *The Dow Story*, and Joseph C. Robert, *Ethyl: A History of the Corporation and the People Who Made It*, Charlottesville, Va.: University Press of Virginia, 1983.

37. Eckes, *Global Struggle*, p. 84.

38. This wartime history is detailed in Eckes, *Global Struggle*, pp. 89–119.

39. See Spitz, *Petrochemicals*, pp. 141–146. Peter Morris provides a more detailed history in *The American Synthetic Rubber Research Program*, Philadelphia: University of Pennsylvania Press, 1989.

40. This transition was in part promoted by government initiative. "In 1931, the U.S. National Bureau of Standards and the American Petroleum Institute began a systematic study of petroleum hydrocarbon synthesis and uses. This culminated after World War II in a fundamental shift of organic chemical production to the petroleum industry, which, along with increased hydrocarbon production from catalytic cracking of petroleum, gave birth to a new petroleum industry." See Samuel S. Epstein, *The Politics of Cancer*, New York: Anchor/Doubleday, 1979, p. 28.

41. Spitz records these early histories in brief detail. See Spitz, *Petrochemicals*, pp. 69–96.

42. Arora and Rosenberg, "Chemicals," p. 94.

43. Chandler et al., "Corporate Capabilities," pp. 422–423.

44. The demand for consumer products was spurred by pentup desires delayed by wartime restraints, but it was also fueled by the growth of advertising and marketing initiatives targeted at the family as the primary consuming entity. For early analyses, see Vance Packard, *The Waste Makers*, New York: David McKay, 1960; and Joseph Seldin, *The Golden Fleece: Selling the Good Life to Americans*, New York: Macmillan, 1963.

45. Peter Spitz quotes Stanford Research Institute figures. See Spitz, *Petrochemicals*, p. 233.

46. Spitz, *Petrochemicals*, p. 291.

47. See Keith Chapman, *The International Petrochemical Industry*, Oxford: Basil Blackwell, 1991, p. 224.

48. U.S. Geological Survey, *Mineral Commodity Summaries, 1998*, Washington, D.C., 1998.

49. Arora and Rosenberg, "Chemicals," p. 71.

Chapter 3

1. Iddo K. Wernick, Robert Herman, Shekhar Govind, and Jesse H. Ausubel, "Materialization and Dematerialization: Measures and Trends," in *Daedalus*, 125 (3), Summer 1996, pp. 171–198. Statistic is on p. 175. The largest share of material movement in the United States is composed of soil erosion, overburden moved during mining, dredged materials, and earth moved during construction. When these "hidden flows" are added in, the total flow of materials in the United States is closer to 22 billion metric tons per year or 84 million tons per person per year. See Albert Adriaanse, Stefan Bringezu, Allen Hammond, Yuichi Moriguchi, Eric Rodenberg, Donald Rogich, and Helmut Schutz, *Resources Flows: The Material Basis of Industrial Economies*, Washington, D.C.: World Resources Institute, 1997.

2. Sadi Carnot in an 1824 book titled *Reflections on the Motive Power of Fire* is credited with offering the thermomechanical insights upon which thermodynamic principles have been built, while Rudolf Clausius codified the principles into laws of physics. See Robert Costanza, John Cumberland, Herman Daly, Robert Goodland, and Richard Norgaard, *An Introduction to Ecological Economics*, Boca Raton, Fla.: St. Lucie Press, 1997.

3. This model has been developed and used by various engineers and economists. See for instance, Allen V. Kneese, Robert U. Ayres, and Ralph C. D'Arge, *Economics and the Environment: A Materials Balance Approach*, Baltimore: John Hopkins University Press for Resources for the Future, 1970.

4. For a fuller exploration of the economics of natural resources, see John Hatwick and Nancy Olewiler, *The Economics of Natural Resource Use*, New York: Harper & Row, 1986; or Tom Tietenberg, *Environmental and Natural Resource Economics*, 4th ed., New York: Harper Collins, 1996.

5. National Mining Association, *Facts About Mining, 1997–1998*, Washington, D.C., 1998.

6. The numbers are from U.S. Geological Survey, *Mineral Commodity Summaries, 1998*, Washington, D.C., 1998. See also Organization for Economic Cooperation and Development, *Mining and Non-Ferrous Metals Policies of the OECD Countries*, Paris, 1994.

7. Robert U. Ayres and Leslie W. Ayres, *Industrial Ecology: Towards Closing the Materials Cycle*, Brookfield, Vt.: Edward Elgar, 1996.

8. L. Harold Bullis and James E. Mielke, *Strategic and Critical Materials*, Boulder, Col.: Westview Press, 1985, p. 29.

9. Keith Chapman, *The International Petrochemical Industry: Evolution and Location*, Oxford: Basil Blackwell, 1991, p. 23.

10. Robert Multhauf, "Industrial Chemistry in the Nineteenth Century," in Melvin Kranzberg and Carrol W. Purcell, Jr., eds., *Technology in Western Civilization*, New York: Oxford University Press, 1967, pp. 468–489.

11. Roderick Eggert, "The Passenger Car Industry: Faithful to Steel," in John E. Tilton, ed., *World Metal Demand: Trends and Prospects*, Baltimore: Johns Hopkins University Press for Resources for the Future, 1990, pp. 161–216.

12. See Donald G. Rogich and staff, "United States Global Material Use Patterns," unpublished paper, Division of Mineral Commodities, U.S. Bureau of Mines, Washington, D.C., September 1993. Population data come from the U.S. Bureau of Census, *Census of Population, 1990*, Washington, D.C., 1991.

13. This wavelike pattern of a technology's life cycle from early development to rapid diffusion and adoption and ultimately to saturation and senescence has long been recognized by economists. See Nicholas D. Kondratiev, "The Long Waves in Economic Life," *Review of Economic Statistics*, 17, 1935, pp. 105–115; and Joseph A. Schumpeter, *Business Cycles: A Theoretical, Historical, and Statistical Analysis of the Capitalist Process*, New York: McGraw-Hill, 1939.

14. Wernick et al., "Materialization and Dematerialization," p. 177.

15. Wernick et al., "Materialization and Dematerialization," p. 175.

16. See Gerald R. Smith, *Materials Flow of Tungsten in the United States*, Washington, D.C.: U.S. Bureau of Mines, 1993; and James H. Holly, "Materials Flow of Zinc in the United States, 1850–1990," *Resources, Conservation and Recycling*, 9 (1–2), 1993, pp. 1–30. The U.S. Geological Survey is continuing a series of materials flow studies begun by the U.S. Bureau of Mines, with twenty-five studies planned by 2001.

17. William Woodbury, Daniel Edelstein, and Stephen Jasinski, "Lead Materials Flow in the United States: 1940–1988," unpublished paper, U.S. Bureau of Mines, Washington, D.C., 1993, p. 6.

18. David Rejeski notes estimates of 93 to 98 percent recycling rates for lead in lead–acid batteries. See "Mars, Materials, and Three Morality Plays: Materials Flows and Environmental Policy," in *Journal of Industrial Ecology*, 1 (4), 1998, p. 14.

19. Robert Socolow and Valerie Thomas review five studies of lead residuals near secondary smelters that demonstrate human exposure. See "The Industrial Ecology of Lead and Electric Vehicles," *Journal of Industrial Ecology*, 1 (1), 1997, pp. 26–27.

20. U.S. Bureau of Mines, *The New Materials Society*, Vol. 3, *Material Shifts in the New Society*, Washington, D.C.: U.S. Government Printing Office, 1991, pp. 1.1–1.3.

21. Most polymers have low densities, with specific gravities below 1.5, compared with the specific gravity of glass at 2.5, aluminum at 2.7, steel at 7.8, and

lead at 11.8. See U.S. Bureau of Mines, *The New Materials Society*, Vol. 1, *New Materials Markets and Issues*, Washington, D.C.: U.S. Government Printing Office, 1990, p. 2.25.

22. Alfred E. Eckes, Jr., *The United States and the Global Struggle for Minerals*, Austin, Texas: University of Texas Press, 1979, pp. 103–104.

23. See Emily Matthews et al., *The Weight of Nations: Material Outflows from Industrial Economies*, Washington, D.C.: World Resources Institute, 2000, p. xi; U.S. Environmental Protection Agency, *National Biennial RCRA Hazardous Waste Report*, 1999; and U.S. Environmental Protection Agency, *Characterization of Municipal Solid Waste in the United States, 1996 Update*, Washington, D.C., 1997.

Chapter 4

1. Donella H. Meadows, Dennis L. Meadows, Jorgen Randers, and William W. Behrens III, *The Limits to Growth*, New York: Universe Books, 1972.

2. Theoretically, this is not a trivial issue. For some time it was not clear how to reconcile the ability of industrial processes to increase the complexity, organization, and refinement of materials with the second law of thermodynamics, which held that all energy (and material) transformations tended to reduce these qualities. The argument about this dilemma is presented by Nicholas Georgescu-Roegen in *The Entropy Law and the Economic Process*, Cambridge, Mass.: Harvard University Press, 1971.

3. Donald Worster offers a detailed history of ecological concepts in *Nature's Economy: A History of Ecological Ideas*, Cambridge: Cambridge University Press, 1977. For a brief review of these concepts in regard to industrial wastes, see Michael. J. Chadwick and Jan Nilsson, "Environmental Quality Objectives—Assimilative Capacity and Critical Load Concepts in Environmental Management," in Tim Jackson, ed., *Cleaner Production Strategies: Developing Preventive Environmental Management in the Industrial Economy*, Boca Raton, Fla.: Lewis, 1993, pp. 29–39. Quotation is on p. 31.

4. See Samuel P. Hays, *Conservation and the Gospel of Efficiency: The Progressive Conservation Movement, 1890–1920*, Cambridge, Mass.: Harvard University Press, 1959; and Stephen Fox, *The American Conservation Movement: John Muir and His Legacy*, Madison, Wis.: University of Wisconsin Press, 1985.

5. U.S. Office of the President, *Proceedings of a Conference of Governors in the White House*, Document No. 48489, Washington, D.C.: U.S. Government Printing Office, 1908.

6. U.S. National Conservation Commission, *Report*, Senate Document No. 676, 60th Congress, Washington, D.C., 1910, p. 2.

7. U.S. National Conservation Commission, *Report of the National Conservation Commission*, 3 volumes, Washington, D.C.: U.S. Government Printing Office, 1909, Vol. 1, p. 109.

8. Gifford Pinchot, *The Fight for Conservation*, New York: Doubleday, Page, 1910, p. 43.

9. Alfred E. Eckes, Jr., *The United States and the Global Struggle for Minerals*, Austin, Texas: University of Texas Press, 1979, p. 8.

10. For a good review of this history, see Grosvenor B. Clarkson, *Industrial America in the World War: The Strategy Behind the Lines, 1917–1918*, Boston: Houghton-Mifflin, 1923.

11. *National Petroleum News*, October 29, 1919, p. 51. Quoted in Daniel Yergin, *The Prize: The Epic Quest for Oil, Money and Power*, New York: Simon & Schasters, 1991, p. 194.

12. Charles K. Leith, *World Minerals and World Politics: A Factual Study of Minerals in their Political and International Relations*, New York: McGraw-Hill, 1931, p. 4.

13. Harold Ickes, "The War and Our Vanishing Resources," in *American Magazine*, 140, December 1945, p. 20.

14. Elmer W. Pehrson, "The Mineral Position of the United States and the Outlook for the Future," in *Annual Report of the Board of Regents of the Smithsonian Institution*, Washington, D.C., 1945, pp. 175–199. Quote is on p. 176.

15. W. E. Wrather, "Mineral Resources and Mineral Resourcefulness," in *Mining and Metallurgy*, 27, August 1946, pp. 427–428. Quote is on p. 428.

16. J. Frederick Dewhurst and Associates, *America's Needs and Resources*, New York: Twentieth Century Fund, 1947.

17. Dewhurst, *America's Needs*, p. 598.

18. For well-documented histories, see Martin V. Melosi, *Garbage in the Cities: Refuse Reform and the Environment, 1880–1980*, College Station, Texas: Texas A&M University Press, 1981; and Joel Tarr, *The Search for the Ultimate Sink: Urban Pollution in Historical Perspective*, Akron, Ohio: University of Akron Press, 1996.

19. See Duane Smith, *Mining America: The Industry and the Environment, 1800–1980*, Lawrence, Kan.: University of Kansas Press, 1987.

20. Duane Smith quotes from various mine field observers in the nineteenth century, including John Muir and Mark Twain. See Smith, *Mining America*. Quote is on p. 54.

21. Smith, *Mining America*, p. 72.

22. Richard V. Francaviglia, "Mining and Landscape Transformation," in Larry M. Dilsaver and Craig E. Colten, *The American Environment: Interpretations of Past Geographies*, Lanham, Md.: Rowman and Littlefield, 1992, pp. 89–114.

23. Smith, *Mining America*, p. 89.

24. Ellis L. Armstrong, Michael C. Robinson, and Suellen M. Hoy, *History of Public Works in the United States, 1776–1976*, Chicago: American Public Works Association, 1976, p. 410.

25. See Craig Colten, "Creating a Toxic Landscape: Chemical Waste Disposal Policy and Practice, 1900–1960," *Environmental History Review*, 18 (1), Spring 1994, pp. 85–116.

26. E. B. Besselievre, "Industrial Waste Disposal as a Chemical Engineering Problem," in *Chemical and Metallurgical Engineering*, 38, 1931, pp. 501–503. Quoted in Colten, "Toxic Landscape," p. 93.

27. J. B. Hill, "Waste Problems in the Petroleum Industry, *Industrial & Engineering Chemistry*, 31, 1939, p. 1363. Quoted in Craig E. Colten and Peter N. Skinner, *The Road to Love Canal: Managing Industrial Waste before EPA*, Austin, Texas: University of Texas Press, 1996, p. 52.

28. U.S. Congress, Committee on Interstate and Foreign Commerce, Subcommittee on Oversight and Investigations, *Waste Disposal Site Survey*, 96th Congress, Committee Report 96-IFC33, October 1979, reported in Colten, "Toxic Landscape," p. 85.

29. These wastes were a combination of domestic trash, organic garbage, coal and wood ashes, street sweepings, and manures. The statistics and analysis come from Melosi, *Garbage in the Cities*, p. 22.

30. Rudolph Hering, "Disposal of Municipal Refuse," in *Transactions of the American Society of Civil Engineers*, 54, 1904, pp. 265–308.

31. Melosi notes the effective role played by civic improvement organizations, many like the influential Ladies Health Protective Association of New York City, in bringing the issue to public attention and promoting its resolution. See Melosi, *Garbage in the Cities*, p. 35; the statistics are on p. 155.

32. For a brief review, see Louis Blumberg and Robert Gottlieb, *War on Waste: Can America Win its Battle with Garbage?*, Washington, D.C.: Island Press, 1989; and Jennifer Seymour Whitaker, *Salvaging the Land of Plenty: Garbage and the American Dream*, New York: William Morrow, 1994.

33. George E. Waring, Jr., *Sewage and Land Drainage*, New York: Van Nostrand, 1889, p. 21, Quoted in Melosi, *Garbage in the Cities*, p. 60.

34. See George E. Waring, Jr. "The Utilization of City Garbage," *Cosmopolitan Magazine*, 24, February 1898, pp. 406–408. Quoted in Melosi, *Garbage in the Cities*, p. 71.

35. John McGaw Woodbury, "The Sanitary Disposal of Municipal Refuse," in *Transactions of the Society of Municipal Engineers*, 50, 1903, p. 104. Quoted in Melosi, *Garbage in the Cities*, p. 101.

36. See Blumberg and Gottlieb, *War on Waste*, pp. 195–199. The quotation is from New Orleans councilor Quitman Kohnke. The survey was reported in "Refuse Collection and Disposal," *Municipal Journal and Engineer*, 39, November 11, 1915, pp. 730–732. See Melosi, *Garbage in the Cities*, p. 176.

37. Reported in Melosi, *Garbage in the Cities*, p. 192.

38. By 1988, mechanical failures had resulted in unscheduled shutdown of over half of these incinerators, and the initial enthusiasm of the investment commu-

nity was cooling over what was seen as a "not yet proven" technology. This history is developed in more detail in Blumberg and Gottlieb, *War on Waste*. The EPA data come from U.S. Environmental Protection Agency, Office of Solid Waste and Emergency Response, *Municipal Waste Combustion Study, Report to Congress*, Washington, D.C., 1987.

39. Kenneth Boulding's often-quoted essay titled "The Economics of the Coming Spaceship Earth" can be found in Kenneth E. Boulding and Henry Jarrett, eds., *Environmental Quality in a Growing Economy: Essays for the Sixth RFF Forum*, Baltimore: John Hopkins University Press, for Resources for the Future, 1966. For a review of the criticism of *Limits to Growth*, see Ferdinand E. Banks, *Scarcity, Energy and Economic Progress*, Lexington, Mass.: Lexington, 1977.

40. Harold J. Barnett and Chandler Morse's study is titled *Scarcity and Growth: The Economics of Natural Resource Availability*, Baltimore: John Hopkins University Press for Resources for the Future, 1965.

41. Robert Heilbroner, *An Inquiry into the Human Prospect*, New York: Norton, 1980. Quote is on p. 47.

42. William Ophuls, *Ecology and the Politics of Scarcity*, San Francisco: W. H. Freeman, 1977. Quotes are on pp. 185 and 9.

43. Barry Commoner, *The Closing Circle*, New York: Knopf, 1971. Quote is on p. 177.

Chapter 5

1. Rachel Carson, *Silent Spring*, New York: Ballantine, 1962.

2. See Henry Carr, *Our Domestic Poisons: Poisonous Effects of Certain Dyes and Colours*, London: Ridgeway, 1883; and Thomas Oliver, *Dangerous Trades*, London: John Murray, 1902.

3. For a brief history of toxicology, see Michael A. Gallo, "History and Scope of Toxicology," in Curtis D. Klaassen, *Casarett and Doull's Toxicology: The Basic Science of Poisons*, 5th ed., New York: McGraw-Hill, 1996, pp. 3–11.

4. For relevant histories of occupational health, see David Rosner and Gerald E. Markowitz, eds., *Dying for Work: Worker Safety and Health in Twentieth Century America*, Bloomington, Ind.: Indiana University Press, 1987; and Jacqueline K. Corn, *Response to Occupational Health Hazards: A Historical Perspective*, New York: Van Nostrand Reinhold, 1992.

5. Don Whitehead, *The Dow Story: The History of the Dow Chemical Company*, New York: McGraw-Hill, 1968, p. 45.

6. See Christopher Sellers, "Factory as Environment: Industrial Hygiene, Professional Collaboration and the Modern Sciences of Pollution," in *Environmental History Review*, 18 (1), Spring 1994, pp. 55–83.

7. Thomas Somerfield, Thomas Oliver, and Felix Putzeys, "List of Industrial Poisons," *Bulletin of the U.S. Bureau of Labor Statistics*, No. 86, January 1910, pp. 147–168.

8. See U.S. Bureau of Labor Statistics, "Occupation Hazards and Diagnostic Signs," *Bulletin of the U.S. Bureau of Labor Statistics*, No. 306, 1922; and "Occupation Hazards and Diagnostic Signs," *Bulletin of the U.S. Bureau of Labor Statistics*, No. 582, 1933.

9. For a good review, see Craig Colten and Peter N. Skinner, *The Road to Love Canal: Managing Industrial Waste before EPA*, Austin, Texas: University of Texas Press, 1996.

10. Alice Hamilton, *Industrial Poisons in the United States*, New York: Macmillan, 1925, p. 1.

11. For histories of these institutions, see Fitzhugh Mullan, *Plagues and Politics: The Story of the United States Public Health Service*, New York: Basic Books, 1989; Victoria Harden, *Inventing the NIH: Federal Biomedical Research Policy, 1887–1937*, Baltimore: Johns Hopkins University Press, 1986; and Elizabeth W. Etheridge, *Sentinels for Health: A History of the Centers for Disease Control*, Berkeley, Calif.: University of California Press, 1992.

12. For a brief history of chemical carcinogenesis, see Henry C. Pitot III and Yvonne P. Dragan, "Chemical Carcinogenesis," in Klaassen, *Casarett and Doull*, pp. 201–267.

13. The connection between chemicals and cancer was not easily accepted. At the beginning of the century many thought that cancer was an infectious disease, and well up into the 1970s many researchers focused on viruses as the causative factor. See Robert A. Weinberg, *Racing to the Beginning of the Road: The Search for the Origin of Cancer*, New York: Harmony, 1996.

14. Elizabeth Miller and James Miller, "The Presence and Significance of Bound Aminoazo Dyes in the Livers of Rats fed *p*-Dimethyllaminoazobenzene," in *Cancer Research*, 7, 1947, pp. 468–480.

15. A good assessment of the current state of knowledge in the 1950s can be found in George Wolf, *Chemical Induction of Cancer*, London: Cassell and Company, 1952.

16. See George R. Hoffmann, "Genetic Toxicology," in Klaassen, *Casarett and Doull*, pp. 269–300.

17. James G. Wilson, *Environment and Birth Defects*, New York: Academic Press, 1973. The thalidomide story is covered in John Elkington's *The Poisoned Womb*, New York: Viking, 1985.

18. Ludwig F. Haber, *The Chemical Industry in the Nineteenth Century*, New York: Oxford University Press, 1958, pp. 206–210. Also see Lee Niedrinhaus Davis, *The Corporate Alchemists: Profit Takers and Problem Makers in the Chemical Industry*, New York: William Morrow, 1984.

19. For a classic history of the public health movement, see George Rosen, *A History of Public Health*, expanded ed. Baltimore: Johns Hopkins University Press, 1993; and John Duffy, *The Sanitarians: A History of American Public Health*, Urbana, Ill.: University of Illinois Press, 1990. For a more specialized review of the history of environmental health, see James Whorton, *Before Silent Spring: Pesticides and Public Health in Pre-DDT America*, Princeton, N.J.: Princeton

University Press, 1974. State and local public health history can be found in Barbara G. Rosenkrantz, *Public Health and the State: Changing Views in Massachusetts, 1842–1936*, Cambridge, Mass.: Harvard University Press, 1972; Stuart Galishoff, *Safeguarding the Public Health: Newark, 1895–1918*, Westport, Conn.: Greenwood, 1975; Judith Walzer Leavitt, *The Healthiest City: Milwaukee and the Politics of Health Reform*, Princeton, N.J.: Princeton University Press, 1982; Martin V. Melosi, *Garbage in the Cities: Refuse, Reform and the Environment, 1880–1980*, College Station, Texas: Texas A&M University Press, 1981; and Martin V. Melosi, ed., *Pollution and Reform in American Cities*, Austin, Texas: University of Texas Press, 1980.

20. Duane Smith quotes an 1889 *Butte* [Montana] *Miner* editorial on smelters: "The thicker the fumes the greater our financial vitality, and Butteites feel best when the fumes are thickest." See Duane Smith, *Mining America: The Industry and the Environment, 1800–1980*, Lawrence, Kan.: University of Kansas Press, 1987, p. 75.

21. Donald Pisani, "Fish Culture and the Dawn of Concern over Water Pollution in the United States," in *Environmental Review*, 8, 1984, pp. 117–131.

22. L. Pearse, ed., *Modern Sewage Disposal*, New York: Federation of Sewage Treatment Works Association, 1938, pp. 340–372. Quoted in Joel Tarr, *The Search for the Ultimate Sink: Urban Pollution in Historical Perspective*, Abron, Ohio: University of Abron Press, p. 345. Physicians and mine superintendents in Butte, Montana, argued that the sulfur-laden smoke from the smelters acted as a "partial disinfectant," destroying "microbes that constitute the germs of disease." See Smith, *Mining America*, p. 77.

23. In a 1916 text on chemical accident prevention, manufacturers were warned not to release volatile industrial wastes into public sewers in order to prevent explosions or fires at waste treatment plants. See D. S. Beyer, *Industrial Accident Prevention*, Boston: Houghton-Mifflin, 1916, p. 353.

24. Wade H. Frost, "Report to the Surgeon General," July, 30, 1913, as reported in Tarr, *Search for the Ultimate Sink*, p. 366.

25. Earle B. Phelps, *Studies of the Treatment and Disposal of Industrial Wastes*, U.S. Public Health Service, Public Health Bulletin, No. 97, Washington, D.C.: U.S. Government Printing Office, 1918.

26. Wade H. Frost, "Report to the Surgeon General," May 28, 1926, as reported in Tarr, *Search for the Ultimate Sink*, p. 368.

27. See Joel A. Tarr's useful history of industrial pollution in "Industrial Wastes, Water Pollution, and Public Health, 1876–1962," and "Searching for a Sink for Industrial Waste," in Tarr, *Search for the Ultimate Sink*, pp. 354–411.

28. "Progress Report of the Committee on Industrial Wastes in Relationship to Water Supplies," in *Journal of the American Water Works Association*, 10, 1923, pp. 415–430.

29. National Resources Committee, Special Advisory Committee on Water Pollution, *Report on Water Pollution*, Washington, D.C., July 1935.

30. The papers from the symposium were published in a special issue of *Industrial & Engineering Chemistry*, 39, May 1947. Quoted in Tarr, *Search for the Ultimate Sink*, p. 374.

31. Samuel Hays argues that "Earth Day was as much a result as a cause. It came after a decade or more of evolution in attitudes and action without which it would not have been possible." See "From Conservation to Environmental Politics in the United States Since World War Two," *Environmental History Review*, 6 (2), Fall 1982, pp. 14–41.

32. These and other early contamination revelations are described in Samuel S. Epstein, Lester O. Brown, and Carl Pope, *Hazardous Waste in America*, San Francisco: Sierra Club Books, 1982.

33. For this history, see Robert Gottlieb, *Forcing the Spring: The Transformation of the American Environmental Movement*, Washington, D.C.: Island Press, 1993; and Mark Dowie, *Losing Ground, American Environmentalism at the Close of the Twentieth Century*, Cambridge, Mass.: MIT Press, 1996.

34. David A. Hounshell and J. Kenley Smith, Jr., *Science and Corporate Strategy: DuPont R&D, 1902–1980*, Cambridge: Cambridge University Press, 1988, pp. 555–563.

35. See various studies published by Du Pont staff scientists such as W. F. von Oettingen and W. Deichmann-Gruebler, including W. F. von Oettingen and W. Deichmann-Gruebler, "The Toxicology and Potential Dangers of Crude 'Duprene'," *Journal of Industrial Hygiene and Toxicology*, 18, 1936, pp. 271–272; W. F. von Oettingen, W. Deichmann-Gruebler, and W. C. Hueper, "Toxicity and Potential Dangers of Phenyl-Hydrazine Zinc Chloride," *Journal of Industrial Hygiene and Toxicology*, 18, 1936, pp. 301–309; F. H. Wiley, W. C. Hueper, and W. F. von Oettingen, "On the Effects of Low Concentrations of Carbon Disulfide," *Journal of Industrial Hygiene and Toxicology*, 18, 1936, pp. 733–740, and W. F. von Oettingen et. al., "The Toxicity and Potential Dangers of Toluene, with Special Reference to Its Maximal Permissible Exposure," U.S. Public Health Service, *Public Health Bulletin*, No. 279, 1942, pp. 1–158.

36. See W. F. von Oettingen, "The Halogenated Hydrocarbons: Their Toxicity and Potential Dangers," *Journal of Industrial Hygiene and Toxicology*, 19, 1937, p. 349; W. F. von Oettingen, "Toxicity and Potential Dangers of Aliphatic and Aromatic Hydrocarbons," U.S. Public Health Service, *Public Health Bulletin*, No. 255, 1940, pp. 135; and W. F. von Oettingen, "The Aliphatic Alcohols: Their Toxicity and Potential Dangers in Relationship to Their Chemical Constitution and Their Fate in Metabolism," U.S. Public Health Service, *Public Health Bulletin*, 281, 1943, pp. 1–253.

37. M. A. Wolf, V. K. Rowe, D. D. McCollister, R. L. Hollinsworth, and F. Oyen, "Toxicological Studies of Certain Alkylated Benzenes and Benzene," *AMA Archives of Industrial Health*, 14, 1956, pp. 387–398.

38. See various studies published by Dow scientists such as H. C. Spenser, D. D. Irish, E. M. Adams, and V. K. Rowe, "Response of Laboratory Animals to Monomeric Styrene," *Journal of Industrial Hygiene and Toxicology*, 24 (10),

1942, pp. 295–301; E. M. Adams, H. C. Spenser, V. K. Rowe, D. D. McCollister, and D. D. Irish, "Vapor Toxicity of Trichloroethane," *Archives of Industrial Hygiene and Occupational Medicine*, 3 (2), 1950, pp. 225–236; E. M. Adams, H. C. Spenser, V. K. Rowe, D. D. McCollister, and D. D. Irish, "Vapor Toxicity of Trichloroethylene," *Archives of Industrial Hygiene and Occupational Medicine*, 4 (4), 1951, pp. 469–481; H. C. Spenser, V. K. Rowe, E. M. Adams, D. D. McCollister, and D. D. Irish, "Vapor Toxicity of Ethylene Dichloride Determined by Experiments in Laboratory Animals," *Archives of Industrial Hygiene and Occupational Medicine*, 4 (5), 1951, pp. 482–493; and E. M. Adams, H. C. Spenser, V. K. Rowe, D. D. McCollister, and D. D. Irish, "Vapor Toxicity of Carbon Tetrachloride," *Archives of Industrial Hygiene and Occupational Medicine*, 6 (1), 1952, pp. 50–66.

39. Reported in Colten and Skinner, *Road to Love Canal*, p. 18.

40. R. Dale Grinder, "The Battle for Clean Air: The Smoke Problem in Post-Civil War America," in Martin V. Melosi, *Pollution and Reform*, pp. 83–104.

41. Arthur C. Stern, "History of Air Pollution Legislation in the United States," *Journal of the Air Pollution Control Association*, 32 (1), January 1982, p. 44.

42. Tarr, *Search for the Ultimate Sink*, pp. 344, 359.

43. Gottlieb, *Forcing the Spring*, pp. 270–282. These early union initiatives are described in Charles Noble's *Liberalism at Work: The Rise and Fall of OSHA*, Philadelphia: Temple University Press, 1986; and the asbestos story is covered by Paul Broudeur in *Outrageous Misconduct: The Asbestos Industry on Trial*, New York: Pantheon, 1985.

44. Quoted in Grinder, "Battle for Clean Air," p. 99.

45. The early physician/alchemist, Aureolus Paracelsus, is credited with noting that "the dose makes the poison"; the scale and duration of exposure to a toxin is almost always a major factor in determining biological outcomes. However, recognizing the importance of exposure does not mean that we can assume that all chemicals are toxic or even equally toxic. The hazardousness of a substance is determined by both chemical structure and the characteristics of exposure.

46. Sellers, "Factory as Environment," pp. 69–70.

47. Edward J. Cleary, "Determining Risks of Toxic Substances in Water," in *Sewage and Industrial Wastes*, 26, February 1954, pp. 203–209.

48. Craig Colten, "Creating a Toxic Landscape: Chemical Waste Disposal and Practice, 1900–1960," *Environmental History Review*, 81, 1991, pp. 215–228, 1994, p. 108.

49. Craig Colten, "A Historical Perspective on Industrial Wastes and Groundwater Contamination," *Geographic Review*, 81, 1991, pp. 215–228.

50. Joel Tarr, "Introduction" in Tarr, *Search for the Ultimate Sink*, footnote 27, p. xlvii.

51. David Ozonoff, "One Man's Meat. Another Man's Poison: Two Chapters in the History of Public Health," in *American Journal of Public Health*, 75 (4), April 1985, p. 339.

Chapter 6

1. John Reiger, "Wildlife, Conservation and the First Forest Reserves," and Harold K. Steen, "The Beginning of the National Forest System," in Char Miller, ed., *American Forests: Nature, Culture, and Politics*, Lawrence, Kan.: University of Kansas Press, 1997, pp. 35–68. Richard Andrews argues that these early forest reservations were the first efforts to differentiate property rights (which were held by the government) from use rights (which could be transferred to private interests by contract, claim, or lease). See Richard N. L. Andrews, *Managing the Environment, Managing Ourselves: A History of American Environmental Policy*, New Haven, Conn.: Yale University Press, 1999, pp. 94–108.

2. Stephan Fox, The *American Conservation Movement: John Muir and His Legacy*, Madison, Wis.: University of Wisconsin Press, 1985, pp. 110–129.

3. General Persifor F. Smith, *Report of General in Command of the Pacific Division*, U.S. Army, October 7, 1849, as reported in Thomas A. Rickard, The *History of American Mining*, New York: McGraw-Hill, 1932, p. 33.

4. For an overview, see Dyan Zaslowsky, *These American Lands*, New York: Henry Holt, 1986, pp. 122–126.

5. This history is well covered in Alfred E. Eckes, Jr., The *United States and the Global Struggle for Materials*, Austin, Texas: University of Texas Press, 1979.

6. Eckes, *Global Struggle*, p. 36.

7. Mining and Metallurgical Society of America, *International Control of Minerals*, New York, 1925, pp. 7–15.

8. See Elmer Pehrson, "What are Strategic and Critical Materials," in *Mining and Metallurgy*, 25, July 1944, pp. 339–341.

9. These included the Rubber Reserve Company, the Metals Reserve Company, Defense Plants Corporation, and Defense Supplies Corporation. See Eckes, *Global Struggle*, p. 94. Eckes notes that these initial efforts were fairly clumsy because the country had little experience with purchasing large amounts of materials and the early estimates of need were substantially low.

10. Indeed, Eckes notes that a loophole in the legislation allowed vast quantities of the war surplus materials to pass into private hands at a hugely discounted price. See Eckes, *Global Struggle*, p. 140.

11. Commission of Organization of the Executive Branch of the Government, *Report on Natural Resources*, Washington, D.C.: U.S. Government Printing Office, 1949, Appendix L, p. 1.

12. Quoted from Harry Truman, "Public Papers of the President, 1951," in Eckes, *Global Struggle*, p. 176.

13. U.S. President's Materials Policy Commission, *Resources for Freedom*, 5 volumes, Washington, D.C.: U.S. Government Printing Office, 1952, Vol. 1, p. 6.

14. President's Materials Policy Commission, *Resources for Freedom*, Vol. 1, p. 6.

15. Resources for the Future, *The Nation Looks at its Resources: Report on the Mid-Century Conference on Resources for the Future*, December 2, 3, 4, 1953, Washington, D.C., 1954.

16. U.S. Congress, *Mining and Minerals Policy Act of 1970*, P.L. 91–631, Washington, D.C., 1970.

17. U.S. Congress, *National Materials Policy Act of 1970*, P.L. 91–512, Washington, D.C. 1970, sect. 202.

18. U.S. National Commission on Materials Policy, *Material Needs and the Environment: Today and Tomorrow*, Washington, D.C.: U.S. Government Printing Office, 1974, pp. 1–5.

19. National Academy of Sciences/National Academy of Engineering, *Elements of a National Materials Policy*, A Report of the National Materials Advisory Board, Washington, D.C.,1972; and National Academy of Sciences/National Academy of Engineering, *Man, Materials and the Environment*, Cambridge, Mass.: MIT Press, 1973.

20. See National Academy of Sciences, Committee on the Survey of Material Sciences and Engineering, *Materials and Man's Needs*, Washington, D.C.: National Academy Press, 1974; National Academy of Sciences, Committee on Mineral Resources and the Environment, *Mineral Resources and the Environment*, Washington, D.C.: National Academy Press, 1975; and U.S. Congress, General Accounting Office, *Federal Materials Research and Development: Modernizing Institutions and Management*, Washington, D.C.: U.S. Government Printing Office, 1975.

21. See U.S. Congress, Senate Committee on Public Works, *Hearings*, Washington, D.C., June 11–13, July 9–11, 15–18, 1974.

22. The recommendations of the Resource Conservation Committee (RCC) followed the status quo, advised against new policies, and urged "further consideration" of the issues. See Resource Conservation Committee, *Choices for Conservation: Final Report to the President and Congress*, Washington, D.C., July 1979. For a critical review of the RCC, see Louis Blumberg and Robert Gottlieb, *War on Waste: Can America Win Its Battle with Garbage?*, Washington, D.C.: Island Press, 1989, pp. 199–203.

23. Craig Colten and Peter N. Skinner, *The Road to Love Canal: Managing Industrial Waste before EPA*, Austin, Texas: University? Texas Press, 1996, p. 19.

24. Joel Tarr, *The Search for the Ultimate Sinic: Urban Pollution in Historical Perspective*, Akron, Ohio: University of Akron Press, 1996, p. 370.

25. See Samuel P. Hays, *Beauty, Health and Permanence: Environmental Politics in the United States, 1955–1985*, New York: Cambridge University Press, 1987, pp. 1–12.

26. For this history, see Andrews, *Managing the Environment*; and Mark K. Landy, Marc J. Roberts, and Stephen R. Thomas, *The Environmental Protection*

Agency: Asking the Wrong Questions, From Nixon to Clinton, New York: Oxford University Press, 1994.

27. U.S. Congress, *Federal Water Pollution Control Act (Clean Water Act) of 1972*, P.L. 92–500, Washington, D.C., 1972, sect. 101.

28. This history is well documented in several historical studies of federal environmental legislation. For example, see Hays, *Beauty, Health and Permanence*; and Andrews, *Managing the Environment*.

29. U.S. Congress, *Toxic Substances Control Act*, P.L. 94–469, Washington, D.C., 1976, sect. 2.

30. See U.S. Environmental Protection Agency, Office of Solid Waste and Emergency Response, *Report to Congress: Minimization of Hazardous Waste*, Washington, D.C., October 1986; and U.S. Congress, Office of Technology Assessment, *Serious Reduction of Hazardous Waste*, Washington, D.C.: U.S. Government Printing Office, 1986.

31. U.S. National Commission on Supplies and Shortages, *Government and the Nation's Resources*, Washington, D.C.: U.S. Government Printing Office, 1976.

32. U.S. Congress, House of Representatives, Committee on Science and Technology, *House Report 97–761*, Part 1, Washington, D.C., August 18, 1982, p. 9.

33. National Academy of Sciences, *Science: The Endless Frontier*, Washington, D.C., 1945. For a review of the early government support for the materials sciences, see Julius Harwood, "Emergence of the Field and Early Hopes," in Rustum Roy, ed., *Materials Science and Engineering in the United States*, University Park, Pa.: Pennsylvania State University Press, 1970.

34. The advocacy for comprehensive national materials policies has not disappeared. Resources for the Future still promotes a national policy, as does the Federation of Materials Societies, a broad coalition of materials trade organizations based in Washington.

Chapter 7

1. This argument is well presented by Paul Hawken, Amory Lovins, and L. Hunter Lovins in a provocative book titled *Natural Capitalism: Creating the Next Industrial Revolution*, Boston: Little, Brown, 1999.

2. National materials accounts typically exclude some subjects: atmospheric oxygen and water are typically excluded and materials are measured in dry weight. Some substances of concern, e.g., dioxin and furans, are not counted because their total annual production by weight is so small. See Iddo K. Wernick and Jesse H. Ausubel, "National Materials Flows and the Environment," *Annual Review of Energy and Environment*, 20, 1995, pp. 463–492.

3. Organization for Economic Cooperation and Development, *Mining and Non-Ferrous Metals Policies of the OECD Countries*, Paris 1994, pp. 221–225.

4. These are the public lands most open to mineral extraction. The actual acreage owned by the federal government (including parkland and military bases) is closer to 740 million acres, or nearly 30 percent of the nation's lands.

5. See Paul C. Hyndman and Harry W. Campbell, *Digital Databases Containing Mining Claim Density Information.... From the BLM Mining Claim Recordation System*, 1996, Washington, D.C.: U.S. Geological Survey, 1999; and Carl J. Mayer and George A Riley, *Public Domain, Private Dominion: A History of Public Mineral Policy in the United States*, San Francisco: Sierra Club Books, 1985, pp. 45–46.

6. The fuel subsidy estimate is from D. Koplow, "Energy Subsidies and the Environment," in Organization for Economic Cooperation and Development, *Subsidies and the Environment: Exploring the Linkages*, Paris, 1996; and the petrochemical product estimate comes from U.S. Bureau of Mines, *The New Materials Society*, Vol. 3, Washington, D.C., 1991, p. I–12.

7. A critique of mining subsidies can be found in David Malin Roodman, *The Natural Wealth of Nations: Harnessing the Market for the Environment*, New York: Norton, 1998; and Thomas Michael Power, *Lost Landscapes and Failed Economies: The Search for a Value of Place*, Washington, D.C.: Island Press, 1996.

8. See GrassRoots Recycling Network, *Welfare for Waste: How Federal Taxpayer Subsidies Waste Resources and Discourage Recycling*, Atlanta, GA, April, 1999. This study paints a more dramatic picture than a 1994 federal EPA study that found that, while federal incentives for virgin materials extraction do not significantly impact the relative prices of virgin and recycled products, the combination of federal and state subsidies and other policies favoring virgin commodity markets can place secondary materials at a disadvantage. See Douglas Koplow and Kevin Dietly, *Federal Disincentives: A Study of Federal Tax Subsidies and Other Programs Affecting Virgin Industries and Recycling*, Washington, D.C.: U.S. Environmental Protection Agency, 1994. For the federal budget numbers, see U.S. Office of Management and Budget, *Budget of the United States Government Fiscal Year 2000, Analytical Perspectives*, Washington, D.C., 1999, table 5.1, p. 107.

9. Thomas J. Misa, "Military Needs, Commercial Realities, and the Development of the Transistor, 1948–1958," in Merritt Roe Smith, *Military Enterprise and Technological Change: Perspectives on the American Experience*, Cambridge, Mass.: MIT Press, 1985, pp. 253–287.

10. See Patricia E. Perkins, *World Metal Markets: The United States Strategic Stockpile and Global Market Influence*, Westport, Conn.: Praeger, 1997.

11. U.S. Environmental Protection Agency, Office of Solid Waste and Emergency Response, *National Biennial RCRA Hazardous Waste Reports, 1989–1993*, Washington, D.C., 1996.

12. Data from the national Toxics Release Inventory show a steady decline in releases of toxic chemicals to air and water and a gradual increase in wastes shipped off-site to treatment facilities.

13. Thomas J. Hilliard, *Golden Patents, Empty Pockets: A Nineteenth Century Law Gives Miners Billions, the Public Pennies*, Washington, D.C.: Mineral Policy Center, 1994.

14. U.S. Environmental Protection Agency, Office of Solid Waste and Emergency Response, *Characterization of Municipal Solid Waste in the United States, 1997 Update*, Washington, D.C., 1998.

15. GrassRoots Recycling Network and Institute for Local Self Reliance (Brenda Platt and Neil Seldman), *Wasting and Recycling in the United States 2000*, Athens, GA, 2000, p. 4; and John Young and Aaron Sachs, *The Next Efficiency Revolution: Creating a Sustainable Materials Economy*, Worldwatch Paper No. 121, Washington, D.C.: Worldwatch Institute, 1994, p. 19.

16. There were roughly 8000 local and municipal landfills in 1988; by 1996, the number had dwindled to 2300, although the increasing size of newer landfills has resulted in an increase in total landfill capacity. See Nora Goldstein, "The State of Garbage in America," *BioCycle*, 38 (4), April 1997, pp. 60–67; and Jim Glenn, "The State of Garbage in America," *BioCycle*, 40 (4), April 1999, pp. 60–71.

17. See T. Clarence Davies and Jan Mazurek, *Pollution Control in the United States: Evaluating the System*, Baltimore: Johns Hopkins University Press for Resources for the Future, 1998, p. 273; National Academy of Sciences, National Research Council, *Toxicity Testing, Strategies to Determine Needs and Priorities*, Washington, D.C.: National Academy Press, 1984; and Environmental Defense Fund, *Toxic Ignorance*, Washington, D.C., 1997.

18. U.S. Environmental Protection Agency, Office of Pollution Prevention and Toxics, *Chemical Hazard Data Availability Study*, Washington, D.C., 1998.

19. Organization for Economic Cooperation and Development, *Environmental Performance Reviews: United States*, Paris, 1996, p. 72.

20. The TSCA New Chemicals Program recieves about 2000 premarket applications a year. Of these, some applications are for limited use or pilot tests and some are withdrawn because of missing data or other concerns. Something like 1000 substances actually enter the market.

21. For analyses that link costs to benefits, see A. Myrick Freeman, *Air and Water Pollution Control: A Benefits-Cost Assessment*, New York: Wiley, 1982; and Allen V. Kneese, *Measuring the Benefits of Clean Air and Water*, Baltimore: Johns Hopkins University Press for Resources for the Future, 1984. For a more general review, see Paul Portney, ed., *Public Policies for Environmental Protection*, Baltimore: Johns Hopkins University Press for Resources for the Resources for the Future, 1990; and Davies and Mazurek, *Pollution Control*.

22. U.S. Environmental Protection Agency, Office of Air Quality Planning and Standards Research, *National Air Pollutant Emissions Trends, 1900–1995*, Research Triangle Park, N.C., 1996.

23. U.S. Environmental Protection Agency, Office of Water, *National Water Quality Inventory, 1994 Report to Congress*, Washington, D.C., 1995.

24. U.S. Environmental Protection Agency, Office of Pollution Prevention and Toxics, *1997 Toxics Release Inventory*, Washington, D.C., 1999.

25. Evan J. Ringquist, *Environmental Protection at the State Level: Politics and Progress in Controlling Pollution*, Armonk, N.Y.: M. E. Sharpe, 1993, pp. 193–194; and Friends of the Earth, *Failing Grades: Clean Water Report Card*, Washington, D.C., March 2000.

26. See Winston Harrington, *The Regulatory Approach to Air Quality Management: A Case Study of New Mexico*, Baltimore: Johns Hopkins University Press for Resources for the Future, 1981; and U.S. General Accounting Office, *Wastewater Discharges Are Not Complying with EPA Pollution Control Permits*, Washington, D.C., December 1983. A more recent analysis of federal enforcement suggests that restraints on EPA's budget often limit the agency's enforcement capacity. See Joel A. Mintz, *Enforcement at the EPA: High Stakes and Hard Choices*, Austin, Texas: University of Texas Press, 1995.

27. Paul W. MacAvoy, "The Record of the Environmental Protection Agency in Controlling Industrial Air Pollution," in R. L. Gordon, H. D. Jacoby, and M. B. Zimmerman, eds., *Energy Markets and Regulation*, Cambridge, Mass.: Ballinger, 1987, pp. 107–137.

28. For a more complete treatment of this assessment, see Charles Noble, *Liberalism at Work: The Rise and Fall of OSHA*, Philadelphia: Temple University Press, 1986; and John Mendeloff, *The Dilemma of Toxic Substance Regulation: How Overregulation Causes Underregulation*, Cambridge, Mass.: MIT Press, 1988.

29. U.S. Congress, Office of Technology Assessment, *Preventing Illness and Injury in the Workplace*, Washington, D.C.: U.S. Government Printing Office, 1985, pp. 34–36 and 48.

30. In particular see Richard Doll and Richard Peto, "Avoidable Risks of Cancer in the U.S.," in *Journal of the National Cancer Institute*, 66 (6), June 1981, pp. 1191–1308.

31. U.S. Congress, Office of Technology Assessment, *Preventing Illness*, p. 268.

32. Fred Hoerger, "Indicators of Exposure Trends," in Richard Peto and Marvin Schneiderman, eds., *Quantification in Occupational Cancer*, Cold Spring Harbor, N.Y.: Cold Spring Laboratory Press, 1981.

33. Sheldon Krimsky, *Hormonal Chaos: The Scientific and Social Origins of the Environmental Endocrine Hypothesis*, Baltimore: John Hopkins University Press, 2000, p. 206. In considering the case of chlorine, Joe Thornton offers an extensive argument against a chemical-by-chemical policy approach; see *Pandora's Poison: Chlorine, Health, and a New Environmental Strategy*, Cambridge, Mass.: MIT Press, 2000. For a critical review of risk assessment, see C. Mark Smith, Karl T. Kelsey, and David Christiani, "Risk Assessment and Occupational Health: Overview and Recommendations," and Daniel Wartenberg and Caron Chess, "The Risk Wars: Assessing Risk Assessment," in Charles Levenstein and John Wooding, eds., *Work, Health, and Environment: Old*

Problems, New Solutions, New York: Guilford, 1997, pp. 236–257 and pp. 258–274, respectively.

34. U.S. Office of Management and Budget, *Budget of the United States Federal Government, Fiscal Years 1980–1993*, Washington, D.C., 1992.

35. U.S. Department of Commerce, Bureau of Economic Analysis, "Pollution Abatement and Control Expenditures," *Survey of Current Business*, Washington, D.C.: U.S. Government Printing Office, September 1996.

36. See Lee Webb and Jeff Tryens, *An Environmental Agenda for the States*, Washington, D.C.: Conference on Alternative State and Local Policies, 1984; and Walter Corson, ed., *The Global Ecology Handbook*, Boston: Beacon Press, 1990.

37. U.S. Environmental Protection Agency, Office of Air and Radiation, *Final Report to Congress on Benefits and Costs of the Clean Air Act, 1970 to 1990*, Washington, D.C., October 1997.

38. See A. Myrick Freeman, "Water Pollution Policy," in Portney, *Public Policies*, pp. 97–150.

39. For a full development of this argument, see Joel S. Hirschorn, "Why the Pollution Prevention Revolution Failed—and Why it Ultimately Will Succeed," *Pollution Prevention Review*, 7 (1), Winter 1997, pp. 11–31.

40. The pivotal data for this argument come from a 1987 study correlating hazardous waste dump sites and the racial composition of nearby neighborhoods. See Charles Lee, *Toxic Waste and Race in the United States, A National Report on the Racial and Socioeconomic Characteristics of Communities with Hazardous Waste Sites*, New York: Commission for Racial Justice, United Church of Christ, 1987. For a broader analysis of these arguments, see Robert D. Bullard, ed., *Environmental Justice and Communities of Color*, San Francisco: Sierra Club Books, 1993.

Chapter 8

1. See the International Union for the Conservation of Nature and Natural Resources, *World Conservation Strategy: Living Resource Conservation for Sustainable Development*, Gland, Switzerland, 1980. For a discussion of the history, see Francis R. Thibodeau and Herman H. Field, eds., *Sustaining Tomorrow: A Strategy for World Conservation and Development*, Hanover, N.H.: University Press of New England, 1984.

2. The so-called "Brundtland Commission Report," named after its chair, Gro Brundtland, the Prime Minister of Norway, was committed to "more rapid growth in both industrial and developing countries," as a means to alleviate poverty because the commission was unwilling to alienate those hostile to the limits-to-growth idea still drifting about during the mid-1980s. See World Commission on Environment and Development, *Our Common Future*, New York: Oxford University Press, 1987. Quotes are on pp. 32–33 and p. 43.

3. See the U.S. President's Council on Sustainable Development, *Sustainable America: A New Consensus*, Washington, D.C.: U.S. Government Printing Office, February 1996; and *Towards a Sustainable America: Advancing Prosperity, Opportunity and a Healthy Environment for the 21st Century*, Washington, D.C., May 1999.

4. Kenneth Boulding used the term *cowboy economics* to characterize the view of the environment as a limitless resource and an infinite sink. See "The Economics of the Coming Spaceship Earth," in Kenneth E. Boulding and Henry Jarrett, ed., *Environmental Quality in a Growing Economy: Essays for the Sixth RFF Forum*, Baltimore: John Hopkins University Press for Resources for the Future, 1966.

5. For an early conceptualization, see Aldo Leopold, *A Sand County Almanac*, New York: Oxford University Press, 1966; and David W. Eherenfeld, *Conserving Life on Earth*, New York: Oxford University Press, 1972. A more recent text is offered by Eugene P. Odum, *Basic Ecology*, Philadelphia: Saunders College Publications, 1983.

6. An early statement of this cradle-to-cradle thinking can be found in Walter R. Stahel and Genevieve Reday-Mulvey, *Jobs for Tomorrow: The Potential for Substituting Manpower for Energy*, New York: Vantage Press, 1981. A more recent version is found in Paul Hawken, Amory Lovins, and L. Hunter Lovins, *Natural Capitalism: Creating the Next Industrial Revolution*, Boston: Little, Brown, 1999, p. 17. "Life-cycle thinking" is covered in Mary Ann Curran, ed., *Environmental Life Cycle Assessment*, New York: McGraw-Hill, 1996.

7. Robert Frosh notes that such a redefinition of wastes as potential commodities encourages the concentration of material in a way that reverses the old adage that dilution is the preferred waste management strategy. See "Towards the End of Waste: Reflections on a New Ecology of Industry," *Daedalus*, 125 (3), Summer 1996, pp. 199–212.

8. For a further development of this broader view of natural capital, see David W. Pearce and R. Kerry Turner, *Economics of Natural Resources and the Environment*, Baltimore: John Hopkins University Press for Resources for the Future, 1990; and Herman E. Daly, *Beyond Growth: The Economics of Sustainable Development*, Boston: Beacon Press, 1996.

9. For the University of Maryland study, see Maureen Roubi, "Fallout from Pricing Earth," *Chemical & Engineering News*, 75, June 30, 1997, p. 38. For an excellent set of essays on natural services, see Gretchen C. Dally, ed., *Nature's Services: Societal Dependence on Natural Ecosystems*, Washington, D.C.: Island Press, 1997.

10. Many authors have used this two-system (system within a system) model with its associated materials flows. For instance, Barry Commoner describes the two systems as the "ecosphere" and the "technosphere." See *Making Peace with the Planet*, New York: Pantheon, 1990.

11. This framework follows that outlined by William McDonough and Michael Braungart, who cleverly describe these two systems as two metabolisms—a bio-

logical metabolism and a technical metabolism. See "The Next Industrial Revolution," *Atlantic Monthly*, 282 (4), October 1998, pp. 82–91.

12. Pearce and Turner, *Economics of Natural Resources*, p. 44. This principle of conservation is quite appealing conceptually, but very difficult to put into operation. The strategy developed here avoids the complexity of measuring natural capital by focusing on reducing material flows, which are at least somewhat more amenable to measurement.

13. Herman Daly and other ecological economists argue that human made capital can be substituted for natural capital only up to a point. Because environmental resources are already so depleted, sustainable development requires the protection of and "reinvestment in" natural capital. See Herman E Daly, *Beyond Growth: The Economics of Sustainable Development*, Boston: Beacon Press, 1996, pp. 75–87.

14. Robert Hermon, Siamak A. Ardekani, and Jesse H. Ausubel, "Dematerialization," in Jesse H. Ausubel and Hedy E. Sladovich, eds., *Technology and Environment*, Washington, D.C.: National Academy Press, 1989, pp. 50–69. Quote is on p. 51.

15. Iddo K. Wernick, Robert Herman, Shekhar Govind, and Jesse H. Ausubel, "Materialization and Dematerialization: Measures and Trends," *Daedalus* 125 (3), Summer 1996, pp. 171–198. Quote is on p. 171.

16. Robert A. Frosch and Nicholas E. Gallopoulos, "Strategies for Manufacturing," *Scientific American*, 261 (3), September 1989, pp. 144–153. For a good history of the concept of industrial ecology see S. Erkman, "Industrial Ecology: An Historical View," *Journal of Cleaner Production*, 5 (1–2), pp. 1–10. See also Barden R. Allenby, *Industrial Ecology: Policy Framework and Implementation*, Upper Saddle River, N.J.: Prentice Hall 1999; and papers from the *Journal of Industrial Ecology*.

17. Barry Commoner, "The Relationship Between Industrial and Ecological Systems," *Journal of Cleaner Production*, 5 (1–2), 1997, pp. 125–129.

18. Herman Daly notes "As stocks of artifacts and people have grown, the throughput necessary for their maintenance has had to grow as well, implying more depletion and more pollution.... Not only has the throughput grown quantitatively, but its qualitative nature has changed. Exotic substances are produced and thrown wholesale into the biosphere—substances with which the world has had no adaptive evolutionary experience, and which are consequently nearly always disruptive." See Herman E. Daly, "Entropy, Growth, and the Political Economy of Scarcity," in V. Kerry Smith, ed., *Scarcity and Growth Reconsidered*, Baltimore: Johns Hopkins University Press for Resources for the Future, 1979, pp. 67–94. Quote is on p. 75.

19. The precautionary principle was first introduced to many in the United States by Konrad von Moltke in *The Vorsorsorgeprinzip in West German Environmental Policy*, Institute for European Environmental Policy, London, 1987. More recent developments of the principle can be found in James Cameron and Timothy O'Riordan, eds., *Interpreting the Precautionary Principle*, London:

Cameron and May, 1994; and Carolyn Raffensperger and Joel Tickner, eds., *Protecting Public Health and the Environment: Implementing the Precautionary Principle*, Washington, D.C.: Island Press, 1999.

20. This is what Mary O'Brien calls alternatives assessment. See *Making Better Environmental Decisions: An Alternative to Risk Assessment*, Cambridge, Mass.: MIT Press, 2000.

Chapter 9

1. For an expansion of this argument, see David Allen and Nasrin Bemanesh, "Wastes as Raw Materials," in Braden Allenby and Deanna J. Richards, eds., *The Greening of Industrial Ecosystems*, Washington, D.C.: National Academy Press, 1994, pp. 69–89.

2. For a social account of waste and recycling, see Susan Stasser, *Waste and Want: A Social History of Trash*, New York: Henry Holt, 1999.

3. Stasser, *Waste and Want*, p. 85.

4. For a history of these charity organizations, see Roy Lubove, *The Professional Altruist: The Emergence of Social Work as a Career, 1880–1930*, Cambridge, Mass: Harvard University Press, 1965. The Sunset Scavenger story is described in Louis Blumberg and Robert Gottlieb, *War on Waste: Can America Win its Battle with Garbage?* Washington, D.C.: Isand Press, 1989, p. 198.

5. These EPA figures are certainly conservative. The survey of state solid waste administrators conducted each year by the magazine *BioCycle* reports that 118 million tons of garbage (31.5 percent) were recycled in 1999. See Jim Glenn, "The State of Garbage in America," *BioCycle*, 40 (4), April 1999, pp. 60–71.

6. David B. Spenser, "Recycling," in Frank Krieth, ed., *Handbook of Solid Waste Management*, New York: McGraw-Hill, 1994, pp. 9.1–9.215, reference is to p. 9.105; and Iddo K. Wernick, Robert Herman, Shekhar Govind, and Jesse H. Ausubel, "Materialization and Dematerialization: Measures and Trends," *Daedalus*, 125 (3) Summer 1996, p. 175.

7. U.S. Geological Survey, *Minerals Yearbook: Metals and Minerals, 1998*, Washington, D.C., U.S. Government Printing Office, 2000, pp. JJJ1; and Scott F. Sibley, William C. Butterman, and staff, "Metals Recycling in the United States," *Resource Conservation and Recycling*, 15, 1995, pp. 259–267.

8. Ambuj D. Sager and Robert Frosch, "A Perspective on Industrial Ecology and Its Applicability to a Metals-Industry Ecosystem," *Journal of Cleaner Production*, 5, 1997, pp. 39–45.

9. Iddo K. Wernick and Nickolas J. Themelis, "Recycling Metals for the Environment," *Annual Review of Energy and the Environment*, 23, 1998, pp. 465–497.

10. Sibley et al., "Metals Recycling," p. 261. For a textbook discussion of scrap, see James F. McDivitt and Gerald Manners, *Minerals and Men*, Baltimore: John's Hopkins University Press for Resources For the Future, 1974 , p. 33.

11. See Wernick and Themelis, "Recycling Metals," pp. 483–484; U.S. Geological Survey, *Minerals Yearbook*, p. JJJ6; and R. S. Kaplan and F. J. Schottman, "Recycling of Metals," in Michael B. Bever, ed., *Concise Encyclopedia of Materials, Economics, Policy and Management*, New York: Pergamon, 1993, pp. 293–302. The new steel minimills can take 100 percent scrap, but these mills can only manufacture a limited range of steel products. See Robert A. Frosch and Nicholas Gallopoulos, "Strategies for Manufacturing," *Scientific American*, 261 (3) September 1989, pp. 144–153. 1989, p. 148.

12. U.S. Geological Survey, *Minerals Yearbook*, p. JJJ3.

13. U.S. Geological Survey, *Minerals Yearbook*, p. JJJ6; and Gary R. Brenniman, Stephen D. Cooper, William H. Hallenbeck, and James M. Lyznicki, "Recycling: Automotive and Household Batteries," in Krieth, *Handbook of Solid Waste*, p. 9.159.

14. For the lead and cadmium data, see U.S. Environmental Protection Agency, Office of Solid Waste, *Characterization of Products Containing Lead and Cadmium in Municipal Solid Waste in the United States, 1970–2000*, Washington, D.C., January 1989; and for the mercury data see U.S. Environmental Protection Agency, Office of Solid Waste, *Characterization of Products Containing Mercury in Municipal Solid Waste in the United States*, Washington, D.C., April 1992.

15. See Wernick and Themelis, "Recycling Metals," p. 479; Kaplan and Schottman, "Recycling of Metals," p. 301; and Robert Frosch and Nicholas Gallopoulos, "Strategies for Manufacturing," p. 149.

16. Mary B. Sikora, "Options for Managing and Marketing Scrap Tires," in Krieth, *Handbook of Solid Waste*, pp. 9.129–9.148.

17. B. D. Lagrone and F. G. Smith, "Reclaiming of Elastomers," in Bever, *Encyclopedia of Materials*, pp. 285–288; and Michael Blumenthal, "What's New with Ground Rubber," *BioCycle*, 39 (3), March 1998, pp. 40–44.

18. The 1985 data are from U.S. Environmental Protection Agency, Office of Solid Waste, *Characterization of the Municipal Solid Waste Stream in the United States: 1990 Update*, Washington, D.C., June 1990, and the 1995 data are from the U.S. Environmental Protection Agency, Office of Solid Waste and Emergency Response, *Characterization of the Municipal Solid Waste Stream in the United States*, 1997.

19. U.S. Environmental Protection Agency, Office of Solid Waste and Emergency Response, *Characterization of Municipal Solid Waste*, 1997, p. 42.

20. Spenser, "Recycling," p. 9.64.

21. Michael Lewis, Russell Clark, Jeffery Vandall, and Neil Seldman, *Reuse Operations: Community Development Through Redistribution of Used Goods*, Washington, D.C.: Institute for Local Self-Reliance, 1995.

22. Brenniman et al., "Recycling," p. 9.153.

23. For the history, see Neil N. Seldman, "Recycling—History in the United States," in Bisio Attilio and Sharon Boots, eds., *Encyclopedia of Energy, Tech-*

nology and the Environment, New York: Wiley, 1995. The municipal recycling and composting rates are from U.S. Environmental Protection Agency, Office of Solid Waste and Emergency Response, *Cutting the Waste Stream in Half: Community Record Setters Show How,* Washington, D.C., June 1999. The *BioCycle* survey found some 9300 curbside recycling programs serving 139.5 million people in 1999. See Jim Glenn, "The State of Garbage in American," *BioCycle* 40 (4) April 1999, p. 63.

24. See Frank Ackerman, *Why Do We Recycle: Markets, Values and Public Policy,* Washington, D.C.: Island Press, 1997.

25. Over 85 percent of all aluminum beverage cans sold in Maine, a bottle bill state, are recycled. See W. U. Chandler, *Materials Recycling: The Virtue of Necessity,* Worldwatch Paper #56, Washington, D.C.: Worldwatch Institute, 1983. The data on the mandatory bottle takeback programs are reported in Spenser, "Recycling," p. 9.19.

26. Cynthia Pollack, "Realizing Recycling's Potential," in Lester R. Brown, Edward C. Wolfe, and Linda Starke, eds., *State of the World—1987,* New York: Norton, 1987, pp. 101–121.

27. Franklin Associates, *The Role of Recycling in Industrial Solid Waste Management in the Year 2000,* Stamford, Conn.: Keep America Beautiful, 1994; and Sound Resource Management Group, *The Economics of Recycling and Recycled Materials,* Seattle: Clean Washington Center, 1993. See also the review by Frank Ackerman in Ackerman, *Why Do We Recycle,* p. 77.

28. William L. Kovacs, "The Coming Era of Conservation and Industrial Utilization of Recyclable Materials," *Ecology Law Quarterly,* 1998, pp. 537–625.

29. John Young and Aaron Sachs, *The Next Efficiency Revolution: Creating a Sustainable Materials Economy,* Worldwatch Paper No. 121, Washington, D.C.: Worldwatch Institute, 1994.

30. See Nelson L. Nemerow, *Zero Pollution for Industry: Waste Minimization Through Industrial Complexes,* New York: Wiley, 1995; and Ernest A. Lowe, "Creating By-Product Resource Exchanges for Eco-Industrial Parks," *Journal of Cleaner Production,* 5 (1–2) 1997, pp. 57–66.

31. Dara O'Rourke, Lloyd Connelly, and Catherine Koshland call this a "stiffness" in their critique of industrial ecology. See "Industrial Ecology: A Critical Review," *International Journal of Environment and Pollution,* 6 (2–3), February 1996, pp. 89–112.

32. U.S. Geological Survey, *Minerals Yearbook,* p. JJJ1. The aluminum data are from Tom Tietenberg, *Environmental and Natural Resource Economies,* 4th ed., New York: Harper Collins, 1996, p. 185. A recent EPA analysis demonstrates the potential reduction in greenhouse gas emissions from recycling, although the effect of source reduction is even more dramatic. See U.S. Environmental Protection Agency, Office of Solid Waste and Emergency Response, *Greenhouse Gas Emissions from Management of Selected Materials in Municipal Solid Waste,* Washington, D.C., September 1999.

33. Sibley et al., "Metals Recycling," p. 266. See also Michael J. McKinley and staff, *Recent Market Trends of Recycled Metals in the United States*, Reston, Va.: U.S. Geological Survey, 1999.

34. This includes about 19,000 gold and silver mining employees, 15,000 copper mining employees, 8700 iron mining employees, 2700 lead mining employees, and various other mineral mining employees. See U.S. Bureau of the Census, *1992 Census of Mining*, Washington, D.C., 1993.

35. For these trends and a similar argument, see Thomas Michael Power, *Lost Landscapes and Failed Economies: The Search for a Value of Place*, Washington, D.C.: Island Press, 1966.

36. U.S. Bureau of Mines, *The New Materials Society*, Vol. 3, *Material Shifts in the New Society*, Washington, D.C.: U.S. Government Printing Office, 1991, p. 1.10.

37. Frank Ackerman argues that the public commitment to recycling is as much about social values, such as responsibility for the environment, public purpose, and frugality, and opposition to waste treatment facilities, as it is about waste management efficiencies. See Ackerman, *Why Do We Recycle*, pp. 184–187.

Chapter 10

1. See Morris Cohen, ed., "Materials Science and Engineering: Its Evolution, Practice and Prospects," special ed., *Materials Science and Engineering*, 37 (1), January 1979.

2. Ivan Amato, *Stuff: The Materials the World Is Made Of*, New York: Avon Books, 1997, p. 88.

3. National Academy of Sciences, *Materials Science and Engineering for the 1990s*, Washington, D.C.: National Academy Press, 1987.

4. George White, "General Introduction," U.S. Bureau of Mines, *The New Materials Society*, Vol. 1, *New Materials Markets and Issues*, Washington, D.C.: U.S. Government Printing Office, 1990, p. 1.2. For a wonderfully rich review of many of these new materials see Philip Ball, *Made to Measure: New Materials for the 21st Century*, Princeton, N.J.: Princeton University Press, 1997.

5. U.S. Congress, Office of Technology Assessment, *Advanced Materials by Design: New Structural Materials Technologies*, Washington, D.C., June 1988.

6. William C. Butterman and Richard S. Gillette, "Markets for New Materials," in U.S. Bureau of Mines, *The New Materials Society*, Vol. 1, pp. 2.7–2.9.

7. Butterman and Gillette, "Markets for New Materials," p. 2.20.

8. Butterman and Gillette, "Markets for New Materials," pp. 2.26–2.27.

9. For a good reference, see Anthony Kelly, ed., *Concise Encyclopedia of Composite Materials*, New York: Pergamon Press, 1989.

10. Butterman and Gillette, "Markets for New Materials," p. 2.29.

11. Harold Bullis and James E. Mielke, *Strategic and Critical Materials*, Boolder, Col.: Westview, 1985, p. 11.

12. A comprehensive review of semiconductor materials can be found in W. O'Mara, B. Herring, and L. Hunt, *Handbook of Semiconductor Silicon Technology*, Westword, N.J.: Noyes Publications, 1990; and Lev I. Berger, *Semiconductor Materials*, Boca Raton, Fla.: CRC Press, 1997.

13. Amato, *Stuff*, p. 172.

14. Ball, *Made to Measure*, pp. 103–142.

15. See Ishrat M. Khan and Joycelyn S. Harrison, eds., *Field Responsive Polymers: Electroresponsive, Photoresponsive, and Responsive Polymers in Chemistry and Biology*, Washington, D.C.: American Chemical Society, 1999.

16. Amato, *Stuff*, p. 174.

17. Kenneth J. Klabunde and Cathy Mohs, "Nanoparticles and Nanostructural Materials," in Leonard V. Interrante and Mark J. Hampden Smith, eds., *Chemistry of Advanced Materials*, New York: Wiley-VCH, 1998, pp. 271–328.

18. Petri Lehmus and Bernard Rieger, "Nanoscale Polymerization Reactors for Polymer Fibers," *Science*, 285, September 24, 1999, pp. 2081–2082.

19. Guojun Liu, "Polymeric Nanostructures," in Hari Singh Nalwa, *Handbook of Nanostructured Materials and Nanotechnology*, San Diego: Academic Press, 2000, pp. 475–500.

20. For a most visionary statement, see Eric Drexler and Chris Peterson, *Unbounding the Future: The Nanotechnology Revolution*, New York: William Morrow, 1991.

21. K. Eric Drexler, "The Coming Era of Nanotechnology," in Tom Forester, ed., *The Materials Revolution: Superconductors, New Materials, and the Japanese Challenge*, Cambridge, Mass.: MIT Press, 1988, pp. 361–370.

22. Butterman and Gillette, "Markets for New Materials," p. 2.25.

23. For a enthusiastic review see John B. Horrigan, Frances H. Irwin and Elizabeth Cook, *Taking a Byte Out of Carbon: Electronics Innovation for Climate Protection*, Washington, D.C.: World Resources Institute, 1998.

24. Much of the public attention to chemical hazards in the semiconductor industry has been driven by the relentless advocacy of the environmental movement, particularly the San Jose-based Silicon Valley Toxics Coalition. See B. Eskanazi, E. B. Gold, S. J. Samuels et al., "Prospective Monitoring of Early Fetal Loss and Clinical Spontaneous Abortion Among Female Semiconductor Workers," *American Journal of Industrial Medicine*, 26, 1995, pp. 833–846; M. B. Schenker, E. B. Gold, J. J. Beaumont et al., "Association of Spontaneous Abortion and Other Reproductive Effects with Work in the Semiconductor Industry," *American Journal of Industrial Medicine*, 28 (6), 1995, pp. 639–659; and Eric A. Hasenbein, "The Semiconductor Industry's Goal of 100% Elimination of 2-Methoxyethanol, 2-Ethoxyethanol and their Acetates by 1995

Compared to Other Industries," *Semiconductor Safety Association Journal*, June 9 (2), 1995, pp. 9–21.

25. For data, see National Safety Council, Environmental Health Center, *Electronic Product Recovery and Recycling Baseline Report*, Washington, D.C., May 1999. This study projects that by 2007, 40 million electronic products will be discarded in the United States per year.

26. Ronald F. Balazik, "Analysis of New Material Issues," in U.S. Bureau of Mines, *The New Materials Society*, Vol. 1, pp. 3.37–3.39.

27. Suppliers of Advanced Composite Materials Association, *Safe Handling of Advanced Composite Materials: Health Information*, Arlington, Va., April 1989.

28. White, "General Introduction," p. 1.3.

29. The journals reviewed include *Advanced Materials, Chemistry of Materials, Materials Research Society Bulletin, Journal of Materials Chemistry, Materials Science and Engineering*, and *Journal of the Minerals, Metals and Materials Societies*. The Massachusetts Institute of Technology offers a course in industrial ecology; no other university programs reviewed offered similar course descriptions.

Chapter 11

1. Judy Stringer, "Biologically Derived Chemicals Find Niches," *Chemical Week*, September 18, 1996, p. 52.

2. William J. Hale, *The Farm Chemurgic*, Boston: Stratford Press, 1934. Hale was also a brother-in-law of Willard Henry Dow, who had succeeded his father, Herbert Dow, as president of the Dow Chemical Company.

3. David L. Lewis, *Henry Ford: An American Folk Hero and His Company*, Detroit: Wayne State University Press, 1976, p. 282.

4. Christy Borth, *The Automobile*, Washington, D.C.: Automobile Manufacturers Association, 1952, pp. 204–205.

5. William J. Hale, *Farmer Victorious: Money, Mart, and Mother Earth*, New York: Coward-McCann, 1949, pp. 105–155.

6. See U.S. Congress, Office of Technology Assessment, *Agricultural Commodities as Industrial Raw Materials*, Washington, D.C.: U.S. Government Printing Office, May 1991; and Lewrence Glaser and Gregory Gajewski, "Agricultural Products for Industry: The Situation and Outlook," *International Journal of Environmentally Conscious Design and Manufacturing*, 3 (2), 1994, pp. 63–66.

7. Beryl B. Simpson and Molly C. Ogorzaly, *Economic Botany: Plants in Our World*, 2nd ed., New York: McGraw-Hill, 1995, p. 363.

8. Glazer and Gajewski, "Agricultural Products," p. 64. Production data are from U.S. Department of Agriculture, Economic Research Service, "Ethanol, Citric Acid, and Lactic Acid Use Corn as a Feedstock," *Industrial Uses of Agri-*

cultural Materials: Situation and Outlook Report, Washington, D.C., July 1997, p. 10.

9. Margaret A. Clarke and Leslie A. Edye, "Sugar Beet and Sugarcane as Renewable Resources," in Glen Fuller, Thomas McKeon, and Donald Bills, eds., *Agricultural Materials as Renewable Resources: Non-Food and Industrial Applications*, Washington, D.C.: American Chemical Society, 1996, pp. 229–247.

10. Glazer and Gajewski, "Agricultural Products," p. 64.

11. American Soybean Association, *Soy Stats, 1999*, St. Louis, Mo., 1999.

12. Marvin O. Bagby, "Products from Vegetable Oils: Two Examples," in Fuller et al., *Agricultural Materials*, pp. 248–257; and Glazer and Gajewski, "Agricultural Products," p. 64.

13. Joseph Roethell, Lewrence Glazer, and Raymond Brigham, *Castor: Assessing the Feasibility of U.S. Production*, U.S. Department of Agriculture, Office of Agricultural Materials, Washington, D.C., May 1991.

14. Stringer, "Biologically Derived Chemicals," pp. 52–53.

15. Stringer, "Biologically Derived Chemicals," p. 52.

16. Robert Braile, "U.S. Starts a Phase Out of Suspected Gas Additive," *Boston Globe*, March 21, 2000, p. A1.

17. See "Fox Fibers," *Industry and Environment* (a publication of the United Nations Environment Program), 20 (1–2), January, June 1997, p. 53. Fox Fibers is reorganizing after filing for bankruptcy protection in 1997 because it overproduced in anticipation of a Levi Straus contract that was cancelled after the cotton had been planted.

18. Simpson and Ororzaly, *Economic Botany*, pp. 520–526. See also Mari Kane, "Growing Pains," *E-Magazine*, 10 (5), September 1997, pp. 37–41; and Bruce Dunford, "Hemp Tested as Possible Crop for Hawaii," *Boston Globe*, December 16, 1999, p. A26.

19. "Fiber Revolution," *The Carbohydrate Economy* (publication of the Institute of Local Self-Reliance, Washington, D.C.), 1 (1), Fall 1997.

20. U.S. Congress, Office of Technology Assessment, *Agricultural Commodities*, p. 82.

21. Kenneth D. Carlson and Donald Van Dyne, eds., *Industrial Uses for High Erucic Acid Oils from Crambe and Rapeseed*, University of Missouri-Columbia, Extension and Agricultural Information Office, October 1992; and U.S. Department of Agriculture, Economic Research Service, "Crambe, Industrial Rapeseed, and Tung Provide Valuable Oils", *Industrial Uses of Agricultural Materials*, Washington, D.C., 1996, p. 17.

22. Joseph C. Roetheli, Kenneth D. Carlson, Robert Kleiman, Anson E. Thompson, David A. Dierig, Lewrence K. Glazer, Melvin G. Blase, and Julia Goodell, *Lesquerella as a Source of Hydroxy Fatty Acids for Industrial Products*,

U.S. Department of Agriculture, Cooperative State Research Service, Washington, D.C., October 1991.

23. Simpson and Ogorzaly, *Economic Botany*, p. 369.

24. Peter H. Spitz quotes a Du Pont manager as saying "In the field of carbon monoxide chemistries [Du Pont] has an answer to anything the oil companies can do with hydrocarbon gases." See Peter Spitz, *Petrochemicals: The Rise of an Industry*, New York: Wiley, 1988, p. 106.

25. David Morris and Irshad Ahmed, *The Carbohydrate Economy: Making Chemicals and Industrial Materials from Plant Matter*, Washington, D.C.: Institute for Local Self-Reliance, 1992, p. 2.

26. Morris and Ahmed, *Carbohydrate Economy*, p. 20.

27. National Academy of Sciences, National Research Council, Committee on Renewable Resources for Industrial Materials, *Renewable Resources for Industrial Materials*, Washington, D.C., 1976; U.S. Congress, Office of Technology Assessment, *Agricultural Commodities*, p. 97; and Morris and Ahmed, *Carbohydrate Economy*, p. 19.

28. U.S. Congress, Office of Technology Assessment, *Agricultural Commodities*, pp. 14–15. Government initiatives at both the federal and state levels have tried to assess the impacts of more vigorous efforts to promote industrial uses of renewable resources. See for example National Research Council, Committee on Renewable Resources for Industrial Materials, *Renewable Resources*; and Texas Department of Agriculture, *Texas Agriculture: Growing A Sustainable Economy, A Greenprint for Action*, Austin, Texas, 1990.

29. U.S. Congress, Office of Technology Assessment, *Agricultural Commodities*, p. 15.

30. National Academy of Sciences, National Research Council, *Jojoba: New Crop for Arid Lands; New Material for Industry*, Washington, D.C.: National Academy Press, 1985.

31. National Academy of Sciences, National Research Council, *Biologically Based Pest Management: New Solutions for a New Century*, Washington, D.C.: National Academy Press, 1996, pp. 23–24.

32. P. A. Matson, W. J. Parton, A. G. Power, and M. J. Smith, "Agricultural Intensification and Ecosystem Properties," *Science*, 277, July 25, 1997, pp. 504–509; and Judith D. Soule and Jon K. Piper, *Farming in Nature's Image: An Ecological Approach to Agriculture*, Washington, D.C.: Island Press, 1992, p. 12.

33. David Pimentel and Wen Dazhong, "Technological Changes in Energy Use in U.S. Agricultural Production," in Ronald Carroll, John H. Vandermeer, and Peter M. Rosset, eds., *Agroecology*, New York: McGraw-Hill, 1990, pp. 150–155.

34. For a good review of agricultural water scarcity, see Sandra Postel, *Water for Agriculture: Facing the Limits*, Worldwatch Paper No. 93, Washington, D.C.: Worldwatch Institute, 1989.

35. Matson et al., "Agricultural Intensification," p. 507; and Soule and Piper, *Farming in Nature's Image*, pp. 31–39.

36. Hale, *Farmer Victorious*, p. 118.

37. See K. H. W. Klages, *Ecological Crop Geography*, New York: Macmillan, 1942, and more recent work such as G. Azzi, *Agricultural Ecology*, London: Constable, 1956; G. E. Dalton, *Study of Agricultural Systems*, London: Applied Sciences, 1975; and Richard Lowrance, Benjamin R. Stinner, and Garfield S. House, *Agricultural Ecosystems: Unifying Concepts*, New York: Wiley, 1984.

38. A recent text that unites theory and practice can be found in Miquel A. Altieri, ed., *Agroecology: The Science of Sustainable Agriculture*, Boulder, Col.: Westview (Harper-Collins), 1989. Quote is on pp. 91–92.

39. See Wes Jackson, *Altars of Unhewn Stone: Science and the Earth*, San Francisco: North Point, 1987. Today's herbaceous perennial grasses yield less seed than annuals, but in terms of net product to resource inputs, the yield is comparable. See Soule and Piper, *Farming in Nature's Image*.

40. See Anton Moser, "Ecological Process Engineering: The Potential of Bioprocessing," in Robert U. Ayres and Paul M. Weaver, *Eco-restructuring: Implications for Sustainable Development*, New York: United Nations University Press, 1998, pp. 77–108; and R. K. Dart and W. B. Betts, "Uses and Potential of Lignocellulose," in W. B. Betts, ed., *Biodegradation: Natural and Synthetic Materials*, London: Springer-Verlag, 1991, pp. 201–217.

41. Morris and Ahmed in their 1992 study note that the amount of "waste bioresources" was roughly three times greater than the tonnage of all organic chemicals produced in the United States that year. Morris and Ahmed, *Carbohydrate Economy*, p. 20.

42. U.S. Environmental Protection Agency, Office of Solid Waste and Emergency Response, *Characterization of Municipal Solid Waste in the United States*, 1996 Update, Washington D.C. 1997, pp. 45–47.

43. The annual volume of sewage waste is from Iddok Wernick, Robert Herman, Shekhar Govind, and Jesse H. Ausubel, "Materialization and Dematerialization: Measures and Trends," *Dedalus* 125 (3), Summer 1966, p. 190. Agricultural needs may dictate that a sizable share of these organic wastes be recycled as soil nutrients, but a reasonable surplus should remain for other uses. See Gary Gardner, *Recycling Organic Waste: From Urban Pollutant to Farm Resource*, Worldwatch Paper No. 135, Washington D.C.: Worldwatch Institute, August 1997.

Chapter 12

1. C. W. Hazeltine, "A Millennium of Fungi, Food, and Fermentation," *Mycologia*, 57, March/April, 1965, pp. 149–197.

2. Luigi Orsenigo, *The Emergence of Biotechnology: Institutions and Markets in Industrial Innovation*, New York: St. Martin's Press, 1989, pp. 33–34.

3. Owen P. Ward, *Fermentation Biotechnology*, Englewood Cliffs, N.J.: Prentice Hall, 1989.

4. Technically detailed descriptions of commercial fermentation can be found in Wulf Crueger and Anneliese Crueger, *Biotechnology: A Textbook of Industrial Microbiology*, 2nd ed., Sunderland, Mass.: Sinauer Associates, 1990; and M. D. Trevan, S. Boffey, K. H. Goulding, and P. Stanbury, *Biotechnology: The Biological Principles*, New York: Taylor and Francis, 1987.

5. Crueger and Crueger, *Biotechnology*, pp. 124–129.

6. Ward, *Fermentation Biotechnology*, pp. 139–142.

7. Crueger and Crueger, *Biotechnology*, pp. 134–149 and Ward, *Fermentation Biotechnology*, pp. 133–139.

8. Ward, *Fermentation Biotechnology*, pp. 143–145.

9. Seth Lubove, "Enzyme-eaten Jeans," *Forbes*, 146 (10), October 29, 1990, pp. 140–141. The world market for industrial enzymes in 1990 was over $650 million per year, and well over 60 percent of this market was dominated by European firms. See U.S. Congress, Office of Technology Assessment, *Biotechnology in a Global Economy*, Washington, D.C.: U.S. Government Printing Office, October 1991, p. 121.

10. See Nora Goldstein and Kevin Gray, "Biosolids Composting in the United States," *BioCycle*, 40 (1), January 1999, pp. 63–75; and Jim Glenn, "The State of Garbage in America," *BioCycle* 40 (4) April 1999, p. 64.

11. Alyson Warhurst, "Metals Biotechnology: Trends and Implications," in Martin Fransman, Gerd Junne, and Annemieke Roobeek, eds., *The Biotechnology Revolution?*, Oxford, UK: Blackwell, 1995, pp. 258–273.

12. J. B. Wood, B. L. Ghosh, and S. N. Sinha, "Solid Substrate Fermentation in the Production of Enzyme Complex for Improved Jute Processing," in Glen Fuller, Thomas McKeon, and Donald Bills, eds., *Agricultural Materials as Renewable Resources: Non-Food and Industrial Applications*, Washington, D.C.: American Chemical Society, 1996, pp. 46–59.

13. This analysis follows the well-constructed framework put forward by Janine M. Benyus in her book, *Biomimicry: Innovation Inspired by Nature*, New York: William Morrow, 1997. See also Mehmet Sarikaya and Ilan A. Aksay, eds., *Biomimetics: Design and Processing of Materials*, New York: American Institute of Physics, 1995.

14. See David Kaplan, W. Wade Adams, Barry Farmer, and Christopher Viney, *Silk Polymers: Materials Science and Biotechnology*, Washington, D.C.: American Chemical Society, 1994; and Karel Grohmann and Michael E. Himmel, "Advances in Protein-Derived Materials," in Halena L. Chum, ed., *Polymers from Biobased Materials*, Park Ridge, N.J.: Noyes Data Corporation, 1991, pp. 144–153.

15. Benyus, *Biomimicry*, pp. 98–101.

16. Benyus, *Biomimicry*, pp. 100–104.

17. Benyus, *Biomimicry*, pp. 120–127.

18. See Ivan Amato, *Stuff: The Materials the World Is Made of*, New York: Avon Books, 1997, pp. 168–169.

19. Helen Ghiradella, Daniel Aneshansley, Thomas Eisner, Robert E. Silberglied, and Howard E. Hinton, "Ultraviolet Reflection of a Male Butterfly," *Science*, 178, 1972, pp. 214–216.

20. For a detailed history of recombinant DNA and the controversy that it generated, see Sheldon Krimsky, *Genetic Alchemy: The Social History of the Recombinant DNA Controversy*, Cambridge, Mass.: MIT Press, 1985.

21. See U.S. Congress, Office of Technology Assessment, *New Developments in Biotechnology: U.S. Investment in Biotechnology*, Washington, D.C.: U.S. Government Printing Office, 1987–1989. The data are from Charles C. Mann, "Biotech Goes Wild," *Technology Review*, 102 (4), July–August 1999, pp. 36–43.

22. U.S. Congress, Office of Technology Assessment, *Biotechnology*, p. 121.

23. So-called "biocompatible polymers" used in medical or health care products are often also called "biopolymers," even though some meet none of these three criteria. For definitions, see U.S. Congress, Office of Technology Assessment, *Biopolymers: Making Materials Nature's Way*, Washington, D.C., U.S. Government Printing Office, September 1993, p. 5.

24. Firms investing in biopolymers include Cargill, Archer-Daniels-Midland, ConAgra, Dow, Du Pont, Union Carbide, Mobile, and Warner-Lambert. See Lawrence Glaser and Gregory Gajewski, "Agricultural Products for Industry: The Situation and Outlook," *International Journal of Environmentally Conscious Design and Manufacturing*, 3 (2), 1994, p. 64.

25. Chistopher Rivard, Michael Himmel, and Karel Grohmann, "Biodegradation of Plastics," in Halena L. Chum, *Polymers from Biobased Materials*, Park Ridge, N.J.: Noyes Data Corporation, 1991, pp. 115–119.

26. See Judy Stringer, "Biologically Derived Materials Find Niches," *Chemical Week*, September 18, 1996, pp. 52–53; and Rivard et al., "Biodegradation of Plastics," pp. 115–119. Zencca ceased production of Biopol in 1998.

27. U.S. Congress, Office of Technology Assessment, *Biopolymers*, p. 21.

28. Jane G. Terrell, Maurille J. Fournier, Thomas L. Mason, and David Terrell, "Biomolecular Materials"; *Chemical & Engineering News*, 72, December 19, 1994, pp. 40–51; and U.S. Congress, Office of Technology Assessment, *Biopolymers*, p. 23.

29. See Stringer, "Biologically Derived Materials"; and Stuart L. Hart and Mark B. Milstein, "Global Sustainability and the Creative Destruction of Industries," *Sloan Management Review*, 41 (1), Fall 1999, pp. 23–33.

30. U.S. Congress, Office of Technology Assessment, *Biotechnology*.

31. Margaret Sharp, "Applications of Biotechnology: An Overview," in Fransman et al., *Biotechnology Revolution*, pp. 163–173.

32. Crueger and Crueger, *Biotechnology*, p. 189.

33. A early review of these difficulties is discussed by Sheldon Krimsky in *Biotechnics and Society: The Rise of Industrial Genetics*. New York, Praeger, 1991, pp. 185–204. More recent coverage can be found in Dan Ferber, "GM Crops in the Cross Hairs," *Science*, 286, November 26, 1999, pp. 1662–1666.

34. Amato, *Stuff*, p. 161.

Chapter 13

1. Orio Giaini, ed., *Dialogue on Wealth and Welfare: An Alternative View of World Capital Formation*, Oxford, U.K.: Pergamon Press, 1980; and Ernst von Weizsacker, Amory B. Lovins, and L. Hunter Lovins, *Factor Four: Doubling Wealth, Halving Resource Use*, London: Earthscan, 1997.

2. Other texts promoting this market-regarding optimism include Stephen Schidheimy and the Business Council for Sustainable Development, *Changing Course: A Global Perspective on Development and the Environment*, Cambridge, Mass.: MIT Press, 1992; Paul Hawken, *The Ecology of Commerce*, New York: Harper Collins, 1993; and Carl Frankel, *In Earth's Company: Business, Environment, and the Challenge of Sustainability*, Babriola Island, Canada: New Society, 1998.

3. Wilfred Malenbaum, *World Demand for Raw Materials in 1985 and 2000*, New York: McGraw-Hill, 1978. See also Nikolai D. Kondratiev, "The Long Waves in Economic Life," *Review of Economic Statistics*, 17, 1935, pp. 105–115; and Simon S. Kuznets, *Toward a Theory of Economic Growth*, New York: Norton, 1968.

4. See Eric D. Larson, Marc H. Ross, and Robert H. Williams, "Beyond the Era of Materials," *Scientific American*, 254, 1986, pp. 34–41; M. Janicke, H. Monch, T. Ranneberg, and U. E. Simonis, "Structural Change and Environmental Impact: Empirical Evidence on Thirty-One Countries in East and West," *Environmental Monitoring and Assessment*, 12, 1989, pp. 99–114; and D. B. Brookes and P. W. Andrews, "Mineral Resources, Economic Growth, and World Population," *Science*, 185, 1974, pp. 13–20.

5. M. Janicke, M. Binder, and H. Monch, "Dirty Industries: Patterns of Change in Industrial Countries," *Environmental and Resource Economics*, 6, 1997, pp. 35–56. Cutler J. Cleveland and Matthias Ruth provide an excellent review of these various studies. See their "Indicators of Dematerialization and the Materials Intensity of Use," *Journal of Industrial Ecology*, 2 (3), 1999, pp. 15–50. Quote is on p. 16.

6. See Iddo K. Wernick, Robert Herman, Shekhar Govind, and Jesse H. Ausubel, "Materialization and Dematerialization: Measures and Trends," *Daedalus* 125 (3), Summer, 1996, pp. 171–198.

7. Wernick et al., "Materialization and Dematerialization," p. 179.

8. Matthias Ruth, "Dematerialization in Five US Metals Sectors: Implications for Energy Use and CO_2 Emissions," *Resource Policy*, 24 (1), 1998, pp. 1–18.

9. Jens Brobech Legarth, "Sustainable Metal Resource Management—The Need for Industrial Development: Efficiency Improvement Demands on Metal Resource Management to Enable a (Sustainable) Supply until 2050," *Journal of Cleaner Production*, 4 (2), 1996, pp. 97–104.

10. Herman Daly, "Toward Some Operational Principles of Sustainable Development," *Ecological Economics*, 2, 1990, pp. 1–6.

11. See Peter Vitousek, Paul Ehrlich, and Pamela Matson, "Human Appropriation of the Products of Photosynthesis," *BioScience*, 34 (6), 1986, pp. 368–373.

12. Beverley Thorpe, *Citizen's Guide to Clean Production*, Lowell Center for Sustainable Development, University of Massachusetts Lowell, 1999.

13. For a good history and analysis of extended producer responsibility by one of the originators of the concept, see Thomas Lindhqvist, "Extended Producer Responsibility in Cleaner Production," unpublished doctoral dissertation, International Institute for Industrial Environmental Economics, Lund University, Lund, Sweden, 2000. For brief coverage of the European proposal, see Stewart Boyle, "Crossed Wires," *Tomorrow*, 10 (3), May–June, 2000, pp. 14–15.

14. See Reid Lifset, "Take it Back: Extended Producer Responsibility as a Form of Incentive-Based Environmental Policy," *Journal of Resource Management and Technology*, 21, 1993, p. 166; Cynthia Pollock Shea, "Give and Take," *Tomorrow*, 10 (1), January 2000, p. 26; and Gary A. Davis, Catherine A. Witt, and Jack N. Barkenbus, "Extended Product Responsibility: A Tool for a Sustainable Economy," *Environment*, 39 (7), September 1997, pp. 10–15, 36–37.

15. See L. Nelson Nemerow, *Zero Pollution for Industry: Waste Minimization Through Industrial Complexes*, New York: Wiley, 1995; and Gunter Pauli, "Zero Emissions: The Ultimate Goal of Cleaner Production," *Journal of Cleaner Production*, 5 (1–2), 1997, pp. 109–113. The Kalunborg story can be found in John Ehrenfeld and Nicholas Gertler's "Industrial Ecology in Practice: The Evolution of Interdependence at Kalunborg," *Journal of Industrial Ecology*, 1 (1), 1997, pp. 67–79.

16. World Business Council for Sustainable Development, *Eco-efficient Leadership for Improved Economic and Environmental Performance*, Geneva, 1995.

17. The Institute for Electronics and Electrical Equipment has provided strong encouragement for design for environmental strategies in a series of conferences on electronic products. Product design has become a central focus of design for the environment, with a sequence of articles appearing in the *Journal of Sustainable Product Design* from 1997 onward. For good texts, see Thomas E. Graedel and Braden R. Allenby, *Design for the Environment*; Upper Saddle River, N.J., Prentice Hall, 1996; and Joseph Fiksel, ed., *Design for the Environment: Creating Eco-Efficient Products*, New York: McGraw-Hill, 1996.

18. Wernick et al., "Materialization and Dematerialization," p. 181.

19. Robert Herman, Siamak A. Ardekani, and Jesse H. Ausubel, "Dematerialization," in Jesse H. Ausubel and Hedy Sladovich, eds., *Technology and Environment*, Washington, D.C.: National Academy Press 1989, p. 53.

20. See Walter Stahel and Genevieve Reday-Mulvey, *Jobs for Tomorrow: the Potential for Substituting Manpower for Energy*, New York: Vantage Press, 1981.

21. Robert T. Lund, "Remanufacturing," *Technology Review*, 87, 1984, pp. 18–23, 28–29.

22. Charles Godfrey Leland, *A Manual of Mending and Repairing, with Diagrams*, New York: Dodd, Mead, 1896, p. xxi, as quoted in Susan Strasser, *Waste and Want: A Social History of Trash*, New York: Henry Holt, 1999, p. 10.

23. Some argue that unlike manufacturing jobs, service jobs are low-skilled, low-paid, and temporary jobs. In fact, service jobs range from such low-skilled, employment to professional and managerial jobs that are highly skilled, excessively compensated, and buttressed by substantial nonwage benefits. See Power's argument in Michael Thomas Power, *Lost Landscapes and Failed Economies: The Search for a Value of Place*, Washington, D.C.: Island Press, 1996.

24. von Weizacker et al., *Factor Four*, 1997, p. 102.

25. See Nevin Cohen, "Greening the Internet: Ten Ways E-Commerce Could Affect the Environment," *Environmental Quality Management*, 9 (1), Autumn 1999, pp. 1–16; and Douglas A. Blackmon, "FedEx CEO Smith Bets His Deal Will Recast the Future of Shipping," *Wall Street Journal*, November 4, 1999.

26. See Robert Braile, "E-commerce May Be Helping the Environment," *Boston Globe*, December 14, 1999 p. c1; and Nevin Cohen, "Keys to the Kingdom," *Tomorrow*, 6 (4), November–December, 1999, pp. 25–30.

27. An expression of this perspective as a self-conscious movement can be found in Diane Elgin, *Voluntary Simplicity: Toward a Way of Life that is Outwardly Simple and Inwardly Rich*, rev. ed., New York: William Morrow, 1991. For a more recent perspective, see Robert Lilienfeld and William Rathje, *Use Less Stuff: Environmental Solutions for Who We Really Are*, New York: Ballantine, 1998.

28. This and other cautions about how dematerialization is measured and whether observed trends are not cyclical are raised by Oliviero Bernardini and Riccardo Galli in "Dematerialization: Long-Term Trends in the Intensity of Use of Materials and Energy," *Futures*, 25 (4), May 1993, pp. 431–448.

29. Friedrich Hinterberger, Fred Luks, and Friedrich Schmidt-Bleek, "Material Flows vs. Natural Capital: What Makes an Economy Sustainable," *Ecological Economics*, 23 (1), October 1997, pp. 1–14.

30. Mathis Wackernagel and William Rees, *Our Ecological Footprint: Reducing Human Impact on the Earth*, Gabriola Island, Canada: New Society Publishers, 1996.

31. Frederich Schmidt-Bleek, *Carnoules Declaration of the Factor Ten Club*, Wuppertal, Germany: Wuppertal Institute for Climate, Environment, and Energy, 1994. Others have argued for a factor of 20. For a review, see Lucas Reijnders, "The Factor X Debate: Setting Targets for Eco-Efficiency," *Journal of Industrial Ecology*, 2 (1), 1998, pp. 13–23.

32. Because such a large share of U.S. industrial metals is imported (100 percent for some metals), this would also require a reduction in the rate of metal imports as well. The specific effects on the global metals markets and various metal-exporting countries need more careful consideration. This is one of those points where the U.S. focus of this study creates frustrating limits.

33. Jim Glenn, "The State of Garbage in America," *BioCycle*, 40 (4), April, 1999, p. 60. If wood and plastic wastes are included, the amount of potentially degradable organic wastes rises to 75 percent of the waste stream, or 160 million tons per year. See U.S. Environmental Protection Agency, Office of Solid Waste and Emergency Response, *Characterization of Municipal Solid Waster in the United States, 1996 Update,* Washington. D.C., 1997. These wastes might have a higher value as soil nutrients for agriculture, although they are little used for this purpose today. See Gary Gardner, *Recycling Organic Waste: From Urban Pollutant to Farm Resource*, Worldwatch Paper No. 135, Washington, D.C.: Worldwatch Institute, August 1997.

34. U.S. Geological Survey, *Minerals Yearbook: Metals and Minerals, 1997*, Washington, D.C.: U.S. Government Printing Office, 1999, p. JJJ5.

35. The Labor Institute in New York advocates a transfer payment of this type for all workers who are negatively affected by environmental protection programs. See Michael Merrill, "Superfund for Workers," *New Solutions*, 1 (3), 1991, pp. 9–12.

Chapter 14

1. See Rachel Carson, *Silent Spring*, New York: Ballantine 1962; and Theo Colborn, Diane Dubanowski, and John Peterson Myers, *Our Stolen Future: Are We Threatening Our Fertility, Intelligence, and Survival?—A Scientific Detective Story*, New York: Dutton (Penguin), 1996.

2. H. Dana Moran, "Substitution—Some Practical Considerations," in U.S. Congress, Office of Technology Assessment, *Engineering Implications of Chronic Materials Scarcity,* Washington, D.C.: U.S. Government Printing Office, 1977.

3. Jason King's, *Separation Processes* (New York: McGraw-Hill, 1971) provides a conventional introduction to these techniques. For other examples, see Lawrence Smith, Jeffery Means, and Adwin Barth, *Recycling and Reuse of Industrial* Wastes, Columbus, Ohio: Battelle Press, 1995.

4. W. Wesley Eckenfelder, Alan Bowers, and John Roth, *Chemical Oxidation, Technologies for the Nineties*, Lancaster, Pa.: Technomic, 1994.

5. See Michael LaGrega, Phillip Buckingham, and Jeffrey Evans, *Hazardous Waste Management*, New York: McGraw-Hill, 1994, pp. 555–570.

6. See various papers in Daphne Stoner, ed., *Biotechnology for the Treatment of Hazardous Waste*, Boca Raton, Fla.: Lewis, 1993.

7. The Commission set limits on vinyl chloride and methylene chloride in propellants, required labels on products containing asbestos, and banned the use of

urea formaldehyde in insulation (although this was overturned by an appellate court). For a brief review, see Robert Gottlieb, *Reducing Toxics: A New Approach to Policy and Industrial Decisionmaking*, Washington, D.C.: Island Press, 1995, pp. 110–119.

8. For example, see Caroline Mayer, "Toys Safe to Melt in the Mouth?", *Washington Post*, December 8, 1999, P. EA.

9. For a good review of labeling, see Monica Hale, "Ecolabeling and Cleaner Production: Principles, Problems, Education and Training in Relation to the Adoption of Environmentally Sound Production Processes," *Journal of Cleaner Production*, 4 (2), 1996, pp. 85–95. For the California program see California Environmental Protection Agency, Office of Environmental Health Hazard Assessment, *The Implementation of Proposition 65: A Progress Report, Sacramento, Calif., July, 1992*; and Bette Fishbein, and Caroline Geld, *Making Less Garbage: A Planning Guide for Communities*, New York: INFORM, 1992, pp. 144–145.

10. See James Salzman, "Informing the Green Consumer," *Journal of Industrial Ecology*, 1 (2), 1997, pp. 11–21.

11. This Danish proposal is identified in Finn Bro-Rasmussen, "Chemical Hazards Are a Serious Matter—Too Serious to be Left for Chemists to Administer," *Journal of Cleaner Production*, 5 (3), 1997, pp. 183–186.

12. The history of the struggle for access to information on workplace chemicals is covered in Tim Morse, "Dying to Know: A Historical Analysis of the Right to Know Movement," *New Solutions*, 8 (1), 1998, pp. 117–145.

13. For example, see Debra Lynn Dadd, *Non-Toxic and Natural: How to Avoid Dangerous Everyday Products and Buy or Make Safer Ones*, Boston: Houghton Mifflin, 1984; and John Elkington and Julia Hailes, *The Green Consumers Supermarket Shopping Guide*, London: Victor Gollancz, 1989.

14. See U.S. Office of the President, Executive Order 12873, "Federal Acquisition, Recycling and Waste Prevention," Washington, D.C., October 20, 1993. See also U.S. Environmental Protection Agency, Office of Pollution Prevention and Toxics, *Selling Environmental Products to the Federal Government*, Washington, D.C., May 1997.

15. Several of the Swedish chemical prohibitions have required special exemptions from European Union directives that are less restrictive. See Swedish Ministry of the Environment, National Chemical Inspectorate, *Swedish Restrictions Benefit the Environment*, Solna, Sweden, October 1997. See also United Nations Environment Program, International Registry of Potentially Toxic Chemicals, *Consolidated List of Products Whose Consumption and/or Sale Have Been Banned, Withdrawn, Severely Restricted, on Not Approved by Governments*, 6th ed., Geneva, 1997.

16. For the Swedish chemical phaseout proposal, see Bo Wahlstrom, "Sunset for Dangerous Chemicals," *Nature*, 341, 1989, p. 276; and Swedish Ministry of the Environment, National Chemical Inspectorate, *Risk Reduction of Chemicals*, Solna, Sweden, 1991.

17. A reflective study of the ban on Alar in apple orchards suggests that reduction in consumer health risks may have been offset by increases in occupational health impacts. See Beth Rosenberg, "Unintended Consequences: Impacts of Pesticides on Industry, Workers, the Public and the Environment," unpublished doctoral dissertation, Department of Work Environment, University of Massachusetts Lowell, 1995

18. States with pollution prevention laws that encourage reduced use of toxics include New Jersey, Massachusetts, Oregon, Washington, Maine, and Minnesota. For example, see the Massachusetts General Court, *Massachusetts Toxics Use Reduction Act of 1989*, Chapter 21I of the Massachusetts General Laws, Boston, Mass., 1989. This law requires firms using large quantities of toxic chemicals to report annually on their use of those chemicals and to periodically plan procedures for reducing the use of those chemicals.

19. Mark Rossi, Michael Ellenbecker, and Kenneth Geiser, "Techniques in Toxics Use Reduction: From Concept to Action," *New Solutions*, 2 (2), Fall 1991, pp. 25–32. For a review of the Massachusetts program, see Monica Becker and Ken Geiser, "Massachusetts Tries to Cut Toxics Chemical Use," *Environmental Science and Technology*, 31 (12), September 1997, pp. 564–567.

20. Although the objectives of programs to reduce use of toxics, prevent pollution, and encourage cleaner production differ somewhat, the techniques are quite similar. For good texts, see Harry M. Freeman, ed., *Industrial Pollution Prevention Handbook*, New York: McGraw-Hill, 1995; and David T. Allen and Kirsten Sinclair Rosselot, *Pollution Prevention for Chemical Processes*, New York: Wiley, 1997.

21. Massachusetts Toxics Use Reduction Institute, "VOC Lacquer Replacement for Wood Finishing," unpublished technical report, No. 19, University of Massachusetts Lowell, 1994.

22. See Joseph E. Paluzzi and Timothy Greiner, "Finding Green in Clean: Progressive Pollution Prevention at Hyde Tool," *Total Quality Environmental Management*, 2 (3), Spring 1993, pp. 283–290; and Massachusetts Toxics Use Reduction Institute, "Cyanide Reduction in Bright Stripping Using an Electrolytic Process," unpublished technical report No. 18, University of Massachusetts Lowell, 1994.

23. See Joseph J. Breen and Michael J. Dellarco, eds., *Pollution Prevention in Industrial Processes: The Role of Process Analytical Chemistry*, Washington, D.C.: American Chemical Society, 1992.

24. See Allen and Rosselot, *Pollution Prevention*, p. 237.

25. U.S. Environmental Protection Agency, Office of Pollution Prevention and Toxics, *Cleaner Technologies Substitutes Assessment: A Methodology and Resource Guide*, Washington, D.C., 1996. The SNAP program is described in "Safe New Alternatives Program (Proposed Rule)," *Federal Register*, 58 (90), May 12, 1993, pp. 28094–28192.

26. For an extensive review, see J. A. Cano-Ruiz and G. J. McRae, "Environmentally Conscious Chemical Process Design," *Annual Review of Energy and Environment*, 23, 1998, pp. 499–536; Allen and Rosselot, *Pollution Prevention*, pp. 383–394; and James W. Mitchell, "Alternative Starting Materials for Industrial Processes," *Proceedings of the National Academy of Sciences*, 89, 1992, pp. 821–826.

27. The most recent Chemical Category is for Persistent, Bioaccumulative, and Toxic Substances.

28. See Stephen C. DeVito and Roger Garrett, eds., *Designing Safer Chemicals: Green Chemistry for Pollution Prevention*, Washington. D.C.: American Chemical Society, 1996; Paul Anastas and Tracy Williamson, eds., *Green Chemistry: Designing Chemistry for the Environment*, Washington, D.C.: American Chemical Society, 1998; and Paul T. Anastas and John C. Warner, *Green Chemistry: Theory and Practice*, New York: Oxford University Press, 1998.

29. Anastas and Warner, *Green Chemistry*, pp. 11 and 30.

30. Joseph J. Bozell, John O. Hoberg, David Claffey, Bonnie R. Hamesm and Donald R. Dimmel, "New Methodology for the Production of Chemicals from Renewable Feedstocks," in Paul T. Anastas and Tracy C. Williamson, eds., *Green Chemistry: Frontiers in Benign Chemical Syntheses and Processes*, New York: Oxford University Press, 1998, pp. 27–45; and Scott Sieburth, "Isosteric Replacement of Carbon with Silicon in the Design of Safer Chemicals," in DeVito and Garrett, *Designing Safer Chemicals*, pp. 74–83.

31. Allen and Rosselot, *Pollution Prevention*, p. 210.

32. Paul T. Anastas and Tracy C. Williamson, "Frontiers in Green Chemistry," in Anastas and Williamson, *Green Chemistry: Frontiers ...*, p. 10.

33. Chris Ryan argues that a factor 20 improvement in materials efficiency should not be beyond reach. See "Designing for Factor 20 Improvements," *Journal of Industrial Ecology*, 2 (2), 1998, pp. 3–5.

34. U.S. Environmental Protection Agency, Office of Pollution Prevention and Toxics, *1997 Toxics Release Inventory*, Washington, D.C., 1999; and Massachusetts Department of Environmental Protection, *1998 Toxics Use Reduction Information Release*, Boston, Mass. Spring 2000.

Chapter 15

1. Stanford Research Institute's *Chemical Economics Handbook* (Palo Alto, Calif.) provides an excellent example of a private data management resource.

2. See Allen Hammond, Albert Adriaanse, Eric Rodenberg, D. Bryant, and R. Woodward, *Environmental Indicators: A Systematic Approach to Measuring and Reporting Environmental Policy Performance in the Context of Sustainable Development*; Washington, D.C.: World Resources Institute, 1995; Bedrich

Moldan, Suzanne Billharz, and Robyn Matravers, eds., *Sustainability Indicators: Report of the Project on Indicators of Sustainable Development*, New York: Wiley, 1997; and U.S. Interagency Working Group on Sustainable Development Indicators, *Sustainable Development in the United States: An Experimental Set of Indicators, Washington*, D.C., December 1998.

3. See Organization for Economic Cooperation and Development, *Environmental Indicators*, Paris, 1994; Albert Adriaanse, Stefan Bringezu, Allen Hammond, Yuichi Moriguchi, Eric Rodenberg, Donald Rogich, and Helmut Schutz, *Resource Flows: The Material Basis of Industrial Economies*, Washington, D.C.: World Resources Institute, 1997; and Emily Matthews et al., *The Weight of Nations: Material Outflows from Industrial Economies*, Washington, D.C.: World Resources Institute, 2000.

4. Stefan Bringezu, "Accounting for the Physical Basis of National Economies: Material Flow Indicators," in Moldan et al., *Sustainability Indicators*, pp. 170–180.

5. World Commission on Environment and Development, *Our Common Future*, New York: Oxford University Press 1987; Robert Repetto, *Wasting Assets*, Washington, D.C.: World Resources Institute, 1989; and Herman E. Daly and John B. Cobb, Jr., *For the Common Good: Redirecting the Economy Toward Community, Environment and a Sustainable Future*, Boston: Beacon Press, 1989.

6. See United Nations, *System of Integrated Environmental and Economic Accounting*, New York: United Nations, 1993; Ismail Serageldin, *Sustainability and the Wealth of Nations: First Steps in an On-going Journey*, Washington, D.C.: World Bank, 1996; and *Expanding the Measures of Wealth: Indicators of Environmentally Sustainable Development*, Washington, D.C.: World Bank, 1997.

7. The quantitative estimate comes from Norman Myers and Jennifer Kent, *Perverse Subsidies: Tax Dollars Undercutting Our Economies and Environments Alike*, Winnipeg: International Institute for Sustainable Development, 1998, p. 60.

8. See Arthur C. Pigou, *The Economics of Welfare*, London: Macmillan, 1920, Allen V. Kneese and Orris C. Herfindahl, *Economic Theory of Natural Resources*, New York: Charles Merrill, 1974; and Albert M. Church, *Taxation of Non-Renewable Resources*, Lexington, Mass.: D. C. Heath, 1981.

9. For a recent review, see Paul Elkins, "European Environmental Taxes and Charges: Recent Experiences, Issues and Trends," *Ecological Economics*, 31, 1999, pp. 39–62.

10. See Robert Repetto, Roger C. Dower, Robin Jenkins, and Jacqueline Geoghegan, *Green Fees: How a Tax Shift Can Work for the Environment and the Economy*, Washington, D.C.: World Resources Institute, 1992; and David Malin Roodman, *The Natural Wealth of Nations: Harnessing the Market for the Environment*, New York: Norton, 1998.

11. See David Malin Roodman, *Getting the Signals Right: Tax Reform to Protect the Environment and the Economy*, Washington, D.C.: Worldwatch Institute, May 1997; and European Environment Agency, *Environmental Taxes: Implementation and Environmental Effectiveness*, Copenhagen, 1996.

12. This tax, which generated $4.1 billion for the federal budget, rose from $3.02 per kilogram in 1990 to $11.80 per kilogram by 1996 when chlorofluorocarbon use was phased out. See Roodman, *Getting the Signals Right*, p. 15.

13. These new approaches are driven by cost, efficiency, reduction of risk and liability, customer demands, and regulatory pressures, many of which are emerging from European governments. They are being promoted by a host of professional advocates, including Paul Hawken, William McDonough, Amory and Hunter Lovins, Ray Anderson of Interface, and the proponents of the "Natural Step," an ecological framework developed and promoted by a Swedish physician, Karl-Henrik Robert. See Paul Hawken, *The Ecology of Commerce*, New York: Harper Collins, 1993; Carl Frankel, *In Earth's Company: Business, Environment, and the Challenge of Sustainability*, Babriola Island, Canada: New Society, 1998; Paul Hawken, Amory Lovins, and L. Hunter Lovins, *Natural Capitalism: Creating the Next Industrial Revolution*, Boston: Little, Brown, 1999.

14. The most popular environmental management system promoted in the United States has been developed by the International Standards Organization (ISO) and is called "ISO 14000." See James L. Lamprecht, *ISO 14000: Issues and Implementation Guidelines for Environmental Management*, New York: American Manufacturing Association, 1997.

15. See Kevin Brady, Patrice Henson, and James A. Fava, "Sustainability, Eco-Efficiency, Life-Cycle Management, and Business Strategy," *Environmental Quality Management*, 8 (3), Spring 1999, pp. 33–40. Life-cycle design has proved similar benefits for IBM, Bristol-Myers Squibb, and Armstrong. See Karen Shapiro and Allen White, "Product Stewardship through Life Cycle Design, *Corporate Environmental Strategy*, 6 (1), Winter 1999, pp. 15–23.

16. See David Heeney and Laura Murphy, "From Waste Management to Knowledge Management," *Environmental Quality Management*, 9 (1), Autumn 1999, pp. 81–85; and David Rejeski, "Organizational Learning and Change: The Forgotten Dimensions of Sustainable Development," *Corporate Environmental Strategy*, 3 (1), Summer 1995, pp. 19–29.

17. See National Academy of Sciences, *Science: The Endless Frontier*, Washington, D.C.: National Science Foundation, 1945. Alternative models are described in Richard Nelson and Sidney Winter, *An Evolutionary Theory of Economic Change*, Cambridge, Mass.: Harvard University Press, 1982; Eric von Hippel, *The Sources of Innovation*, New York: Oxford University Press, 1988; and James M. Utterback, *Mastering the Dynamics of Innovation*, Boston: Harvard Business School Press, 1996.

18. For further development, see Lewis M. Branscomb, ed., *Empowering Technology: Implementing a U.S. Strategy,* Cambridge, Mass.: MIT Press, 1993.

19. Nicholas Ashford and George R. Heaton, "Regulation and Technology Innovation in the Chemical Industry," *Law and Contemporary Problems,* 46 (3), 1983, pp. 109–157.

20. See Swedish Ministry of the Environment, Chemicals Policies Committee, *Towards a Sustainable Chemicals Policy,* Stockholm, 1997; and Wayne M. Shinya, "Canada's New Minerals and Metals Policy," *Resources Policy,* 24 (2), 1998, pp. 95–104. For the Dutch and Austrian plans, see Lucas Reijnders, "The Factor X Debate: Setting Targets for Eco-Efficiency." *Journal of Industrial Ecology,* 21 (1), 1998, p. 14.

21. A 1993 World Resources Institute study warns that public sector institutions, even more than private institutions, may undercut progress toward a sustainable economy. See *Focus Group Report: Institutions,* Washington, D.C.: World Resources Institute 1993. For a critical look at the EPA's limits see Marc K. Landy, Marc J. Roberts, and Stephen R. Thomas, *The Environmental Protection Agency: Asking the Wrong Questions, From Nixon to Clinton,* 2nd ed., New York: Oxford University Press, 1994.

22. This is what Wendell Berry has meant by the new "home economics." For a forceful argument, see David Orr, *Ecological Literacy,* Albany, N.Y.: State University of New York Press, 1992.

23. Early versions of these citizens' advisory committees were inspired by similar committees set up to advise on federal urban renewal and community anti-poverty programs during the 1960s. See Sherry Arnstein, "A Ladder of Citizen Participation," *Journal of the American Institute of Planners,* 35, 1969, pp. 216–224. For an overview, see Francis Lynn and Jack Kartez, "The Redemption of Citizen Advisory Committees: A Perspective from Critical Theory," in Ortin Renn, Thomas Webler, and Peter Wiedemann, eds., *Fairness and Competence in Citizen Participation: Evaluating Models for Environmental Discourse,* Dordrecht, The Netherlands: Kluwer Academic Press, 1995, pp. 87–101.

24. Ned Crosby, "Citizen Juries: One Solution for Difficult Environmental Questions," in Renn et al. *Fairness and Competence,* pp. 157–174. A caution is offered by Klaus Peter Rippe and Peter Schaber, who argue that citizen panels will never be as effective as simply encouraging more direct democracy. See "Democracy and Environmental Decision Making," *Environmental Values,* 8 (1), 1999, pp. 75–88.

25. U.S. President's Materials Policy Commission, *Resources for Freedom,* 5 vols., Washington, D.C.: U.S. Government Printing Office. 1952, Vol. 1, p. 2.

Bibliography

The text relies on a broad array of secondary materials from a variety of different fields. Therefore a bibliography must mix together many quite diverse citations. In order to facilitate a review of the references, they are presented in two sections. The first covers the materials used in preparing the more historical chapters, specifically chapters 2, 4, 5, and 6. The second covers the more contemporary literature cited in the remaining chapters.

Historical Chapters (chapters 2, 4, 5, and 6)

Adams, E. M., H. C. Spenser, V. K. Rowe, D. D. McCollister and D. D. Irish, "Vapor Toxicity of Trichloroethane," *Archives of Industrial Hygiene and Occupational Medicine*, 1950, pp. 225–236.

Adams, E. M., H. C. Spenser, V. K. Rowe, D. D. McCollister, and D. D. Irish, "Vapor Toxicity of Trichloroethylene," *Archives of Industrial Hygiene and Occupational Medicine*, 4 (4), 1951, pp. 469–481.

Adams, E. M., H. C. Spenser, V. K. Rowe, D. D. McCollister, and D. D. Irish, "Vapor Toxicity of Carbon Tetrachloride," *Archives of Industrial Hygiene and Occupational Medicine*, 6 (1), 1952, pp. 50–66.

Amato, Ivan, *Stuff: The Materials the World Is Made Of*, New York: Avon, 1997.

Anderson, Nils, Jr., and Mark W. DeLawyer, *Chemicals, Metals and Men: Gas, Chemicals and Coke: A Bird's Eye View of the Materials that Make the World Go Around*, New York: Vantage, 1995.

Andrews, Richard N. L., *Managing the Environment, Managing Ourselves: A History of American Environmental Policy*, New Haven Conn.: Yale University Press, 1999.

Armstrong, Ellis C., Michael Robinson, and Suellen Hoy, *History of Public Works in the United States, 1776–1976*, Chicago: American Public Works Association, 1976.

Arora, Ashish, and Nathan Rosenberg, "Chemicals: A U.S. Success Story," in Ashish Arora, Ralph Landeau and Nathan Rosenberg, *Chemicals and Long-Term Economic Growth: Insights from the Chemical Industry*, New York: Wiley, 1998, pp. 71–102.

Arora, Ashish, Ralph Landeau, and Nathan Rosenberg, *Chemicals and Long Term Economic Growth: Insights from the Chemical Industry*, New York: Wiley, 1998.

Banks, Ferdinand E., *Scarcity, Energy and Economic Progress*, Lexington, Mass.: Lexington, 1977.

Barnett, Harold, and Chandler Morse, *Scarcity and Growth: The Economics of Natural Resource Availability*, Baltimore: John Hopkins University Press, for Resources for the Future, 1965.

Beyer, D. S., *Industrial Accident Prevention*, Boston: Houghton-Mifflin, 1916.

Bijker, Wiebe E., *Of Bicycles, Bakelites, and Bulbs: Toward a Theory of Sociotechnical Change*, Cambridge, Mass.: MIT Press, 1997.

Blumberg, Louis, and Robert Gottlieb, *War on Waste: Can America Win its Battle with Garbage?*, Washington, D.C.: Island Press, 1989.

Boulding, Kenneth, "The Economics of the Coming Spaceship Earth" in Kenneth Boulding and Henry Jarrett, eds., *Environmental Quality in a Growing Economy: Essays for the Sixth RFF Forum*, Baltimore: John Hopkins University Press, for Resources for the Future, 1966, pp. 3–14.

Boulding, Kenneth, and Henry Jarrett, eds., *Environmental Quality in a Growing Economy: Essays for the Sixth RFF Forum*, Baltimore: John Hopkins University Press, for Resources for the Future, 1966.

Broudeur, Paul, *Outrageous Misconduct: The Asbestos Industry on Trial*, New York: Pantheon, 1985.

Campbell, Murray, and Harrison Hatton, *Herbert H. Dow: Pioneer in Creative Chemistry*, New York: Appleton-Century-Crofts, 1951.

Carr, Henry, *Our Domestic Poisons: Poisonous Effects of Certain Dyes and Colours*, London: Ridgeway, 1883.

Chadwick, Michael J., and Jan Nilsson, "Environmental Quality Objectives—Assimilative Capacity and Critical Load Concepts in Environmental Management," in Tim Jackson, ed., *Clean Production Strategies: Developing Preventive Environmental Management in the Industrial Economy*, Boca Raton, Fla.: Lewis, 1993, pp. 29–39.

Chandler, Alfred D., Jr., and Stephen Salsbury, *Pierre S. du Pont and the Making of the Modern Corporation*, New York Harper & Row, 1971.

Chandler, Alfred, Jr., Takashi Hikino, and David Mowery, "The Evolution of Corporate Capabilities and Corporate Strategy and Structure within the World's Largest Chemical Firms: The Twentieth Century in Perspective," in Ashish Arora, Ralph Landau and Nathan Rosenberg, *Chemicals and Long-Term Eco-*

nomic Growth: Insights from the Chemical Industry, New York: Wiley, 1998, pp. 415–457.

Chapman, Keith, *The International Petrochemical Industry*, Oxford: Basil Blackwell, 1991.

Clarkson Grosvenor B., *Industrial America in the World War: The Strategy Behind the Lines, 1917–1918*, Boston: Houghton-Mifflin, 1923.

Cleary, Edward J., "Determining Risks of Toxic Substances in Water," *Sewage and Industrial Wastes*, 26, February 1954, pp. 203–209.

Colten, Craig, "A Historial Perspective on Industrial Wastes and Groundwater Contamination," *Geographic Review*, 81, 1991, pp. 215–228.

Colten, Craig, "Creating a Toxic Landscape: Chemical Waste Disposal Policy and Practice, 1900–1960," *Environmental History Review*, 18 (1), Spring 1994, pp. 85–116.

Colten, Craig E., and Peter N. Skinner, *The Road to Love Canal: Managing Industrial Waste before EPA*, Austin, Texas: University of Texas Press, 1996.

Commission on Organization of the Executive Branch of the Government, *Report on Natural Resources*, Washington: U.S. Government Printing Office, 1949.

Commoner, Barry, *The Closing Circle*, New York: Knopf, 1971.

Corn, Jacqueline K., *Response to Occupational Health Hazards: A Historical Perspective*, New York: Van Nostrand Reinhold, 1992.

Davis, Lee Niedrinhaus, *The Corporate Alchemists: Profit Takers and Problem Makers in the Chemical Industry*, New York: William Morrow, 1984.

Dewhurst, J. Frederick, and Associates, *America's Needs and Resources*, New York: Twentieth Century Fund, 1947.

Dilsaver, Larry M., and Craig E. Colten, eds., *The American Environment: Interpretations of Past Geographies*, Lanham, Md.: Rowman and Littlefield, 1992.

Dowie, Mark, *Losing Ground, American Environmentalism at the Close of the Twentieth Century*, Cambridge, Mass.: MIT Press, 1996.

Duffy, John, *The Sanitarians: A History of American Public Health*, Urbana, Ill.: University of Illinois Press, 1990.

Eckes, Alfred E., Jr., *The United States and the Global Struggle for Materials*, Austin, Texas: University of Texas Press, 1979.

Elkington, John, *The Poisoned Womb*, New York: Viking, 1985.

Epstein, Samuel S., *The Politics of Cancer*, New York: Anchor/Doubleday, 1979.

Epstein, Samuel S., Lester O. Brown, and Carl Pope, *Hazardous Waste in America*, San Francisco: Sierra Club Books, 1982.

Etheridge, Elizabeth, W., *Sentinels for Health: A History of the Centers for Disease Control*, Berkeley, Calif.: University of California Press, 1992.

Forrestal, Dan, *Faith, Hope and $5,000: The Story of Monsanto: The Trials and Triumphs of the First 75 Years*, New York: Simon & Schuster, 1977.

Fox, Stephen, *The American Conservation Movement: John Muir and His Legacy*, Madison, Wisc., University of Wisconsin Press, 1985.

Francaviglia, Richard V., "Mining and Landscape Transformation," in Larry M. Dilsaver and Craig E. Colten, *The American Environment: Interpretations of Past Geographies*, Lanham, Md.: Rowman and Littlefield, 1992, pp. 89–114.

Galishoff, Stuart, *Safeguarding the Public Health: Newark, 1895–1918*, Westport, Conn.: Greenwood, 1975.

Gallo, Michael A., "History and Scope of Toxicology," in Curtis D. Klassen, ed., *Casarett and Doull's Toxicology: The Basic Science of Poisons*, 5th ed., New York: McGraw-Hill, 1996, pp. 3–11.

Georgescu-Roegen, Nicholas, *The Entropy Law and the Economic Process*, Cambridge, Mass.: Harvard University Press, 1971.

Gottlieb, Robert, *Forcing the Spring: The Transformation of the American Environmental Movement*, Washington, D.C.: Island Press, 1993.

Grinder, R. Dale, "The Battle for Clean Air: The Smoke Problem in Post-Civil War America," in Martin V. Melosi, ed., *Pollution and Reform in American Cities*, Austin, Texas: University of Texas Press, 1980, pp. 83–104.

Haber, Ludwig F., *The Chemical Industry in the Nineteenth Century*, New York: Oxford University Press, 1958.

Hamilton, Alice, *Industrial Poisons in the United States*, New York: Macmillan, 1925.

Halloway, Steven Kendall, *The Aluminum Multinationals and the Bauxite Cartel*, London: Macmillan, 1988.

Harden, Victoria, *Inventing the NIH: Federal Biomedical Research Policy, 1887–1937*, Baltimore: Johns Hopkins University Press, 1986.

Harwood, Julius, "Emergence of the Field and Early Hopes," in Rustum Roy, ed., *Materials Science and Engineering in the United States*, University Park, Pa.: Pennsylvania State University Press, 1970.

Hays, Samuel P., *Conservation and the Gospel of Efficiency*, Cambridge, Mass.: Harvard University Press, 1959.

Hays, Samuel P., "From Conservation to Environmental Politics in the United States Since World War Two," *Environmental History Review*, 6 (2), Fall 1982, pp. 14–41.

Hays, Samuel P., *Beauty, Health and Permanence: Environmental Politics in the United States, 1955–1985*, New York: Cambridge University Press, 1987.

Haynes, William, *American Chemical Industry*, Vols. 1–6, New York: Van Nostrand, 1945–1954.

Haynes, William, and Edward Goody, *Chemical Industry's Contribution to the Nation; 1635–1935, A Record of Chemical Accomplishments, with an Index of the Chemicals Made in America*, supplement to *Chemical Industries*, May 1935.

Heilbroner, Robert, *An Inquiry into the Human Prospect*, New York: Norton, 1980.

Hering, Rudolph, "Disposal of Municipal Refuse," *Transactions of the American Society of Civil Engineers*, 54, 1904, pp. 265–308.

Hoffmann, George R., "Genetic Toxicology," in Curtis D. Klaassen, ed., *Casarett and Doull's Toxicology: The Basic Science of Poisons*, 5th ed., New York: McGraw-Hill, 1996, pp. 269–300.

Hogan, William T., *Economic History of the Iron and Steel Industry in the United States*, Vol. 1, Lexington, Mass: D.C. Heath, 1971.

Hounshell, David A., and J. Kenley Smith, Jr., *Science and Corporate Strategy: Du Pont R&D, 1902–1980*, Cambridge, Mass.: Cambridge University Press, 1988.

Ickes, Harold, "The War and Our Vanishing Resources," *American Magazine*, 140, December 1945, p. 20.

Jackson, Tim, ed., *Cleaner Production Strategies: Developing Preventive Environmental Management in the Industrial Economy*, Boca Raton, Fla.: Lewis, 1993.

Klaassen, Curtis D. ed., *Casarett and Doull's Toxicology: The Basic Science of Poisons*, 5th ed., New York: McGraw-Hill, 1996.

Landy, Mark K., Marc J. Roberts, and Stephen R. Thomas, *The Environmental Protection Agency: Asking the Wrong Questions, From Nixon to Clinton*, New York: Oxford University Press, 1994.

Leavitt, Judith Walzer, *The Healthiest City: Milwaukee and the Politics of Health Reform*, Princeton, N.J.: Princeton University Press, 1982.

Leith, Charles K., *World Minerals and World Politics: A Factual Study of Minerals in their Political and International Relations*, New York: McGraw-Hill, 1931.

Meadows, Donella H., Dennis L. Meadows, Jorgen Randers, and William H. Behrens III, *The Limits to Growth*, New York: Universe Books, 1972.

Melosi, Martin V., *Garbage in the Cities: Refuse Reform and the Environment, 1880–1980*, College Station, Texas: Texas A&M University Press, 1981.

Melosi, Martin V., ed., *Pollution and Reform in American Cities*, Austin, Texas: University of Texas Press, 1980.

Miller, Char, ed., *American Forests: Nature, Culture, and Politics*, Lawrence, Kan.: University of Kansas Press, 1997.

Miller, Elizabeth, and James Miller, "The Presence and Significance of Bound Aminoazo Dyes in the Livers of Rats fed p-Dimethyllaminoazobenzene," *Cancer Research*, 7, 1947, pp. 468–480.

Mining and Metallurgical Society of America, *International Control of Minerals*, New York, 1925.

Morris, Peter, *The American Synthetic Rubber Research Program*, Philadelphia: University of Pennsylvania Press, 1989.

Mossman, Susan T., "Parkesine and Celluloid," in Susan T. Mossman and Peter J. Morris, eds., *The Development of Plastics*, London: Royal Society of Chemistry, 1994, pp. 10–25.

Mossman, Susan T., and Peter J. Morris, eds., *The Development of Plastics*, London: Royal Society of Chemistry, 1994.

Mullan, Fitzhugh, *Plagues and Politics: The Story of the United States Public Health Service*, New York: Basic Books, 1989.

National Academy of Sciences, *Science: The Endless Frontier*, Washington, D.C., 1945.

National Academy of Sciences, Committee on Mineral Resources and the Environment, *Mineral Resources and the Environment*, Washington, D.C.: National Academy Press, 1975.

National Academy of Sciences, Committee on the Survey of Material Sciences and Engineering, *Materials and Man's Needs*, Washington, D.C.: National Academy Press, 1974.

National Academy of Sciences/National Academy of Engineering, *Elements of a National Materials Policy*, A Report of the National Materials Advisory Board, Washington, D.C., 1972.

National Academy of Sciences/National Academy of Engineering, *Man, Materials and the Environment*, Cambridge, Mass.: MIT Press, 1973.

National Resources Committee, Special Advisory Committee on Water Pollution, *Report on Water Pollution*, Washington, D.C., July 1935.

Noble, Charles, *Liberalism at Work: The Rise and Fall of OSHA*, Philadelphia: Temple University Press, 1986.

Noble, David, *America by Design: Science, Technology and the Rise of Corporate Capitalism*, New York: Knopf, 1977.

Nriagu, J. O., "Global Metal Pollution—Poisoning the Bioshere," *Environment*, 32 (7), 1990, pp. 7–32.

Oliver, Thomas, *Dangerous Trades*, London: John Murray, 1902.

Ophuls, William, *Ecology and the Politics of Scarcity*, San Francisco: W. H. Freeman, 1977.

Ozonoff, David, "One Man's Meat, Another Man's Poison: Two Chapters in the History of Public Health," *American Journal of Public Health*, 75 (4), April 1985, pp. 338–340.

Packard, Vance, *The Waste Makers*, New York: David McKay, 1960.

Pehrson, Elmer, "What are Strategic and Critical Materials?", *Mining and Metallurgy*, 25, July 1944, pp. 339–341.

Pehrson, Elmer, "The Mineral Position of the United States and the Outlook for the Future," *Annual Report of the Board of Regents of the Smithsonian Institution*, Washington, D.C., 1945, pp. 175–199.

Phelps, Earle B., *Studies of the Treatment and Disposal of Industrial Wastes*, U.S. Public Health Service, *Public Health Bulletin, No. 97*, Washington, D.C.: U.S. Government Printing Office, 1918.

Pinchot, Gifford, *The Fight for Conservation*, New York: Doubleday, 1910.

Pisani, Donald, "Fish Culture and the Dawn of Concern over Water Pollution in the United States," *Environmental Review*, 8, 1984, pp. 117–131.

Pitot, Henry C. III, and Yvonne P. Dragan, "Chemical Carcinogenesis," in Curtis D. Klassen, ed., *Casarett and Doull's Toxicology: The Basic Science of Poisons*, 5th ed., New York: McGraw-Hill, 1996, pp. 201–267.

"Progress Report of the Committee on Industrial Wastes in Relationship to Water Supplies," *Journal of the American Water Works Association*, 10, 1923, pp. 415–430.

Resource Conservation Committee, *Choices for Conservation: Final Report to the President and Congress*, Washington, D.C., July 1979.

Resources for the Future, *The Nation Looks at its Resources: Report on the Mid-Century Conference on Resources for the Future*, December 2, 3, 4, 1953, Washington, D.C., 1954.

Rickard, Thomas A., *The History of American Mining*, New York: McGraw-Hill, 1932.

Rosen, George, *A History of Public Health*, expanded ed. Baltimore: Johns Hopkins University Press, 1993.

Rosenkrantz, Barbara G., *Public Health and the State: Changing Views in Massachusetts, 1842–1936*, Cambridge, Mass.: Harvard University Press, 1972.

Rosner, David, and Gerald E. Markowitz, eds., *Dying for Work: Worker Safety and Health in Twentieth Century America*, Bloomington, Ind.: Indiana University Press, 1987.

Seldin, Joseph, *The Golden Fleece: Selling the Good Life to Americans*, New York: Mcmillan, 1963.

Sellers, Christopher, "Factory as Environment: Industrial Hygiene, Professional Collaboration and the Modern Sciences of Pollution," *Environmental History Review*, 18 (1) Spring 1994, pp. 55–83.

Smith, Duane, *Mining America: The Industry and the Environment, 1800–1980*, Lawrence, Kan.: University of Kansas Press, 1987.

Somerfield, Thomas, Thomas Oliver, and Felix Putzeys, "List of Industrial Poisons," *Bulletin of the U.S. Bureau of Labor, No. 86*, January 1910, pp. 147–168.

Spenser, H. C., D. D. Irish, E. M. Adams, and V. K. Rowe, "Response of Laboratory Animals to Monomeric Styrene," *Journal of Industrial Hygiene and Toxicology*, 24 (10), 1942, pp. 295–301.

Spenser, H. C., V. K. Rowe, E. M. Adams, D. D. McCollister, and D. D. Irish, "Vapor Toxicity of Ethylene Dichloride Determined by Experiments in Laboratory Animals," *Archives of Industrial Hygiene and Occupational Medicine*, 4 (5), 1951, pp. 482–493.

Spitz, Peter, *Petrochemicals: The Rise of an Industry*, New York: Wiley, 1988.

Stern, Arthur C., "History of Air Pollution Legislation in the United States," *Journal of the Air Pollution Control Association*, 32 (1), January 1982, pp. 44–61.

Tarr Joel, *The Search for the Ultimate Sink: Urban Pollution in Historical Perspective*, Akron, Ohio: University of Akron Press, 1996.

Tarr, Joel, "Industrial Wastes, Water Pollution, and Public Health, 1876–1962," in Joel Tarr, *The Search for the Ultimate Sink: Urban Pollution in Historical Perspective*, Akron, Ohio: University of Akron Press, 1996, pp. 354–384.

Trescott, Martha Moore, *The Rise of the American Electrochemicals Industry, Studies in the American Technological Environment*, Westport, Conn.: Greenwood, 1981.

U.S. Bureau of Labor Statistics, "Occupation Hazards and Diagnostic Signs," *Bulletin of the U.S. Bureau of Labor Statistics*, No. 306, 1922; and "Occupation Hazards and Diagnostic Signs," *Bulletin of the U.S. Bureau of Labor Statistics*, No. 582, 1933.

U.S. Congress, *Mining and Mineral Policy Act of 1970*, P.L. 91-631, Washington, D.C., 1970.

U.S. Congress, *National Materials Policy Act of 1970*, P.L. 91-152, Washington, D.C., 1970.

U.S. Congress, *Federal Water Pollution Control Act (Clean Water Act) of 1972*, P.L. 92-500, Washington, D.C., 1972.

U.S. Congress, *Toxic Substances Control Act*, P.L. 94-469, Washington, D.C., 1976.

U.S. Congress, General Accounting Office, *Federal Materials Research and Development: Modernizing Institutions and Management*, Washington, D.C.: U.S. Government Printing Office, 1975.

U.S. Congress, House of Representatives, Committee on Science and Technology, *House Report 97-761*, Part 1, Washington, D.C., August 18, 1982.

U.S. Congress, Office of Technology Assessment, *Serious Reduction of Hazardous Waste*, Washington, D.C.: U.S. Government Printing Office, 1986.

U.S. Congress, Senate Committee on Public Works, *Hearings*, Washington, D.C., June 11–13, July 9–11, 15–18, 1974.

U.S. Environmental Protection Agency, Office of Solid Waste and Emergency Response, *Report to Congress: Minimization of Hazardous Waste*, Washington, D.C., October 1986.

U.S. Environmental Protection Agency, Office of Solid Waste and Emergency Response, *Municipal Waste Combustion Study, Report to Congress*, Washington, D.C., 1987.

U.S. Geological Survey, *Mineral Commodity Summaries, 1998*, Washington, D.C., 1998.

U.S. National Commission on Materials Policy, *Material Needs and the Environment: Today and Tomorrow*, Washington, D.C.: U.S. Government Printing Office, 1974.

U.S. National Commission on Supplies and Shortages, *Government and the Nation's Resources*, Washington, D.C.: U.S. Government Printing Office, 1976.

U.S. National Conservation Commission, *Report of the National Conservation Commission*, 3 vols., Washington, D.C.: U.S. Government Printing Office, 1909.

U.S. National Conservation Commission, *Report*, Senate Document No. 676, 60th Congress, Washington, D.C., 1910.

U.S. Office of the President, *Proceedings of a Conference of Governors in the White House*, Document No. 48489, Washington, D.C.: U.S. Government Printing Office, 1908.

U.S. President's Materials Policy Commission, *Resources for Freedom*, 5 vols., Washington, D.C.: U.S. Government Printing Office, 1952.

von Oettingen, W. F., "The Halogenated Hydrocarbons: Their Toxicity and Potential Dangers," *Journal of Industrial Hygiene and Toxicology*, 19, 1937, pp. 342–349.

von Oettingen, W. F., "Toxicity and Potential Dangers of Aliphatic and Aromatic Hydrocarbons," U.S. Public Health Service, *Public Health Bulletin*, No. 255, 1940, pp. 1–135.

von Oettingen, W. F., "The Aliphatic Alcohols: Their Toxicity and Potential Dangers in Relationship to Their Chemical Constitution and Their Fate in Metabolism," U.S. Public Health Service, *Public Health Bulletin*, No. 281, 1943, pp. 1–253.

von Oettingen, W. F., and W. Deichmann-Gruebler, "The Toxicology and Potential Dangers of Crude 'Duprene'," *Journal of Industrial Hygiene and Toxicology*, 18, 1936, pp. 271–272.

von Oettingen, W. F., W. Deichmann-Gruebler, and W. C. Hueper, "Toxicity and Potential Dangers of Phenyl-Hydrazine Zinc Chloride," *Journal of Industrial Hygiene and Toxicology*, 18, 1936, pp. 301–309.

von Oettingen, W. F., et al., "The Toxicity and Potential Dangers of Toluene, with Special Reference to Its Maximal Permissible Exposure," U.S. Public Health Service, *Public Health Bulletin*, No. 279, 1942, pp. 1–158.

Weinberg, Robert A., *Racing to the Beginning of the Road: The Search for the Origin of Cancer*, New York: Harmony, 1996.

Whitaker, Jennifer Seymour, *Salvaging the Land of Plenty: Garbage and the American Dream*, New York: William Morrow, 1994.

Whitehead, Don, *The Dow Story: The History of the Dow Chemical Company*, New York: McGraw-Hill, 1968.

Whorton, James, *Before Silent Spring: Pesticides and Public Health in Pre-DDT America*, Princeton, N.J.: Princeton University Press, 1974.

Wiley, F. H., W. C. Hueper, and W. F. von Oettingen, "On the Effects of Low Concentrations of Carbon Disulfide," *Journal of Industrial Hygiene and Toxicology*, 18, 1936, pp. 733–740.

Wilson, Charles Morrow, *Trees and Test Tubes: The Story of Rubber*, New York: Henry Holt, 1943.

Wilson, James G., *Environment and Birth Defects*, New York: Academic Press, 1973.

Wolf, George, *Chemical Induction of Cancer*, London: Cassell and Company, 1952.

Wolf, M. A., V. K. Rowe, D. D. McCollister, R. L. Hollinsworth, and F. Oyen, "Toxicological Studies of Certain Alkylated Benzenes and Benzene," *AMA Archives of Industrial Health*, 14, 1956, pp. 387–398.

Worster, Donald, *Nature's Economy: A History of Ecological Ideas*, Cambridge: Cambridge University Press, 1977.

Wrather, W. E., "Mineral Resources and Mineral Resourcefulness," *Mining and Metallurgy*, 27, August 1946, pp. 427–428.

Yergin, Daniel, *The Prize: The Epic Quest for Oil, Money and Power*, New York: Simon & Schuster, 1991.

Zaslowsky, Dyan, *These American Lands*, New York: Henry Holt, 1986.

Contemporary Chapters (chapters 1, 3, 7, 8, 9, 10, 11, 12, 13, 14, and 15)

Ackerman Frank, *Why do We Recycle: Markets, Values and Public Policy*, Washington, D.C.: Island Press, 1997.

Adriaanse, Albert, Stefan Bringezu, Allen Hammond, Yuichi Moriguchi, Eric Rodenberg, Donald Rogich, and Helmut Schutz, *Resource Flows: The Material Basis of Industrial Economies*, Washington, D.C.: World Resources Institute, 1997.

Allen, David, and Nasrin Bemanesh, "Wastes as Raw Materials," in Barden R. Allenby and Deanna J. Richards, eds., *The Greening of Industrial Ecosystems*, Washington, D.C.: National Academy Press, 1994, pp. 69–89.

Allen, David T., and Kirsten Sinclair Rosselot, *Pollution Prevention for Chemical Processes*, New York: Wiley, 1997.

Allenby, Barden R., *Industrial Ecology: Policy Framework and Implementation*, Upper Saddle River, N.J.: Prentice Hall, 1999.

Allenby, Barden R., and Deanna J. Richards, eds., *The Greening of Industrial Ecosystems*, Washington, D.C.: National Academy Press, 1994.

Altieri, Miquel A., ed., *Agroecology: The Science of Sustainable Agriculture*, Boulder, Col.: Westview, 1989.

Amato, Ivan, *Stuff: The Materials the World Is Made Of*, New York: Avon Books, 1997.

American Soybean Association, *Soy Stats, 1999*, St. Louis, Mo., 1999.

Anastas, Paul T., and John C. Warner, *Green Chemistry: Theory and Practice*, New York: Oxford University Press. 1998.

Anastas, Paul T., and Tracy C. Williamson, "Frontiers in Green Chemistry," in Paul T. Anastas and Tracy C. Williamson, eds., *Green Chemistry: Frontiers in Benign Chemical Syntheses and Processes*, New York: Oxford University Press, 1998, pp. 1–25.

Anastas, Paul T., and Tracy C. Williamson, eds., *Green Chemistry: Frontiers in Benign Chemical Syntheses and Processes*, New York: Oxford University Press, 1998.

Anastas, Paul T., and Tracy Williamson, eds., *Green Chemistry: Designing Chemistry for the Environment*, Washington, D.C.: American Chemical Society, 1998.

Arnstein, Sherry, "A Ladder of Citizen Participation," *Journal of the American Institute of Planners*, 35, 1969, pp. 216–224.

Ashford, Nicholas, and George R. Heaton, "Regulation and Technology Innovation in the Chemical Industry," *Law and Contemporary Problems*, 46 (3), 1983, pp. 109–157.

Attilio, Bisio, and Sharon Boots, eds., *Encyclopedia of Energy, Technology and the Environment*, New York: Wiley, 1995.

Ausubel, Jesse H., and Hedy E. Sladovich, eds., *Technology and Environment*, Washington, D.C.: National Academy Press, 1989.

Ayres, Robert U., and Leslie W. Ayres, *Industrial Ecology: Towards Closing the Materials Cycle*, Brookfield, Vt: Edward Elgar, 1996.

Ayres, Robert U., and Paul M. Weaver, eds., *Eco-restructuring: Implications for Sustainable Development*, New York: United Nations University Press, 1998.

Azzi, G., *Agricultural Ecology*, London: Constable, 1956.

Bagby, Marvin O., "Products from Vegetable Oils: Two Examples," in Glen Fuller, Thomas McKeon, and Donald Bills, eds., *Agricultural Materials as Renewable Resources: Non-Food and Industrial Applications*, Washington, D.C.: American Chemical Society, 1996, pp. 248–257.

Balazik, Ronald F., "Analysis of New Material Issues," in U.S. Bureau of Mines, *The New Materials Society*. Vol. 1: *New Materials, Markets and Issues*, Washington, D.C.: U.S. Government Printing Office, 1990, pp. 3.37–3.39.

Ball, Phillip, *Made to Measure: New Materials for the 21st Century*, Princeton, N.J.: Princeton University Press, 1997.

Becker, Monica, and Ken Geiser, "Massachusetts Tries to Cut Toxics Chemical Use," *Environmental Science and Technology*, 31 (12), September 1997, pp. 564–567.

Berger, Lev I., *Semiconductor Materials*, Boca Raton, Fla.: CRC Press, 1997.

Bernardini, Oliviero, and Riccardo Galli, "Dematerialization: Long-Term Trends in the Intensity of Use of Materials and Energy," *Futures*, 25 (4), May 1993, pp. 431–448.

Benyus, Janine M., *Biomimicry: Innovation Inspired by Nature*, New York: William Morrow, 1997.

Betts, W. B., ed., *Biodegradation: Natural and Synthetic Materials*, London: Springer-Verlag, 1991.

Bever, Michael B., ed., *Concise Encyclopedia of Materials, Economics, Policy and Management*, New York: Pergamon, 1993.

Blackmon, Douglas A., "FedEx CEO Smith Bets His Deal Will Recast the Future of Shipping," *Wall Street Journal*, November 4, 1999.

Blumenthal, Michael, "What's New With Ground Rubber," *BioCycle*, 39 (3), March 1998, pp. 40–44.

Borth, Christy, *The Automobile*, Washington, D.C.: Automobile Manufacturers Association, 1952.

Boulding, Kenneth, "The Economics of the Coming Spaceship Earth," in Kenneth E. Boulding and Henry Jarrett, eds., *Environmental Quality in a Growing Economy: Essays for the Sixth RFF Forum*, Baltimore: John Hopkins University Press, for Resources for the Future, 1966, pp. 3–14.

Boyle, Stewart, "Crossed Wires," *Tomorrow*, 10 (3), May–June 2000, pp. 14–15.

Bozell, Joseph J., John O. Hoberg, David Claffey, Bonnie R. Hamesm, and Donald R. Dimmel, "New Methodology for the Production of Chemicals from Renewable Feedstocks," in Paul T. Anastas and Tracy C. Williamson, eds., *Green Chemistry: Frontiers in Benign Chemical Syntheses and Processes*, New York: Oxford University Press, 1998, pp. 27–45.

Brady, Kevin, Patrice Henson, and James A. Fava, "Sustainability, Eco-Efficiency, Life-Cycle Management, and Business Strategy," *Environmental Quality Management*, 8 (3), Spring 1999, pp. 33–40.

Braile, Robert, "E-commerce May Be Helping the Environment," *Boston Globe*, December 14, 1999, p. C1.

Braile, Robert, "U.S. Starts a Phase Out of Suspected Gas Additive," *Boston Globe*, March 21, 2000, p. A1.

Branscomb, Lewis M., ed., *Empowering Technology: Implementing a U.S. Strategy*. Cambridge, Mass.: MIT Press, 1993.

Breen, Joseph J., and Michael J. Dellarco, eds., *Pollution Prevention in Industrial Processes: The Role of Process Analytical Chemistry*, Washington, D.C.: American Chemical Society, 1992.

Brenniman, Gary R., Stephen D. Cooper, William H. Hallenbeck, and James M. Lyznicki, "Recycling: Automotive and Household Batteries," in Frank Krieth, ed., *Handbook of Solid Waste Management*, New York: McGraw-Hill, 1994, pp. 9.149–9.178.

Bringezu, Stefan, "Accounting for the Physical Basis of National Economies: Material Flow Indicators," in Bedrich Moldan, Suzanne Billharz, and Robyn

Matravers, eds., *Sustainability Indicators: Report on the Project on Indicators of Sustainable Development*, New York: Wiley, 1997, pp. 170–180.

Bullard, Robert D., ed., *Environmental Justice and Communities of Color*, San Francisco: Sierra Club Books, 1993.

Bullis, L. Harold, and James E. Mielke, *Strategic and Critical Materials*, Boulder, Col.: Westview P, 1985.

Butterman, William C., and Richard S. Gillette, "Markets for New Materials," in U.S. Bureau of Mines, *The New Materials Society. Vol. 1: New Materials, Markets and Issues*, Washington, D.C.: U.S. Government Printing Office, 1990, pp. 2.7–2.9

Brookes, D. B., and P. W. Andrews, "Mineral Resources, Economic Growth, and World Population," *Science*, 185, 1974, pp. 13–20.

Bro-Rasmussen, Finn, "Chemical Hazards Are a Serious Matter—Too Serious to be Left for Chemists to Administer," *Journal of Cleaner Production*, 5 (3), 1997, pp. 183–186.

Brown, Lester R., Edward C. Wolfe, and Linda Starke, eds., *State of the World—1987*, New York: W. W. Norton, 1987.

California Environmental Protection Agency, Office of Environmental Health Hazard Assessment, *The Implementation of Proposition 65: A Progress Report*, Sacramento, Calif., July 1992.

Cameron, James, and Timothy O'Riordan, eds., *Interpreting the Precautionary Principle*, London: Cameron and May, 1994.

Cano-Ruiz, J. A., and G. J. McRae, "Environmentally Conscious Chemical Process Design," *Annual Review of Energy and Environment*, 23, 1998, pp. 499–536.

Carlson, D. Kenneth, and Donald Van Dyne, eds., *Industrial Uses for High Erucic Acid Oils from Crambe and Rapeseed*, University of Missouri-Columbia, Extension and Agricultural Information Office, October, 1992.

Carroll, Ronald, John H. Vandermeer, and Peter M. Rosset, eds., *Agroecology*, New York: McGraw-Hill, 1990.

Carson, Rachel, *Silent Spring*, New York: Ballantine, 1962.

Chandler, W. U., *Materials Recycling: The Virtue of Necessity*, Worldwatch Paper No. 56, Washington, D.C.: Worldwatch Institute, 1983.

Chapman, Keith, *The International Petrochemical Industry: Evolution and Location*, Oxford: Basil Blackwell, 1991.

Chum, Halena L., ed., *Polymers from Biobased Materials*, Park Ridge, N.J.: Noyes Data Corporation, 1991.

Church, Albert M., *Taxation of Non-Renewable Resources*, Lexington, Mass.: D.C. Heath, 1981.

Clarke, Margaret A., and Leslie A. Edye, "Sugar Beet and Sugarcane as Renewable Resources," in Glen Fuller, Thomas McKeon and Donald Bills, eds.,

Agricultural Materials as Renewable Resources: Now-Food and Industrial Applications, Washington, D.C.: American Chemical Society, 1996, pp. 229–247.

Cleveland, Cutler J., and Matthias Ruth, "Indicators of Dematerialization and the Materials Intensity of Use," *Journal of Industrial Ecology*, 2 (3), 1999, pp. 15–50.

Cohen, Morris, ed., "Materials Science and Engineering: Its Evolution, Practice and Prospects," special ed., *Materials Science and Engineering*, 37 (1), January, 1979.

Cohen, Nevin, "Greening the Internet: Ten Ways E-Commerce Could Affect the Environment," *Environmental Quality Management*, 9 (1), Autumn 1999, pp. 1–16.

Cohen, Nevin, "Keys to the Kingdom," *Tomorrow*, 6 (4), November–December 1999, pp. 25–30.

Colborn, Theo, Diane Dubanowski, and John Peterson Myers, *Our Stolen Future: Are We Threatening Our Fertility, Intelligence, and Survival?—A Scientific Detective Story*, New York: Dutton, 1996.

Commoner, Barry, *Making Peace with the Planet*, New York: Pantheon, 1990.

Commoner, Barry, "The Relationship Between Industrial and Ecological Systems," *Journal of Cleaner Production*, 5 (1–2), 1997, pp. 125–129.

Corson, Walter, ed., *The Global Ecology Handbook*, Boston: Beacon Press, 1990.

Costanza, Robert, John Cumberland, Herman Daly, Robert Goodland, and Richard Norgaard, *An Introduction to Ecological Economics*, Boca Raton, Fla.: St. Lucie Press, 1997.

Crosby, Ned, "Citizen Juries: One Solution for Difficult Environmental Questions," in Ortin Renn, Thomas Webler, and Peter Wiedemann, eds., *Fairness and Competence in Citizen Participation: Evaluating Models for Environmental Discourse*, Dordrecht, Netherlands: Kluwer Academic Press, 1995, pp. 157–174.

Crueger, Wulf, and Anneliese Crueger, *Biotechnology: A Textbook of Industrial Microbiology*, 2nd ed., Sunderland, Mass.: Sinauer Associates, 1990.

Curran, Mary Ann, ed., *Environmental Life Cycle Assessment*, New York: McGraw-Hill, 1996.

Dadd, Debra Lynn, *Non-Toxic and Natural: How to Avoid Dangerous Everyday Products and Buy or Make Safer Ones*, Boston: Houghton Mifflin, 1984.

Dally, Gretchen C., ed., *Nature's Services: Societal Dependence on Natural Ecosystems*, Washington, D.C.: Island Press, 1997.

Dalton, G. E., *Study of Agricultural Systems*, London: Applied Sciences, 1975.

Daly, Herman E., "Entropy, Growth, and the Poltical Economy of Scarcity," in V. Kerry Smith, ed., *Scarcity and Growth Reconsidered*, Baltimore: Johns Hopkins University Press for Resources for the Future, 1979, pp. 67–94.

Daly, Herman E., "Toward Some Operational Principles of Sustainable Development," *Ecological Economics*, 2, 1990, pp. 1–6.

Daly, Herman E., *Beyond Growth: The Economics of Sustainable Development*, Boston: Beacon Press, 1996.

Daly, Herman E., and John B. Cobb, Jr., *For the Common Good: Redirecting the Economy Toward Community, Environment and a Sustainable Future*, Boston: Beacon Press, 1989.

Dart, R. K., and W. B. Betts, "Uses and Potential of Lignocellulose," in W. B. Betts, ed., *Biodegradation: Natural and Snythetic Materials*, London: Springer-Verlag, 1991, pp. 201–217.

Davies, J. Clarence, and Jan Mazurek, *Pollution Control in the United States: Evaluating the System*, Baltimore: Johns Hopkins University Press for Resources for the Future, 1998.

Davis, Gary A., Catherine A. Witt, and Jack N. Barkenbus, "Extended Product Responsibility: A Tool for a Sustainable Economy," *Environment*, 39 (7), September 1997, pp. 10–15, 36–37.

DeVito, Stephen C., and Roger Garrett, eds., *Designing Safer Chemicals: Green Chemistry for Pollution Prevention*, Washington. D.C.: American Chemical Society, 1996.

Doll, Richard, and Richard Peto, "Avoidable Risks of Cancer in the U.S.," *Journal of the National Cancer Institute*, 66 (6), June 1981, pp. 1191–1308.

Drexler, K. Eric, "The Coming Era of Nanotechnology," in Tom Forester, ed., *The Materials Revolution: Superconductors, New Materials and the Japanese Challenge*, Cambridge, Mass.: MIT Press, 1988, pp. 361–370.

Drexler, K. Eric, and Chris Peterson, *Unbounding the Future: The Nanotechnology Revolution*, New York: William Morrow, 1991.

Dunford, Bruce, "Hemp Tested as Possible Crop for Hawaii," *Boston Globe*, December 16, 1999, p. A26.

Eckenfelder, W. Wesley, Alan Bowers, and John Roth, *Chemical Oxidation, Technologies for the Nineties*, Lancaster, Pa.: Technomic, 1994.

Eckes, Alfred, E., Jr., *The United States and the Global Struggle for Minerals*, Austin, Texas: University of Texas Press, 1979.

Eggert, Roderick, "The Passenger Car Industry: Faithful to Steel," in John E. Tilton, ed., *World Metal Demand: Trends and Prospects*, Baltimore: Johns Hopkins University Press for Resources for the Future, 1990, pp. 161–216.

Eherenfeld, David W., *Conserving Life on Earth*, New York: Oxford University Press, 1972.

Ehrenfeld, John, and Nicholas Gertler, "Industrial Ecology in Practice: The Evolution of Interdependence at Kalunborg," *Journal of Industrial Ecology*, 1 (1), 1997, pp. 67–79.

Elgin, Diane, *Voluntary Simplicity: Toward a Way of Life that is Outwardly Simple and Inwardly Rich*, rev. ed., New York: William Morrow, 1991.

Elkington, John, and Julia Hailes, *The Green Consumers Supermarket Shopping Guide*, London: Victor Gollancz, 1989.

Elkins, Paul, "European Environmental Taxes and Charges: Recent Experiences, Issues and Trends," *Ecological Economics*, 31, 1999, pp. 39–62.

Environmental Defense Fund, *Toxic Ignorance*, Washington, D.C., 1997.

Erkman, S., "Industrial Ecology: An Historical View," *Journal of Cleaner Production*, 5 (1–2), pp. 1–10.

Eskanazi, B., E. B. Gold, S. J. Samuels, et al., "Prospective Monitoring of Early Fetal Loss and Clinical Spontaneous Abortion Among Female Semiconductor Workers," *American Journal of Industrial Medicine*, 26, 1995, pp. 833–846.

European Environment Agency, *Environmental Taxes: Implementation and Environmental Effectiveness*, Copenhagen, Denmark, 1996.

Ferber, Dan, "GM Crops in the Cross Hairs," *Science*, 286, November 26, 1999, pp. 1662–1666.

"Fiber Revolution," *The Carbohydrate Economy* (a publication of the Institute of Local Self-Reliance, Washington, D.C.), 1 (1), Fall 1997, p. 1, 13–14.

Fiksel, Joseph, ed., *Design for the Environment: Creating Eco-Efficient Products*, New York: McGraw-Hill, 1996.

Fishbein, Bette, and Caroline Geld, *Making Less Garbage: A Planning Guide for Communities*, New York: INFORM, 1992.

Forester, Tom, ed., *The Materials Revolution: Superconductors, New Materials, and the Japanese Challenge*, Cambridge, Mass.: MIT Press, 1988.

Forrestal, Dan J., *Faith, Hope and $5,000: The Story of Monsanto*, New York: Simon & Schuster, 1977.

"Fox Fibers," *Industry and Environment* (a publication of the United Nations Environment Program), 20 (1–2), January–June 1997, p. 53.

Frankel, Carl, *In Earth's Company: Business, Environment, and the Challenge of Sustainability*, Babriola Island, Canada: New Society Publishers, 1998.

Franklin Associates, *The Role of Recycling in Industrial Solid Waste Management in the Year 2000*, Stamford, Conn.: Keep America Beautiful, 1994.

Fransman, Martin, Gerd Junne, and Annemieke Roobeek, eds., *The Biotechnology Revolution?*, Oxford, U.K.: Blackwell, 1995.

Freeman, A. Myrick, *Air and Water Pollution Control: A Benefits-Cost Assessment*, New York: Wiley, 1982.

Freeman, Harry M., ed., *Industrial Pollution Prevention Handbook*, New York: McGraw-Hill, 1995.

Friends of the Earth, *Failing Grades: Clean Water Report Card*, Washington, D.C., March 2000.

Frosh, Robert A., "Towards the End of Waste: Reflections on a New Ecology of Industry," *Daedalus*, 125 (3), Summer 1996, pp. 199–212.

Frosch, Robert A., and Nicholas E. Gallopoulos, "Strategies for Manufacturing," *Scientific American*, 261 (3), September 1989, pp. 144–153.

Fuller, Glen, Thomas McKeon, and Donald Bills, eds., *Agricultural Materials as Renewable Resources: Non-Food and Industrial Applications*, Washington, D.C.: American Chemical Society, 1996.

Gardner, Gary, *Recycling Organic Waste: From Urban Pollutant to Farm Resource*, Worldwatch Paper No. 135, Washington D.C.: Worldwatch Institute, August 1997.

Gardner, Gary, and Payal Sampat, *Mind Over Matter: Recasting the Role of Materials in Our Lives*, Worldwatch Paper No. 144, Washington, D.C.: Worldwatch Institute, December 1998.

Ghiradella, Helen, Daniel Aneshansley, Thomas Eisner, Robert E. Silberglied, and Howard E. Hinton, "Ultraviolet Reflection of a Male Butterfly," *Science*, 178, 1972, pp. 214–216.

Giaini, Orio, ed., *Dialogue on Wealth and Welfare: An Alternative View of World Capital Formation*, Oxford, U.K.: Pergamon Press, 1980.

Glaser, Lewrence, and Gregory Gajewski, "Agricultural Products for Industry: The Situation and Outlook," *International Journal of Environmentally Conscious Design and Manufacturing*, 3 (2), 1994, pp. 63–66.

Glenn, Jim, "The State of Garbage in America," *BioCycle*, 40 (4), April 1999, pp. 60–71.

Goldstein, Nora, "The State of Garbage in America," *BioCycle*, 38 (4), April 1997, pp. 60–67.

Goldstein, Nora, and Kevin Gray, "Biosolids Composting in the United States," *BioCycle*, 40 (1), January 1999, pp, 63–75.

Gottlieb, Robert, *Forcing the Spring: The Transformation of the American Environmental Movement*, Washington, D.C.: Island Press, 1993.

Gottlieb, Robert, *Reducing Toxics: A New Approach to Policy and Industrial Decisionmaking*, Washington, D.C.: Island Press, 1995.

Graedel, Thomas E., and Braden R. Allenby, *Design for the Environment*; Upper Saddle River, N.J.: Prentice Hall, 1996.

GrassRoots Recycling Network, *Welfare for Waste: How Federal Taxpayer Subsidies Waste Resources and Discourage Recycling*, Atlanta, Ga., April 1999.

GrassRoots Recycling Network and Institute for Local Self Reliance (Brenda Platt and Neil Seldman), *Wasting and Recycling in the United States 2000*, Athens, Ga., 2000.

Grohmann, Karel, and Michael E. Himmel, "Advances in Protein-Derived Materials," in Helena L. Chum, ed., *Polymers from Biobased Materials*, Park Ridge, N.J.: Noyes Data Corporation, 1991, pp. 144–153.

Hale, Monica, "Ecolabeling and Cleaner Production: Principles, Problems, Education and Training in Relation to the Adoption of Environmentally Sound Production Processes," *Journal of Cleaner Production*, 4 (2), 1996, pp. 85–95.

Hale, William J., *The Farm Chemurgic*, Boston: Stratford Press, 1934.

Hale, William J., *Farmer Victorious: Money, Mart, and Mother Earth*, New York: Coward-McCann, 1949.

Hammond, A., A. Adriaanse, E. Rodenberg, D. Bryant, and R. Woodward, *Environmental Indicators: A Systematic Approach to Measuring and Reporting Environmental Policy Performance in the Context of Sustainable Development*; Washington, D.C.: World Resources Institute, 1995.

Harrington, Winston, *The Regulatory Approach to Air Quality Management: A Case Study of New Mexico*, Baltimore: Johns Hopkins University Press for Resources for the Future, 1981.

Hart, Stuart L., and Mark B. Milstein, "Global Sustainability and the Creative Destruction of Industries," *Sloan Management Review*, 41 (1), Fall 1999, pp. 23–33.

Hasenbein, Eric A., "The Semiconductor Industry's Goal of 100% Elimination of 2-Methoxyethanol, 2-Ethoxyethanol and their Acetates by 1995 Compared to Other Industries," *Semiconductor Safety Association Journal*, 9 (2), June 1995, pp. 9–21.

Hatwick, John, and Nancy Olewiler, *The Economics of Natural Resource Use*, New York: Harper & Row, 1986.

Hawken, Paul, *The Ecology of Commerce*, New York: Harper Collins, 1993.

Hawken, Paul, Amory Lovins, and L. Hunter Lovins, *Natural Capitalism: Creating the Next Industrial Revolution*, Boston: Little, Brown, 1999.

Hazeltine, C. W., "A Millenium of Fungi, Food, and Fermentation," *Mycologia*, 57, March/April 1965, pp. 149–197.

Herman, Robert, Siamak A. Ardekani, and Jesse H. Ausubel, "Dematerialization," in Jesse H. Ausubel and Hedy E. Sladovich, eds., *Technology and Environment*, Washington, D.C.: National Academy Press, 1989, pp. 50–69.

Heeney, David, and Laura Murphy, "From Waste Management to Knowledge Management," *Environmental Quality Management*, 9 (1), Autumn 1999, pp. 81–85.

Hilliard, Thomas J., *Golden Patents, Empty Pockets: A Nineteenth Century Law Gives Miners Billions, the Public Pennies*, Washington, D.C.: Mineral Policy Center, 1994.

Hinterberger, Fredrich, Fred Luks, and Friedrich Schmidt-Bleek, "Material Flows vs. Natural Capital: What Makes an Economy Sustainable," *Ecological Economics*, 23 (1), October 1997, pp. 1–14.

Hirschorn, Joel S., "Why the Pollution Prevention Revolution Failed—and Why it Ultimately Will Succeed," *Pollution Prevention Review*, 7 (1), Winter 1997, pp. 11–31.

Hoerger, Fred, "Indicators of Exposure Trends," in Richard Peto and Marvin Schneiderman, eds, *Quantification in Occupational Cancer*, Cold Springs Harbor, N.Y.: Cold Springs Laboratory Press, 1981.

Holly, James H., "Materials Flow of Zinc in the United States, 1850–1990," *Resources, Conservation and Recycling*, 9 (1–2), 1993, pp. 1–30.

Horrigan, John B., Frances H. Irwin and Elizabeth Cook, *Taking a Byte Out of Carbon: Electronics Innovation for Climate Protection*, Washington, D.C.: World Resources Institute, 1998.

Hutzinger, Otto, Stephen Safe, and V. Zitko, *The Chemistry of PCBs*, Cleveland, Ohio: CRC, 1974.

Hyndman, Paul C., and Harry W. Campbell, *Digital Databases Containing Mining Claim Density Information ... From the BLM Mining Claim Recordation System*, 1996, Washington, D.C.: U.S. Geological Survey, 1999.

Interrante, Leonard V., and Mark J. Hampden Smith, eds., *Chemistry of Advanced Materials*, New York: Wiley-VCH, 1998.

Jackson, Tim, *Material Concerns, Pollution, Profit and Quality of Life*, London: Routledge, 1996.

Jackson, Wes, *Altars of Unhewn Stone: Science and the Earth*, San Francisco: North Point, 1987.

Janicke, M., M. Binder, and H. Monch, "Dirty Industries: Patterns of Change in Industrial Countries," *Environmental and Resource Economics*, 6, 1997, pp. 35–36.

Janicke, M., H. Monch, T. Ranneberg, and U. E. Simonis, "Structural Change and Environmental Impact: Empirical Evidence on Thirty-One Countries in East and West," *Environmental Monitoring and Assessment*, 12, 1989, pp. 99–114.

Jarrett, Henry, ed., *Environmental Quality in a Growing Economy: Essays for the Sixth RFF Forum*, Baltimore: John Hopkins University Press for Resources for the Future, 1966.

Johnsen, Soren, "Report of a New Chemical Hazard," *New Scientist*, 32, 1966, p. 612.

Kane, Mari, "Growing Pains," *E-Magazine*, 10 (5), September 1997, pp. 37–41.

Kaplan, David, W. Wade Adams, Barry Farmer, and Christopher Viney, *Silk Polymers: Materials Science and Biotechnology*, Washington, D.C.: American Chemical Society, 1994.

Kaplan, R. S., and F. J. Schottman, "Recycling of Metals," in Michael B. Bever, ed., *Concise Encyclopedia of Materials, Economics, Policy and Management*, New York: Pergamon, 1993, pp. 293–302.

Kelly, Anthony, ed., *Concise Encyclopedia of Composite Materials*, New York: Pergamon Press, 1989.

Khan, Ishrat M., and Joycelyn S. Harrison, eds., *Field Responsive Polymers: Electroresponsive, Photoresponsive, and Responsive Polymers in Chemistry and Biology*, Washington, D.C.: American Chemical Society, 1999.

King, Jason, *Separation Processes*, New York: McGraw-Hill, 1971.

Klabunde, Kenneth J., and Cathy Mohs, "Nanoparticles and Nanostructural Materials," in Leonard V. Interrante and Mark J. Smith, eds., *Chemistry of Advanced Materials*, New York: Wiley-VCH, 1998, pp. 271–328.

Klages, K. H. W., *Ecological Crop Geography*, New York: Macmillan, 1942.

Kneese, Allen V., *Measuring the Benefits of Clean Air and Water*, Baltimore: Johns Hopkins University Press for Resources for the Future, 1984.

Kneese, Allen V., and Orris C. Herfindahl, *Economic Theory of Natural Resources*, New York: Charles Merrill, 1974.

Kneese, Allen V., Robert U. Ayres, and Ralph C. D'Arge, *Economics and the Environment: A Materials Balance Approach*, Baltimore: Johns Hopkins University Press for Resources for the Future, 1970.

Kondratiev, Nikolai D., "The Long Waves in Economic Life," *Review of Economic Statistics*, 17, 1935, pp. 105–115.

Koplow, Douglas, "Energy Subsidies and the Environment," in Organization for Economic Cooperation and Development, *Subsidies and the Environment: Exploring the Linkages*, Paris, 1996.

Koplow, Douglas, and Kevin Dietly, *Federal Disincentives: A Study of Federal Tax Subsidies and Other Programs Affecting Virgin Industries and Recycling*, Washington, D.C.: U.S. Environmental Protection Agency, 1994.

Kovacs, William L., "The Coming Era of Conservation and Industrial Utilization of Recyclable Materials," *Ecology Law Quarterly*, 1998, pp. 537–625.

Krieth, Frank, ed., *Handbook of Solid Waste Management*, New York: McGraw-Hill, 1994.

Krimsky, Sheldon, *Genetic Alchemy: The Social History of the Recombinant DNA Controversy*, Cambridge, Mass.: MIT Press, 1985.

Krimsky, Sheldon, *Biotechics and Society, the Rise of Industrial Genetics*, New York: Praeger, 1991.

Krimsky, Sheldon, *Hormonal Chaos: The Scientific and Social Origins of the Environmental Endrocrine Hypothesis*, Baltimore: John Hopkins University Press, 2000.

Kuznets, Simon S., *Toward a Theory of Economic Growth*, New York: Norton, 1968.

LaGrega, Michael, Phillip Buckingham, and Jeffrey Evans, *Hazardous Waste Management*, New York: McGraw-Hill, 1994.

Lagrone, B. D., and F. G. Smith, "Reclaiming of Elastomers," in Michael B. Bever, ed., *Concise Encyclopedia of Materials, Economics, Policy and Management*, New York: Pergamon, 1993, pp. 285–288.

Lamprecht, James L., *ISO 14000: Issues and Implementation Guidelines for Environmental Management*, New York: American Manufacturing Association, 1997.

Landis, Wayne G., and Ming-Ho Yu, *Introduction to Environmental Toxicology: Impacts of Chemicals Upon Ecological Systems*, 2nd ed., Boca Raton, Fla.: Lewis, 1999.

Landy, Marc K., Marc J. Roberts, and Stephen R. Thomas, *The Environmental Protection Agency: Asking the Wrong Questions, From Nixon to Clinton*, 2nd ed., New York: Oxford University Press, 1994.

Larson, Eric D., Marc H. Ross, and Robert H. Williams, "Beyond the Era of Materials," *Scientific American*, 254 (6), 1986, pp. 34–41.

Lee, Charles, *Toxic Waste and Race in the United States, A National Report on the Racial and Socioeconomic Characteristics of Communities with Hazardous Waste Sites*, New York: Commission for Racial Justice, United Church of Christ, 1987.

Legarth, Jens Brobech, "Sustainable Metal Resource Management—The Need for Industrial Development: Efficiency Improvement Demands On Metal Resource Management to Enable a (Sustainable) Supply until 2050," *Journal of Cleaner Production*, 4 (2), 1996, pp. 97–104.

Lehmus, Petri, and Bernard Rieger, "Nanoscale Polymerization Reactors for Polymer Fibers," *Science*, 285, September 24, 1999, pp. 2081–2082.

Leopold, Aldo, *A Sand County Almanac*, New York: Oxford University Press, 1966.

Levenstein, Charles, and John Wooding, eds., *Work, Health, and Environment: Old Problems, New Solutions*, New York: Guilford, 1997.

Lewis, David L., *Henry Ford: An American Folk Hero and His Company*, Detroit: Wayne State University Press, 1976.

Lewis, Michael, Russell Clark, Jeffery Vandall, and Neil Seldman, *Reuse Operations: Community Development Through Redistribution of Used Goods*, Washington, D.C.: Institute for Local Self-Reliance, 1995.

Lifset, Reid, "Take it Back: Extended Producer Responsibility as a Form of Incentive-Based Environmental Policy," *The Journal of Resource Management and Technology*, 21, 1993, p. 166.

Lilienfeld, Robert, and William Rathje, *Use Less Stuff: Environmental Solutions for Who We Really Are*, New York: Ballantine, 1998.

Lindhqvist, Thomas, "Extended Producer Responsibility in Cleaner Production," unpublished doctoral disssertation, International Institute for Industrial Environmental Economics, Lund University, Lund, Sweden, 2000.

Liu, Guojun, "Polymeric Nanostructures," in Hari Single Nalwa, *Handbook of Nanostructured Materials and Nanotechnology*, San Diego: Academic Press, 2000, pp. 475–500.

Lowe, Ernest A., "Creating By-Product Resource Exchanges for Eco-Industrial Parks," *Journal of Cleaner Production*, 5 (1–2), 1997, pp. 57–66.

Lowrance, Richard, Benjamin R. Stinner, and Garfield S. House, *Agricultural Ecosystems: Unifying Concepts*, New York: Wiley, 1984.

Lubove, Roy, *The Professional Altruist: The Emergence of Social Work as a Career, 1880–1930*, Cambridge, Mass.: Harvard University Press, 1965.

Lubove, Seth, "Enzyme-eaten Jeans," *Forbes*, 146 (10), October 29, 1990, pp. 140–141.

Lund, Robert T., "Remanufacturing," *Technology Review*, 87, 1984, pp. 18–23, 28–29.

Lynn, Francis, and Jack Kartez, "The Redemption of Citizen Advisory Committees: A Perspective from Critical Theory," in Ortin Renn, Thomas Webler, and Peter Wiedemann, eds., *Fairness and Competence in Citizen Participation: Evaluating Models for Environmental Discourse*, Dordrecht, Germany: Kluwer Academic Press, 1995, pp. 87–101.

MacAvoy, Paul W., "The Record of the Environmental Protection Agency in Controlling Industrial Air Pollution," in R. L. Gordon, H. D. Jacoby, and M. B. Zimmerman, eds., *Energy Markets and Regulation*, Cambridge, Mass.: Ballinger, 1987, pp. 107–137.

Malenbaum, Wilfred, *World Demand for Raw Materials in 1985 and 2000*, New York: McGraw-Hill, 1978.

Mann, Charles C., "Biotech Goes Wild," *Technology Review*, 102 (4), July–August, 1999, pp. 36–43.

Massachusetts Department of Environmental Protection, *1998 Toxics Use Reduction Information Release*, Boston, Mass., Spring 2000.

Massachusetts General Court, *Massachusetts Toxics Use Reduction Act of 1989*, Chapter 21I of the Massachusetts General Laws, Boston, Mass., 1989.

Massachusetts Toxics Use Reduction Institute, "Cyanide Reduction in Bright Stripping Using an Electrolytic Process," unpublished Technical Report No. 18, University of Massachusetts Lowell, 1994.

Massachusetts Toxics Use Reduction Institute, "VOC Lacquer Replacement for Wood Finishing," unpublished Technical Report, No. 19, University of Massachusetts Lowell, 1994.

Matos, Grecia, and Lori Wagner, "Consumption of Materials in the United States, 1900–1995," *Annual Review of Energy and Environment*, 23, 1998, pp. 107–122.

Matson, P. A., W. J. Parton, A. G. Power, and M. J. Smith, "Agricultural Intensification and Ecosystem Properties," *Science*, 277, July 25, 1997, pp. 504–509.

Matthews, Emily et al., *The Weight of Nations: Material Outflows from Industrial Economies*, Washington, D.C.: World Resources Institute, 2000.

Mayer, Carl J., and George A. Riley, *Public Domain, Private Dominiom: A History of Public Mineral Policy in the United States*, San Francisco: Sierra Club Books, 1985.

Mayer, Caroline, "Toys Safe to Melt in the Mouth?," *Washington Post*, December 8, 1999, p. E4.

McDivitt, James F., and Gerald Manners, *Minerals and Men*, Baltimore: John's Hopkins University Press for Resources for the Future, 1974.

McDonough, William, and Michael Braungart, "The Next Industrial Revolution," *Atlantic Monthly*, 282 (4), October 1998, pp. 82–91.

McKinley, Michael J., and staff, *Recent Market Trends of Recycled Metals in the United States*, Reston, Va: U.S. Geological Survey, 1999.

McNeill, J. R., *Something New Under the Sun: An Environmental History of the Twentieth Century World*, New York: Norton, 2000.

Mendeloff, John, *The Dilemma of Toxic Substance Regulation: How Overregulation Causes Underregulation*, Cambridge, Mass.: MIT Press, 1988.

Merrill, Michael, "Superfund for Workers," *New Solutions*, 1 (3), 1991, pp. 9–12.

Mintz, Joel A., *Enforcement at the EPA: High Stakes and Hard Choices*, Austin, Texas: University of Texas Press, 1995.

Misa, Thomas J., "Military Needs, Commercial Realities, and the Development of the Transistor, 1948–1958," in Merritt Roe Smith, ed., *Military Enterprise and Technological Change: Perspectives on the American Experience*, Cambridge, Mass.: MIT Press, 1985, pp. 253–287.

Mitchell, James W., "Alternative Starting Materials for Industrial Processes," *Proceedings of the National Academy of Sciences*, 89, 1992, pp. 821–826.

Moldan, Bedrich, Suzanne Billharz, and Robyn Matravers, eds., *Sustainability Indicators: Report of the Project on Indicators of Sustainable Development*, New York: Wiley, 1997.

Molina, Mario, and F. Sherwood Rowland, "Stratospheric Sink of Chlorofluoromethanes: Chlorine Atom-Catalyzed Destruction of Ozone," *Nature*, 249, June 1974, pp. 810–812.

Moran, Dana H., "Substitution—Some Practical Considerations," in U.S. Congress, Office of Technology Assessment, *Engineering Implications of Chronic Materials Scarcity*, Washington, D.C.: U.S. Government Printing Office, 1977.

Morris, David, and Irshad Ahmed, *The Carbohydrate Economy: Making Chemicals and Industrial Materials from Plant Matter*, Washington, D.C.: Institute for Local Self-Reliance, 1992.

Morse, Tim, "Dying to Know: A Historical Analysis of the Right to Know Movement," *New Solutions*, 8 (1), 1998, pp. 117–145.

Moser, Anton, "Ecological Process Engineering: The Potential of Bioprocessing," in Robert W. Ayres and Paul M. Weaver, eds., *Eco-restructuring: Implications for Sustainable Development*, New York: United Nations University Press, 1998, pp. 77–108.

Multhauf, Robert, "Industrial Chemistry in the Nineteenth Century," in Melvin Kranzberg and Carrol W. Purcell, Jr., eds., *Technology in Western Civilization*, New York: Oxford University Press, 1967, pp. 468–489.

Myers, Norman, and Jennifer Kent, *Perverse Subsidies: Tax Dollars Undercutting Our Economies and Environments Alike*, Winnipeg: International Institute for Sustainable Development, 1998.

Nalwa, Hari Singh, *Handbook of Nanostructured Materials and Nanotechnology*, San Diego: Academic Press, 2000.

National Academy of Sciences, *Science: The Endless Frontier*, Washington, D.C.: National Science Foundation, 1945.

National Academy of Sciences, *Materials Science and Engineering for the 1990s*, Washington, D.C.: National Academy Press, 1987.

National Academy of Sciences, National Research Council, Committee on Renewable Resources for Industrial Materials, *Renewable Resources for Industrial Materials*, Washington, D.C.: National Academy Press, 1976.

National Academy of Sciences, National Research Council, *Toxicity Testing, Strategies to Determine Needs and Priorities*, Washington, D.C.: National Academy Press, 1984.

National Academy of Sciences, National Research Council, *Jojoba: New Crop for Arid Lands; New Material for Industry*, Washington, D.C.: National Academy Press, 1985.

National Academy of Sciences, National Research Council, *Biologically Based Pest Management: New Solutions for a New Century*, Washington, D.C.: National Academy Press, 1996.

National Mining Association, *Facts About Mining, 1997–1998*, Washington, D.C., 1998.

National Safety Council, Environmental Health Center, *Electronic Product Recovery and Recycling Baseline Report*, Washington, D.C., May 1999.

Nelson, Richard, and Sidney Winter, *An Evolutionary Theory of Economic Change*, Cambridge, Mass.: Harvard University Press, 1982.

Nemerow, L. Nelson, *Zero Pollution for Industry: Waste Minimization Through Industrial Complexes*, New York: Wiley, 1995.

Noble, Charles, *Liberalism at Work: The Rise and Fall of OSHA*, Philadelphia: Temple University Press, 1986.

O'Brien, Mary, *Making Better Environmental Decisions: An Alternative to Risk Assessment*, Cambridge, Mass.: MIT Press, 2000.

Odum, Eugene P., *Basic Ecology*, Philadelphia: Saunders College Publications, 1983.

O'Mara, W., B. Herring, and L. Hunt, *Handbook of Semiconductor Silicon Technology*, Park Ridge, N.J.: Noyes Publications, 1990.

Organization for Economic Cooperation and Development, *Mining and Non-Ferrous Metals Policies of the OECD Countries*, Paris, 1994.

Organization for Economic Cooperation and Development, *Environmental Indicators*, Paris, 1994.

Organization for Economic Cooperation and Development, *Environmental Performance Reviews: United States*, Paris, 1996.

O'Rourke, Dara, Lloyd Connelly, and Catherine Koshland, "Industrial Ecology: A Critical Review," *International Journal of Environment and Pollution*, 6 (2–3), February 1996, pp. 89–112.

Orr, David, *Ecological Literacy*, Albany, N.Y.: State University of New York Press, 1992.

Orsenigo, Luigi, *The Emergence of Biotechnology: Institutions and Markets in Industrial Innovation*, New York: St. Martin's Press, 1989.

Paluzzi, Joseph E., and Timothy Greiner, "Finding Green in Clean: Progressive Pollution Prevention at Hyde Tool," *Total Quality Environmental Management*, 2 (3), Spring 1993, pp. 283–290.

Pauli, Gunter, "Zero Emissions: The Ultimate Goal of Cleaner Production," *Journal of Cleaner Production*, 5 (1–2), 1997, pp. 109–113.

Pearce, David W., and R. Kerry Turner, *Economics of Natural Resources and the Environment*, Baltimore: John Hopkins University Press for Resources for the Future, 1990.

Perkins, Patricia E., *World Metal Markets: The United States Strategic Stockpile and Global Market Influence*, Westport, Conn.: Praeger, 1997.

Pigou, Arthur C., *The Economics of Welfare*, London: Macmillan, 1920.

Pimentel, David, and Wen Dazhong, "Technological Changes in Energy Use in U.S. Agricultural Production," in Ronald Carroll, John H. Vandermeer, and Peter M. Rosset, eds., *Agroecology*, New York: McGraw-Hill, 1990, pp. 150–155.

Pollack, Cynthia, "Realizing Recycling's Potential," in Lester R. Brown, Edward C. Wolfe, and Linda Starke, eds., *State of the World—1987*, New York: Norton, 1987, pp. 101–121.

Portney, Paul, ed., *Public Policies for Environmental Protection*, Baltimore: John's Hopkins University Press for Resources for the Future, 1990.

Postel, Sandra, *Water for Agriculture: Facing the Limits*, Worldwatch Paper No. 93, Washington, D.C.: Worldwatch Institute, 1989.

Power, Thomas Michael, *Lost Landscapes and Failed Economies: The Search for a Value of Place*, Washington, D.C.: Island Press, 1996.

Raffensperger, Carolyn, and Joel Tickner, eds., *Protecting Public Health and the Environment: Implementing the Precautionary Principle*, Washington, D.C.: Island Press, 1999.

Reijnders, Lucas, "The Factor X Debate: Setting Targets for Eco-Efficiency," *Journal of Industrial Ecology*, 2 (1), 1998, pp. 13–23.

Rejeski, David, "Organizational Learning and Change: The Forgotten Dimensions of Sustainable Development," *Corporate Environmental Strategy*, 3 (1), Summer 1995, pp. 19–29.

Rejeski, David, "Mars, Materials, and Three Morality Plays: Materials Flows and Environmental Policy," *Journal of Industrial Ecology*, 1 (4), 1998, p. 13–18.

Renn, Ortin, Thomas Webler, and Peter Wiedemann, eds., *Fairness and Competence in Citizen Participation: Evaluating Models for Environmental Discourse*, Dordrecht, the Netherlands: Kluwer Academic Press, 1995.

Repetto, Robert, *Wasting Assets*, Washington, D.C.: World Resources Institute, 1989.

Repetto, Robert, Roger C. Dower, Robin Jenkins, and Jacqueline Geoghegan, *Green Fees: How a Tax Shift Can Work for the Environment and the Economy*, Washington, D.C.: World Resources Institute, 1992.

Ringquist, Evan J., *Environmental Protection at the State Level: Politics and Progress in Controlling Pollution*, Armonk, N.Y.: M. E. Sharpe, 1993.

Rippe, Klaus Peter, and Peter Schaber, "Democracy and Environmental Decision Making," *Environmental Values*, 8 (1), 1999, pp. 75–88.

Rivard, Christopher, Michael Himmel, and Karel Grohmann, "Biodegradation of Plastics," in Halena L. Chum, ed., *Polymers from Biobased Materials*, Park Ridge, N.J.: Noyes Data Corporation, 1991, pp. 115–119.

Robert, Joseph C., *Ethyl: A History of the Corporation and the People Who Made It*, Charlottesville, Va: University Press of Virginia, 1983.

Roetheli, Joseph, Lawrence K. Glazer, and Raymond Brigham, *Castor: Assessing the Feasibility of U.S. Production*, U.S. Department of Agriculture, Office of Agricultural Materials, May 1991.

Roetheli, Joseph C., Kenneth D. Carlson, Robert Kleiman, Anson E. Thompson, David A. Dierig, Lawrence K. Glazer, Melvin G. Blase, and Julia Goodell, *Lesquerella as a Source of Hydroxy Fatty Acids for Industrial Products*, U.S. Department of Agriculture, Cooperative State Research Service, Washington, D.C., October 1991.

Rogich, Donald G., and staff, "United States Global Material Use Patterns," unpublished paper, Division of Mineral Commodities, U.S. Bureau of Mines, Washington, D.C., September 1993.

Roodman, David Malin, *Getting the Signals Right: Tax Reform to Protect the Environment and the Economy*, Washington, D.C.: Worldwatch Institute, May 1997.

Roodman, David Malin, *The Natural Wealth of Nations: Harnessing the Market for the Environment*, New York: Norton, 1998.

Rosen, George, *A History of Public Health*, rev. ed., Baltimore: John Hopkins University Press, 1993.

Rosenberg, Beth, "Unintended Consequences: Impacts of Pesticides on Industry, Workers, the Public and the Environment," unpublished doctoral dissertation, Department of Work Environment, University of Massachusetts Lowell, 1995.

Rosner, David, and Gerald Markowitz, "A 'Gift of God'?: The Public Health Controversy Over Leaded Gasoline During the 1920s," *American Journal of Public Health*, 75 (4), April 1985, pp. 344–352.

Rossi, Mark, Michael Ellenbecker, and Kenneth Geiser, "Techniques in Toxics Use Reduction: From Concept to Action," *New Solutions*, 2 (2), Fall 1991, pp. 25–32.

Roubi, Maureen, "Fallout from Pricing Earth," *Chemical & Engineering News*, 75, June 30, 1997, p. 38.

Ruth, Matthias, "Dematerialization in Five US Metals Sectors: Implications for Energy Use and CO_2 Emissions," *Resource Policy*, 24 (1), 1998, pp. 1–18.

Ryan, Chris, "Designing for Factor 20 Improvements," *Journal of Industrial Ecology*, 2 (2), 1998, pp. 3–5.

"Safe New Alternatives Program (Proposed Rule)," *Federal Register*, 58 (90), May 12, 1993, pp. 28094–28192.

Sager, Ambuj D., and Robert Frosch, "A Perspective on Industrial Ecology and Its Applicability to a Metals-Industry Ecosystem," *Journal of Cleaner Production*, 5, 1997, pp. 39–45.

Salzman, James, "Informing the Green Consumer," *Journal of Industrial Ecology*, 1 (2), 1997, pp. 11–21.

Sarikaya, Mehmet, and Ilan A. Aksay, eds., *Biomimetics: Design and Processing of Materials*, New York: American Institute of Physics, 1995.

Schenker, M. B., E. B. Gold, J. J. Beaumont, et al., "Association of Spontaneous Abortion and Other Reproductive Effects with Work in the Semiconductor Industry," *American Journal of Industrial Medicine*, 28 (6), 1995, pp. 639–659.

Schidheimy, Stephen, and the Business Council for Sustainable Development, *Changing Course: A Global Perspective on Development and the Environment*, Cambridge, Mass.: MIT Press, 1992.

Schmidt-Bleek, Frederich, *Carnoules Declaration of the Factor Ten Club*, Wuppertal, Germany: Wuppertal Institute for Climate, Environment, and Energy, 1994.

Schumpeter, Joseph A., *Business Cycles: A Theoretical, Historical, and Statistical Analysis of the Capitalist Process*, New York: McGraw-Hill, 1939.

Seldman, Neil N., "Recycling–History in the United States," in Bisio Attilio and Sharon Boots, eds., *Encyclopedia of Energy, Technology and the Environment*, New York: Wiley 1995, pp. 2352–2368.

Serageldin, Ismail, *Sustainability and the Wealth of Nations: First Steps in an On-going Journey*, Washington, D.C.: World Bank, 1996.

Serageldin, Ismail, *Expanding the Measures of Wealth: Indicators of Environmentally Sustainable Development*, Washington, D.C.: World Bank, 1997.

Shabecoff, Phillip, "Study Shows Significant Decline in Ozone Layer," *New York Times*, March 16, 1988, p. A25.

Shapiro, Karen, and Allen White, "Product Stewardship through Life Cycle Design, *Corporate Environmental Strategy*, 6 (1), Winter 1999, pp. 15–23.

Sharp, Margaret, "Applications of Biotechnology: An Overview," in Martin Fransman, Gerd Junne, and Annemieke Roobeek, eds., *The Biotechnology Revolution?*, Oxford, U.K.: Blackwell, 1995, pp. 163–173.

Shea, Cynthia Pollock, "Give and Take," *Tomorrow*, 10 (1), January 2000, p. 26.

Sibley, Scott F., William C. Butterman and staff, "Metals Recycling in the United States," *Resource Conservation and Recycling*, 15, 1995, pp. 259–267.

Sieburth, Scott, "Isosteric Replacement of Carbon with Silicon in the Design of Safer Chemicals," in Stephen C. DeVito and Roger Garrett, eds., *Designing Safer Chemicals: Green Chemistry for Pollution Prevention*, Washington, D.C.: American Chemical Society, 1996, pp. 74–83.

Sikora, Mary B., "Options for Managing and Marketing Scrap Tires," in Frank Krieth, ed., *Handbook of Solid Waste Management*, New York: McGraw-Hill, 1994, p. 9.129–9.148.

Simpson, Beryl B., and Molly C. Ogorzaly, *Economic Botany: Plants in Our World*, 2nd ed., New York: McGraw-Hill, 1995, p. 363.

Shinya, William M., "Canada's New Minerals and Metals Policy," *Resources Policy*, 24 (2), 1998, pp. 95–104.

Smith, C. Mark, Karl T. Kelsey, and David Christiani, "Risk Assessment and Occupational Health: Overview and Recommendations," in Charles Levenstein and John Wooding, eds., *Work, Health and Environment: Old Problems, New Solutions*, New York: Guilford, 1997, pp. 236–257.

Smith, Gerald R., *Materials Flow of Tungsten in the United States*, Washington, D.C.: U.S. Bureau of Mines, 1993.

Smith, Lawrence, Jeffery Means, and Adwin Barth, *Recycling and Reuse of Industrial Wastes*, Columbus: Battelle Press, 1995.

Smith, Merritt Roe, ed., *Military Enterprise and Technological Change: Perspectives on the American Experience*, Cambridge, Mass.: MIT Press, 1985.

Socolow, Robert, and Valerie Thomas, "The Industrial Ecology of Lead and Electric Vehicles," *Journal of Industrial Ecology*, 1 (1), 1997, pp. 26–27.

Soule, Judith D., and Jon K. Piper, *Farming in Nature's Image: An Ecological Approach to Agriculture*, Washington, D.C.: Island Press, 1992.

Sound Resource Management Group, *The Economics of Recycling and Recycled Materials*, Seattle: Clean Washington Center, 1993.

Spenser, David B., "Recycling," in Frank Krieth, ed., *Handbook of Solid Waste Management*, New York: McGraw-Hill, 1994, pp. 9.1–9.129.

Stahel, Walter R., and Genevieve Reday-Mulvey, *Jobs for Tomorrow: The Potential for Substituting Manpower for Energy*, New York: Vantage Press, 1981.

Stanford Research Institute, *Chemical Economics Handbook*, Palo Alto, Calif.

Stoner, Daphne, ed., *Biotechnology for the Treatment of Hazardous Waste*, Boca Raton, Fla.: Lewis, 1993.

Strasser, Susan, *Waste and Want: A Social History of Trash*, New York: Henry Holt, 1999.

Stringer, Judy, "Biologically Derived Chemicals Find Niches," *Chemical Week*, September 18, 1996, pp. 52–53.

Suppliers of Advanced Composite Materials Association, *Safe Handling of Advanced Composite Materials: Health Information*, Arlington, Va: April 1989.

Swedish Ministry of the Environment, Chemicals Policies Committee, *Towards a Sustainable Chemicals Policy*, Stockholm, 1997.

Swedish Ministry of the Environment, National Chemical Inspectorate, *Risk Reduction of Chemicals*, Solna, Sweden, 1991.

Swedish Ministry of the Environment, National Chemical Inspectorate, *Swedish Restrictions Benefit the Environment*, Solna, Sweden, October 1997.

Terrell, Jane G., Maurille J. Fournier, Thomas L. Mason, and David Terrell, "Biomolecular Materials," *Chemical & Engineering News*, 72, December 19, 1994, pp. 40–51.

Texas Department of Agriculture, *Texas Agriculture: Growing A Sustainable Economy, A Greenprint for Action*, Austin, Texas, 1990.

Thibodeau, Francis R., and Herman H. Field, eds., *Sustaining Tomorrow: A Strategy for World Conservation and Development*, Hanover, N.H.: University Press of New England, 1984.

Thornton, Joe, *Pandora's Poison: Chlorine, Health, and a New Environmental Strategy*, Cambridge, Mass.: MIT Press, 2000.

Thorpe, Beverley, *Citizen's Guide to Clean Production*, Lowell Center for Sustainable Development, University of Massachusetts Lowell, 1999.

Tietenberg, Tom, *Environmental and Natural Resource Economics*, 4th ed., New York: Harper Collins, 1996.

Trevan, M. D., S. Boffey, K. H. Goulding, and P. Stanbury, *Biotechnology: The Biological Principles*, New York: Taylor and Francis, 1987.

Union for the Conservation of Nature and Natural Resources, *World Conservation Strategy: Living Resource Conservation for Sustainable Development*, Gland, Switzerland, 1980.

United Nations, *System of Integrated Environmental and Economic Accounting*, New York: United Nations, 1993.

United Nations Environment Program, International Registry of Potentially Toxic Chemicals, *Consolidated List of Products Whose Consumption and/or Sale Have Been Banned, Withdrawn, Severely Restricted or Not Approved by Governments*, 6th ed., Geneva, 1997.

U.S. Bureau of the Census, *Census of Population, 1990*, Washington, D.C., 1991.

U.S. Bureau of the Census, *1992 Census of Mining*, Washington, D.C., 1993.

U.S. Bureau of Mines, *The New Materials Society*, Vol. 1: *New Materials, Markets and Issues*, Washington, D.C.: U.S. Government Printing Office, 1990.

U.S. Bureau of Mines, *The New Materials Society*, Vol. 3: *Material Shifts in the New Society*, Washington, D.C.: U.S. Government Printing Office, 1991.

U.S. Congress, Office of Technology Assessment, *Engineering Implications of Chronic Materials Scarcity*, Washington, D.C.: U.S. Government Printing Office, 1977.

U.S. Congress, Office of Technology Assessment, *Preventing Illness and Injury in the Workplace*, Washington, D.C., U.S. Government Printing Office, 1985.

U.S. Congress, Office of Technology Assessment, *Advanced Materials by Design: New Structural Materials Technologies*, Washington, D.C.: U.S. Government Printing Office, June 1988.

U.S. Congress, Office of Technology Assessment, *New Developments in Biotechnology: U.S. Investment in Biotechnology*, Washington, D.C.: U.S. Government Printing Office, 1987–1989.

U.S. Congress, Office of Technology Assessment, *Agricultural Commodities as Industrial Raw Materials*, Washington, D.C.: U.S. Government Printing Office, May 1991.

U.S. Congress, Office of Technology Assessment, *Biotechnology in a Global Economy*, Washington, D.C.: U.S. Government Printing Office, October 1991.

U.S. Congress, Office of Technology Assessment, *Green Products by Design: Choices for a Cleaner Environment*, Washington, D.C.: U.S. Government Printing Office, 1992.

U.S. Congress, Office of Technology Assessment, *Biopolymers: Making Materials Nature's Way*, Washington, D.C.: U.S. Government Printing Office, September 1993.

U.S. Department of Agriculture, Economic Research Service, "Ethanol, Critic Acid, and Lactic Acid Use Corn as a Feedstock," *Industrial Uses of Agricultural Materials: Situation and Outlook Report*, Washington, D.C., July 1997, p. 10–12.

U.S. Department of Agriculture, Economic Research Service, "Crambe, Industrial Rapeseed, and Tury Provide Valuable Oils" *Industrial Uses of Agricultural Materials: Situation and Outlook*, Washington, D.C., August, 1996, pp. 17–23.

U.S. Department of Commerce, Bureau of Economic Analysis, "Pollution Abatement and Control Expenditures," in *Survey of Current Business*, Washington, D.C.: U.S. Government Printing Office, September 1996.

U.S. Environmental Protection Agency, Office of Air and Radiation, *Final Report to Congress on Benefits and Costs of the Clean Air Act, 1970 to 1990*, Washington, D.C., October 1997.

U.S. Environmental Protection Agency, Office of Air Quality Planning and Standards Research, *National Air Pollutant Emissions Trends, 1900–1995*, Research Triangle Park, N.C., 1996.

U.S. Environmental Protection Agency, Office of Pollution Prevention and Toxics, *Cleaner Technologies Substitutes Assessment: A Methodology and Resource Guide*, Washington, D.C., 1996.

U.S. Environmental Protection Agency, Office of Pollution Prevention and Toxics, *Selling Environmental Products to the Federal Government*, Washington, D.C., May 1997.

U.S. Environmental Protection Agency, Office of Pollution Prevention and Toxics, *Chemical Hazard Data Availability Study: What Do We Really Know About the Safety of High Production Volume Chemicals?*, Washington, D.C., April 1998.

U.S. Environmental Protection Agency, Office of Pollution Prevention and Toxics, *1997 Toxics Release Inventory*, Washington, D.C., April 1999.

U.S. Environmental Protection Agency, Office of Solid Waste, *Characterization of Products Containing Lead and Cadmium in Municipal Solid Waste in the United States, 1970–2000*, Washington, D.C., January 1989.

U.S. Environmental Protection Agency, Office of Solid Waste, *Characterization of the Municipal Solid Waste Stream in the United States: 1990 Update*, Washington, D.C., June 1990,

U.S. Environmental Protection Agency, Office of Solid Waste, *Characterization of Products Containing Mercury in Municipal Solid Waste in the United States*, Washington, D.C., April 1992.

U.S. Environmental Protection Agency, Office of Solid Waste and Emergency Response, *National Biennial RCRA Hazardous Waste Reports, 1989–1993*, Washington, D.C., 1996.

U.S. Environmental Protection Agency, Office of Solid Waste and Emergency Response, *Characterization of Municipal Solid Waste in the United States, 1996 Update*, Washington, D.C., 1997.

U.S. Environmental Protection Agency, Office of Solid Waste and Emergency Response, *National Biennial RCRA Hazardous Waste Report*, Washington, D.C., August 1997.

U.S. Environmental Protection Agency, Office of Solid Waste and Emergency Response, *Characterization of Municipal Solid Waste in the United States, 1997 Update*, Washington, D.C., 1998.

U.S. Environmental Protection Agency, Office of Solid Waste and Emergency Response, *Cutting the Waste Stream in Half: Community Record Setters Show How*, Washington, D.C., June 1999.

U.S. Environmental Protection Agency, Office of Solid Waste and Emergency Response, *Greenhouse Gas Emissions from Management of Selected Materials in Municipal Solid Waste*, Washington, D.C., September 1999.

U.S. Environmental Protection Agency, Office of Water, *National Water Quality Inventory, 1994 Report to Congress*, Washington, D.C., 1995.

U.S. General Accounting Office, *Wastewater Discharges Are Not Complying with EPA Pollution Control Permits*, Washington, D.C., December 1983.

U.S. Geological Survey, *Mineral Commodity Summaries, 1998*, Washington, D.C., 1998.

U.S. Geological Survey, *Minerals Yearbook: Metals and Minerals, 1997*, Washington, D.C., U.S. Government Printing Office, 1999.

U.S. Interagency Working Group on Sustainable Development Indicators, *Sustainable Development in the United States*, Washington, D.C., December 1998.

U.S. Office of Management and Budget, *Budget of the United States Federal Government, Fiscal Years 1980–1993*, Washington, D.C., 1992.

U.S. Office of Management and Budget, *Budget of the United States Government Fiscal Year 2000, Analytical Perspectives*, Washington, D.C., 1999.

U.S. Office of the President, Executive Order 12873, "Federal Aquisition, Recycling and Waste Prevention," Washington, D.C., October 20, 1993.

U.S. President's Council on Sustainable Development, *Sustainable America: A New Consensus*, Washington, D.C.: U.S. Government Printing Office, February 1996; and *Towards a Sustainable America: Advancing Prosperity, Opportunity and a Healthy Environment for the 21st Century*, Washington, D.C., May 1999.

U.S. Senate, Committee on Commerce, Subcommittee on the Environment, *The Toxic Substances Control Act of 1971 and Amendments, Part 1*, August 3 and 4, 1971, Washington, D.C., 1972.

Utterback, James M., *Mastering the Dynamics of Innovation*, Boston: Harvard Business School Press, 1996.

Vitousek, Peter, Paul Ehrlich, and Pamela Matson, "Human Appropriation of the Products of Photosynthesis," *BioScience*, 34 (6), 1986, pp. 368–373.

von Hippel, Eric, *The Sources of Innovation*, New York: Oxford University Press, 1988.

von Moltke, Konrad, *The Vorsorsorgeprinzip in West German Environmental Policy*, Institute for European Environmental Policy, London, 1987.

von Weizsacker, Ernst, Amory B. Lovins, and L. Hunter Lovins, *Factor Four: Doubling Wealth, Halving Resource Use*, London: Earthscan, 1997.

Wackernagel, Mathis, and William Rees, *Our Ecological Footprint: Reducing Human Impact on the Earth*, Gabriola Island, Canada: New Society Publishers, 1996.

Wahlstrom, Bo, "Sunset for Dangerous Chemicals," *Nature*, 341, 1989, p. 276.

Ward, Owen P., *Fermentation Biotechnology*, Englewood Cliffs, N.J.: Prentice Hall, 1989.

Warhurst, Alyson, "Metals Biotechnology: Trends and Implications," in Martin Fransman, Gerd Junne and Annemieke Roobeek, eds., *The Biotechnology Revolution?*, Oxford, U.K.: Blackwell 1995, pp. 258–273.

Wartenberg, Daniel, and Caron Chess, "The Risk Wars: Assessing Risk Assessment," in Charles Levenstein and John Wooding, eds., *Work Health and Environment: Old Problems, New Solutions*, New York: Guilford, 1997, pp. 258–274,

Webb, Lee, and Jeff Tryens, *An Environmental Agenda for the States*, Washington, D.C.: Conference on Alternative State and Local Policies, 1984.

Wernick, Iddo K., and Jesse H. Ausubel, "National Materials Flows and the Environment," *Annual Review of Energy and Environment*, 20, 1995, pp. 463–492.

Wernick, Iddo K., and Nickolas J. Themelis, "Recycling Metals for the Environment," *Annual Review of Energy and the Environment*, 23, 1998, pp. 465–497.

Wernick, Iddo K., Robert Herman, Shekhar Govind, and Jesse H. Ausubel, "Materialization and Dematerialization: Measures and Trends," *Daedalus*, 125 (3), Summer 1996, pp. 171–198.

White, George, "General Introduction," U.S. Bureau of Mines, *The New Materials Society*, Vol. 1: *New Materials, Markets and Issues*, Washington, D.C.: U.S. Government Printing Office, 1990, pp. 1.1–1.11.

Widmark, Gunnar, "Possible Interference by Chlorinated Biphenyl," *Journal of the Association of Official Analytical Chemists*, 50, 1967, p. 1069.

Willcox, P. Jeanene, Samuel P. Guido, Wayne Mueller, and David L. Kaplan, "Evidence of Cholesteric Liquid Crystalline Phase in Natural Silk Spinning Processes," *Macromolecules*, 29 (15), 1996, pp. 5106–5110.

Wood, J. B., B. L. Ghosh, and S. N. Sinha, "Solid Substrate Fermentation in the Production of Enzyme Complex for Improved Jute Processing," in Glen Fuller, Thomas McKeon and Donald Bills, eds., *Agricultural Materials as Renewable Resources: Non-Food and Industrial Applications*, Washington, D.C.: American Chemical Society, 1996, pp. 46–59.

Woodbury, William, Daniel Edelstein, and Stephen Jasinski, "Lead Materials Flow in the United States: 1940–1988," unpublished paper, U.S. Bureau of Mines, Washington, D.C., 1993.

World Business Council for Sustainable Development, *Eco-efficient Leadership for Improved Economic and Environmental Performance*, Geneva, 1995.

World Commission on Environment and Development, *Our Common Future*, New York: Oxford University Press, 1987.

World Resources Institute and Future Studies Group, *Focus Group Report: Institutions*, Washington, D.C., 1993.

Young, John, and Aaron Sachs, *The Next Efficiency Revolution: Creating a Sustainable Materials Economy*, Worldwatch Paper No. 121, Washington, D.C.: Worldwatch Institute, September 1994.

Index